A first course in general relativity

A first course in

general relativity

BERNARD F. SCHUTZ

Professor, Applied Mathematics and Astronomy, University College, Cardiff

CAMBRIDGE
UNIVERSITY PRESS

Published by the Press Syndicate of the University of Cambridge
The Pitt Building, Trumpington Street, Cambridge CB2 1RP
40 West 20th Street, New York NY 10011-4211, USA
10 Stamford Road, Oakleigh, Melbourne 3166, Australia

First published 1985
Reprinted 1986 (with corrections), 1988, 1989, 1990, 1992, 1993, 1994

Printed in Great Britain by
Athenæum Press Ltd., Newcastle upon Tyne

Library of Congress catalogue card number: 83-23205

British Library cataloguing in publication data

Schutz, Bernard F.
A first course in relativity.
1. Relativity (Physics)
I. Title
530.1'1   QC173.55

ISBN 0 521 27703 5 paperback

AS

To Siân

Contents

Preface

This book has evolved from lecture notes for a full-year undergraduate course in general relativity which I taught from 1975 to 1980, an experience which firmly convinced me that general relativity is not significantly more difficult for undergraduates to learn than the standard undergraduate-level treatments of electromagnetism and quantum mechanics. The explosion of research interest in general relativity in the past 20 years, largely stimulated by astronomy, has not only led to a deeper and more complete understanding of the theory; it has also taught us simpler, more physical ways of understanding it. Relativity is now in the mainstream of physics and astronomy, so that no theoretical physicist can be regarded as broadly educated without some training in the subject. The formidable reputation relativity acquired in its early years (Interviewer: 'Professor Eddington, is it true that only three people in the world understand Einstein's theory?' Eddington: 'Who is the third?') is today perhaps the chief obstacle that prevents it being more widely taught to theoretical physicists. The aim of this textbook is to present general relativity at a level appropriate for undergraduates, so that the student will understand the basic physical concepts and their experimental implications, will be able to solve elementary problems, and will be well prepared for the more advanced texts on the subject.

In pursuing this aim, I have tried to satisfy two competing criteria: first, to assume a minimum of prerequisites; and second, to avoid watering down the subject matter. Unlike most introductory texts, this one does not assume that the student has already studied electromagnetism in its manifestly relativistic formulation, the theory of electromagnetic waves, or fluid dynamics. The necessary fluid dynamics is developed in the relevant chapters. The main consequence of not assuming a familiarity with electromagnetic waves is that gravitational waves have to be introduced slowly: the wave equation is studied from scratch. A full list of prerequisites appears below.

The second guiding principle, that of not watering down the treatment, is very subjective and rather more difficult to describe. I have tried to introduce differential geometry fully, not being content to rely only on analogies with curved surfaces, but I have left out subjects that are not essential to general relativity at this level, such as nonmetric manifold theory, Lie derivatives, and fiber bundles.[1] I have introduced the full nonlinear field equations, not just those of linearized theory, but I solve them only in the plane and spherical cases, quoting and examining, in addition, the Kerr solution. I study gravitational waves mainly in the linear approximation, but go slightly beyond it to derive the energy in the waves and the reaction effects in the wave emitter. I have tried in each topic to supply enough foundation for the student to be able to go to more advanced treatments without having to start over again at the beginning.

The first part of the book, up to Ch. 8, introduces the theory in a sequence which is typical of many treatments: a review of special relativity, development of tensor analysis and continuum physics in special relativity, study of tensor calculus in curvilinear coordinates in Euclidean and Minkowski spaces, geometry of curved manifolds, physics in a curved spacetime, and finally the field equations. The remaining four chapters study a few topics which I have chosen because of their importance in modern astrophysics. The chapter on gravitational radiation is more detailed than usual at this level because the observation of gravitational waves may be one of the most significant developments in astronomy in the next decade. The chapter on spherical stars includes, besides the usual material, a useful family of exact compressible solutions due to Buchdahl. A long chapter on black holes studies in some detail the physical nature of the horizon, going as far as the Kruskal coordinates,

---

1   The treatment here is therefore different in spirit from that in my book *Geometrical Methods of Mathematical Physics* (Cambridge University Press 1980*b*), which may be used to supplement this one.

then exploring the rotating (Kerr) black hole, and concluding with a simple discussion of the Hawking effect, the quantum mechanical emission of radiation by black holes. The concluding chapter on cosmology derives the homogeneous and isotropic metrics and briefly studies the physics of cosmological observation and evolution. There is an appendix summarizing the linear algebra needed in the text, and another appendix containing hints and solutions for selected exercises. One subject I have decided not to give as much prominence to as other texts traditionally have is experimental tests of general relativity and of alternative theories of gravity. Points of contact with experiment are treated as they arise, but systematic discussions of tests now require whole books (Will 1981). Physicists today have far more confidence in the validity of general relativity than they had a decade or two ago, and I believe that an extensive discussion of alternative theories is therefore almost as out of place in a modern elementary text on gravity as it would be in one on electromagnetism.

The student is assumed already to have studied: special relativity, including the Lorentz transformation and relativistic mechanics; Euclidean vector calculus; ordinary and simple partial differential equations; thermodynamics and hydrostatics; Newtonian gravity (simple stellar structure would be useful but not essential); and enough elementary quantum mechanics to know what a photon is.

The notation and conventions are essentially the same as in Misner *et al.*, *Gravitation* (W. H. Freeman 1973), which may be regarded as one possible follow-on text after this one. The physical point of view and development of the subject are also inevitably influenced by that book, partly because Thorne was my teacher and partly because *Gravitation* has become such an influential text. But because I have tried to make the subject accessible to a much wider audience, the style and pedagogical method of the present book are very different.

Regarding the use of the book, it is designed to be studied sequentially as a whole, in a one-year course, but it can be shortened to accommodate a half-year course. Half-year courses probably should aim at restricted goals. For example, it would be reasonable to aim to teach gravitational waves and black holes in half a year to students who have already studied electromagnetic waves, by carefully skipping some of Chs. 1–3 and most of Chs. 4, 7, and 10. Students with preparation in special relativity and fluid dynamics could learn stellar structure and cosmology in half a year, provided they could go quickly through the first four chapters and then skip Chs. 9 and 11. A graduate-level course can, of course, go much more quickly, and it should be possible to cover the whole text in half a year.

Each chapter is followed by a set of exercises, which range from trivial ones (filling in missing steps in the body of the text, manipulating newly introduced mathematics) to advanced problems that considerably extend the discussion in the text. Some problems require programmable calculators or computers. I cannot overstress the importance of doing a selection of problems. The easy and medium–hard ones in the early chapters give essential practice, without which the later chapters will be much less comprehensible. The medium–hard and hard problems of the later chapters are a test of the student's understanding. It is all too common in relativity for students to find the conceptual framework so interesting that they relegate problem solving to second place. Such a separation is false and dangerous: a student who can't solve problems of reasonable difficulty doesn't really understand the concepts of the theory either. There are generally more problems than one would expect a student to solve; several chapters have more than 30. The teacher will have to select them judiciously. Another rich source of problems is the *Problem Book in Relativity and Gravitation*, Lightman *et al.* (Princeton University Press 1975).

I am indebted to many people for their help, direct and indirect, with this book. I would like especially to thank my undergraduates at University College, Cardiff, whose enthusiasm for the subject and whose patience with the inadequacies of the early lecture notes encouraged me to turn them into a book. And I am certainly grateful to Suzanne Ball, Jane Owen, Margaret Vallender, Pranoat Priesmeyer and Shirley Kemp for their patient typing and retyping of the successive drafts.

BFS
1984

1
Special relativity

1.1 Fundamental principles of special relativity theory (SR)

The way in which special relativity is taught at an elementary undergraduate level – the level which the reader is assumed competent at – is usually close in spirit to the way it was first understood by physicists. This is an algebraic approach, based on the Lorentz transformation (§ 1.7 below). At this basic level, one learns how to use the Lorentz transformation to convert between one observer's measurements and another's, to verify and understand such remarkable phenomena as time dilation and Lorentz contraction, and to make elementary calculations of the conversion of mass into energy.

This purely algebraic point of view began to change, to widen, less than four years after Einstein proposed the theory.[1] Minkowski pointed out that it is very helpful to regard (t, x, y, z) as simply four coordinates in a four-dimensional space which we now call spacetime. This was the beginning of the geometrical point of view which led directly to general relativity in 1914–16. It is this geometrical point of view on special relativity which we must study before all else.

---

1 Einstein's original paper was published in 1905, while Minkowski's discussion of the geometry of spacetime was given in 1908. Einstein's and Minkowski's papers are reprinted (in English translation) in *The Principle of Relativity* by A. Einstein, H. A. Lorentz, H. Minkowski & H. Weyl (Dover).

As we shall see, special relativity can be deduced from two fundamental postulates:

(1) *Principle of relativity* (Galileo): No experiment can measure the absolute velocity of an observer; the results of any experiment performed by an observer do not depend on his speed relative to other observers who are not involved in the experiment.

(2) *Universality of the speed of light* (Einstein): The speed of light relative to any unaccelerated observer is $c = 3 \times 10^8 \, \mathrm{m \, s^{-1}}$, regardless of the motion of the light's source relative to the observer. Let us be quite clear about this postulate's meaning: two different unaccelerated observers measuring the speed of the *same photon* will each find it to be moving at $3 \times 10^8 \, \mathrm{m \, s^{-1}}$ relative to themselves, regardless of their state of motion relative to each other.

As noted above, the principle of relativity is not at all a modern concept: it goes back all the way to Galileo's hypothesis that a body in a state of uniform motion remains in that state unless acted upon by some external agency. It is fully embodied in Newton's second law, which contains only accelerations, not velocities themselves. Newton's laws are, in fact, all invariant under the replacement

$$v(t) \rightarrow v'(t) = v(t) - V,$$

where V is any *constant* velocity. This equation says that a velocity $v(t)$ relative to one observer becomes $v'(t)$ when measured by a second observer whose velocity relative to the first is V. This is called the Galilean law of addition of velocities.

By saying that Newton's laws are *invariant* under the Galilean law of addition of velocities, we are making a statement of a sort which we will often make in our study of relativity, so it is well to start by making it very precise. Newton's first law, that a body moves at a constant velocity in the absence of external forces, is unaffected by the replacement above, since if $v(t)$ is really a constant, say v_0, then the new velocity $v_0 - V$ is also a constant. Newton's second law,

$$F = ma = m \, dv/dt,$$

is also unaffected, since

$$a' = dv'/dt = d(v - V)/dt = dv/dt = a.$$

Therefore the second law will be valid according to the measurements of both observers, provided that we add to the Galilean transformation law the statement that F and m are themselves invariant, i.e. the same regardless of which of the two observers measures them. Newton's third law, that the force exerted by one body on another is equal and opposite to that exerted by the second on the first, is clearly unaffected by the change of observers, again because we assume the forces to be invariant.

So there is no absolute velocity. Is there an absolute acceleration? Newton argued that there was. Suppose, for example, that I am in a train on a perfectly smooth track,[2] eating a bowl of soup in the dining car. Then if the train moves at constant speed the soup remains level, thereby offering me no information about what my speed is. But if the train changes its speed then the soup climbs up one side of the bowl, and I can tell by looking at it how large and in what direction the acceleration is.[3]

Therefore, it is reasonable and useful to single out a class of preferred observers: those who are unaccelerated. They are called *inertial observers*, and each one has a constant velocity with respect to any other one. These inertial observers are fundamental in special relativity, and when we use the term 'observer' from now on we will mean an inertial observer.

The postulate of the universality of the speed of light was Einstein's great and radical contribution to relativity. It smashes the Galilean law of addition of velocities because it says that if v has magnitude c then so does v', regardless of V. Einstein felt that the postulate was forced on him by, among other things, the Michelson–Morley experiment. The counter-intuitive predictions of special relativity all flow from this postulate, and they are amply confirmed by experiment. In fact it is probably fair to say that special relativity has a firmer experimental basis than any other of our laws of physics, since it is tested every day in all the giant particle accelerators, which send particles nearly to the speed of light.

Although the concept of relativity is old, it is customary to refer to Einstein's theory simply as 'relativity'. The adjective 'special' is applied in order to distinguish it from Einstein's theory of gravitation, which acquired the name 'general relativity' because it permits one to describe physics from the point of view of both accelerated and inertial observers and is in that respect a more general form of relativity. But the real physical distinction between these two theories is that special relativity (SR) is capable of describing physics only in the absence of gravitational fields, while general relativity (GR) extends SR to describe gravitation

---

2 Physicists frequently have to make such idealizations, which often are far removed from common experience!

3 For Newton's discussion of this point, see the excerpt from his *Principia* in Williams (1968).

itself.[4] One can only wish that an earlier generation of physicists had chosen more appropriate names for these theories!

1.2   Definition of an inertial observer in SR

It is important to realize that an 'observer' is in fact a huge information-gathering system, not simply one man with binoculars. In fact, we shall remove the human element entirely from our definition, and say that an inertial observer is simply a coordinate system for spacetime, which makes an observation simply by recording the location (x, y, z) and time (t) of any event. This coordinate system must satisfy the following three properties to be called *inertial*:

(1) The distance between point P_1 (coordinates x_1, y_1, z_1) and point P_2 (coordinates x_2, y_2, z_2) is independent of time.

(2) The clocks that sit at every point ticking off the time coordinate t are synchronized and all run at the same rate.

(3) The geometry of space at any constant time t is Euclidean.

Notice that this definition does not mention whether the observer accelerates or not. That will come later. It will turn out that only an unaccelerated observer can keep his clocks synchronized. But we prefer to start out with this geometrical definition of an inertial observer. It is a matter for experiment to decide whether such an observer can exist: it is not self-evident that any of these properties *must* be realizable, although we would probably expect a 'nice' universe to permit them! However, we will see later in the course that a gravitational field does in fact make it impossible to construct such a coordinate system, and this is why GR is required. But let us not get ahead of the story. At the moment we are assuming that we *can* construct such a coordinate system (that, if you like, the gravitational fields around us are so weak that they do not really matter). One can envision this coordinate system, rather fancifully, as a lattice of rigid rods filling space, with a clock at every intersection of the rods. We shall now define how we use this coordinate system to make observations.

An *observation* made by the inertial observer is the act of assigning to any event the coordinates x, y, z of the location of its occurrence, and

---

4   It is easy to see that gravitational fields cause problems for SR. If an astronaut in orbit about Earth holds a bowl of soup, does the soup climb up the side of the bowl in response to the gravitational 'force' which holds the spacecraft in orbit? Two astronauts in different orbits accelerate relative to one another, but neither *feels* an acceleration. Problems like this make gravity special, and we will have to wait until Ch. 5 to resolve them. Until then, the word 'force' will refer to a nongravitational force.

the time read by the clock at (x, y, z) when the event occurred. It is *not* the time t on the wrist watch worn by a scientist located at $(0, 0, 0)$ when he first learns of the event. A *visual* observation is of this second type: the eye regards as simultaneous all events it *sees* at the same time; an inertial observer regards as simultaneous all events that *occur* at the same time as recorded by the clock nearest them when they occurred. This distinction is important and must be borne in mind. Sometimes we will say 'an observer sees...' but this will only be shorthand for 'measures'. We will never mean a *visual* observation unless we say so explicitly.

An inertial observer is also called an *inertial reference frame*, which we will often abbreviate to 'reference frame' or simply 'frame'.

1.3 New units

Since the speed of light c is so fundamental, we shall from now on adopt a new system of units for measurements in which c simply has the value 1! It is perfectly okay for slow-moving creatures like engineers to be content with the SI units: m, s, kg. But it seems silly in SR to use units in which the fundamental constant c has the ridiculous value 3×10^8. The SI units evolved historically. Meters and seconds are not fundamental; they are simply convenient for human use. What we shall now do is adopt a new unit for time, the meter. One meter of time is the time it takes light to travel one meter. (You are probably more familiar with an alternative approach: a year of distance – called a 'light year' – is the distance light travels in one year.) The speed of light in these units is:

$$c = \frac{\text{distance light travels in any given time interval}}{\text{the given time interval}}$$

$$= \frac{1 \text{ m}}{\text{the time it takes light to travel one meter}}$$

$$= \frac{1 \text{ m}}{1 \text{ m}} = 1.$$

So if we consistently measure time in meters, then c is not merely 1, it is also dimensionless! In converting from SI units to these 'natural' units, you can use any of the following relations:

$$3 \times 10^8 \text{ m s}^{-1} = 1,$$

$$1 \text{ s} = 3 \times 10^8 \text{ m},$$

$$1 \text{ m} = \frac{1}{3 \times 10^8} \text{ s}.$$

The SI units contain many 'derived' units, such as joules and newtons, which are defined in terms of the basic three: m, s, kg. By converting from s to m these units simplify considerably: energy and momentum are measured in kg, acceleration in m^{-1}, force in $kg\,m^{-1}$, etc. Do the exercises on this. With practice, these units will seem as natural to you as they do to most modern theoretical physicists.

1.4 Spacetime diagrams

A very important part of learning the geometrical approach to SR is mastering the spacetime diagram. In the rest of this chapter we will derive SR from its postulates by using spacetime diagrams, because they provide a very powerful guide for threading one's way among the many pitfalls SR presents to the beginner. Fig. 1.1 below shows a two-dimensional slice of spacetime, the t–x plane, in which are illustrated

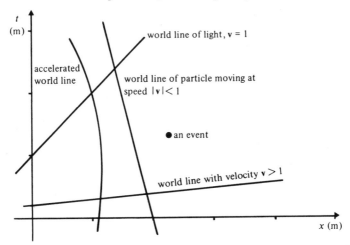

Fig. 1.1   A spacetime diagram in natural units.

the basic concepts. A single point in this space[5] is a point of fixed x *and* fixed t, and is called an *event*. A line in the space gives a relation $x = x(t)$, and so can represent the position of a particle at different times. This is called the particle's *world line*. Its slope is related to its velocity,

$$\text{slope} = dt/dx = 1/v.$$

Notice that a light ray (photon) always travels on a 45° line in this diagram.

---

5   We use the word 'space' in a more general way than you may be used to. We do not mean a Euclidean space in which Euclidean distances are necessarily physically meaningful. Rather, we mean just that it is a set of points which is continuous (rather than discrete, as a lattice is). This is the first example of what we will define in Ch. 5 to be a 'manifold'.

We shall adopt the following notational conventions:

(1) *Events* will be denoted by cursive capitals, e.g. $\mathcal{A}$, $\mathcal{B}$, $\mathcal{P}$. However, the letter $\mathcal{O}$ is reserved to denote observers.

(2) The *coordinates* will be called (t, x, y, z). Any quadruple of numbers like $(5, -3, 2, 10^{16})$ denotes an event whose coordinates are $t = 5$, $x = -3$, $y = 2$, $z = 10^{16}$. Thus, we always put t first. All coordinates are measured in meters.

(3) It is often convenient to refer to the coordinates (t, x, y, z) as a whole, or to each indifferently. That is why we give them the *alternative names* (x^0, x^1, x^2, x^3). These superscripts are *not* exponents, but just labels, called indices. Thus $(x^3)^2$ denotes the square of coordinate 3 (which is z), not the square of the cube of x. *Generically*, the coordinates x^0, x^1, x^2, and x^3 are referred to as x^α. A *Greek index* (e.g. α, β, μ, ν) will be assumed to take a value from the set $(0, 1, 2, 3)$. If α is not given a value, then x^α is *any* of the four coordinates.

(4) There are occasions when we want to distinguish between t on the one hand and (x, y, z) on the other. We use *Latin indices* to refer to the spatial coordinates alone. Thus a Latin index (e.g. a, b, i, j, k, l) will be assumed to take a value from the set $(1, 2, 3)$. If i is not given a value, then x^i is *any* of the three spatial coordinates. Our conventions on the use of Greek and Latin indices are by no means universally used by physicists. Some books reverse them, using Latin for $\{0, 1, 2, 3\}$ and Greek for $\{1, 2, 3\}$; others use a, b, c, ... for one set and i, j, k for the other. Students should always check the conventions used by whatever work they are reading.

1.5 Construction of the coordinates used by another observer

Since any observer is simply a coordinate system for spacetime, and since all observers look at the same events (the same spacetime), it should be possible to draw the coordinate lines of one observer on the spacetime diagram drawn by another observer. To do this we have to make use of the postulates of SR.

Suppose an observer $\mathcal{O}$ uses the coordinates t, x as above, and that another observer $\bar{\mathcal{O}}$, with coordinates $\bar{t}$, $\bar{x}$, is moving with velocity v in the x direction relative to $\mathcal{O}$. Where do the coordinate axes for $\bar{t}$ and $\bar{x}$ go in the spacetime diagram of $\mathcal{O}$?

$\bar{t}$ *axis*: This is the locus of events at constant $\bar{x} = 0$ (and $\bar{y} = \bar{z} = 0$, too, but we shall ignore them here), which is the locus of the origin of $\bar{\mathcal{O}}$'s spatial coordinates. This is $\bar{\mathcal{O}}$'s world line, and it looks like that shown in Fig. 1.2.

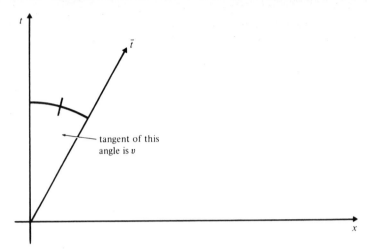

Fig. 1.2   The time-axis of a frame whose velocity is v.

$\bar{x}$ *axis*: To locate this we make a construction designed to determine the locus of events at $\bar{t} = 0$, i.e. those that $\bar{O}$ measures to be simultaneous with the event $\bar{t} = \bar{x} = 0$.

Consider the picture in $\bar{O}$'s spacetime diagram, shown in Fig. 1.3. The events on the $\bar{x}$ axis all have the following property: A light ray emitted at event $\mathscr{E}$ from $\bar{x} = 0$ at, say, time $\bar{t} = -a$ will reach the $\bar{x}$ axis at $\bar{x} = a$ (we call this event $\mathscr{P}$); if reflected, it will return to the point $\bar{x} = 0$ at $\bar{t} = +a$, called event $\mathscr{R}$. The $\bar{x}$ axis can be *defined*, therefore, as the locus of events that reflect light rays in such a manner that they return to the

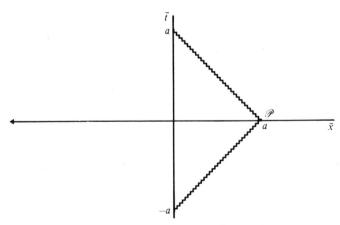

Fig. 1.3   Light reflected at a, as measured by $\bar{O}$.

$\bar{t}$ axis at $+a$ if they left it at $-a$, for any a. Now look at this in the spacetime diagram of $\mathcal{O}$, Fig. 1.4.

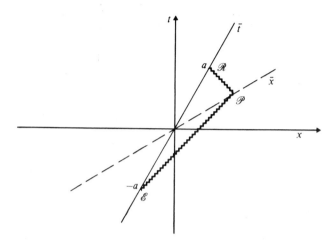

Fig. 1.4   The reflection in Fig. 1.3, as measured $\mathcal{O}$.

We know where the $\bar{t}$ axis lies, since we constructed it in Fig. 1.2. The events of emission and reception, $\bar{t} = -a$ and $\bar{t} = +a$, are shown in Fig. 1.4. Since a is arbitrary, it does not matter where along the negative $\bar{t}$ axis we place event $\mathcal{E}$, so no assumption need yet be made about the calibration of the $\bar{t}$ axis relative to the t axis. All that matters for the moment is that the event $\mathcal{R}$ on the $\bar{t}$ axis must be as far from the origin as event $\mathcal{E}$. Having drawn them in Fig. 1.4, we next draw in the same light beam as before, emitted from $\mathcal{E}$, and traveling on a 45° line in this diagram. The reflected light beam must arrive at $\mathcal{R}$, so it is the 45° line with negative slope through $\mathcal{R}$. The intersection of these two light beams must be the event of reflection $\mathcal{P}$. This establishes the location of $\mathcal{P}$ in our diagram. The line joining it with the origin – the dashed line – must be the $\bar{x}$ axis: it does *not* coincide with the x axis. If you compare this diagram with the previous one you will see why: in both diagrams light moves on a 45° line, while the t and $\bar{t}$ axes change slope from one diagram to the other. This is the embodiment of the second fundamental postulate of SR: that the light beam in question has speed $c = 1$ (and hence slope $= 1$) with respect to *every* observer. When we apply this to these geometrical constructions we immediately find that the events simultaneous to $\bar{\mathcal{O}}$ (the line $\bar{t} = 0$, his x axis) are not simultaneous to $\mathcal{O}$ (are not parallel to the line $t = 0$, the x axis). This *failure of simultaneity* is inescapable.

The following diagrams (Fig. 1.5) represent the same physical situation. The one on the left is the spacetime diagram O, in which $\bar{O}$ moves to the right. The one on the right is drawn from the point of view of $\bar{O}$, in which O moves to the left. The four angles are all equal to arc tan $|v|$, where $|v|$ is the relative speed of O and $\bar{O}$.

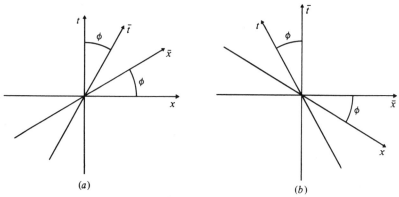

(a)   (b)

Fig. 1.5   Spacetime diagrams of O (left) and $\bar{O}$ (right).

1.6   Invariance of the interval

We have, of course, not quite finished the construction of $\bar{O}$'s coordinates. We have the position of his axes but not the length scale along them. We shall find this scale by proving what is probably the most important theorem of SR, the invariance of the interval.

Consider two events on the world line of the same light beam, such as $\mathscr{E}$ and $\mathscr{P}$ in Fig. 1.4. The differences between the coordinates of $\mathscr{E}$ and $\mathscr{P}$ in some frame O ($\Delta t, \Delta x, \Delta y, \Delta z$) satisfy the relation $(\Delta x)^2 + (\Delta y)^2 + (\Delta z)^2 - (\Delta t)^2 = 0$, since the speed of light is 1. But by the universality of the speed of light, the coordinate differences between the same two events in the coordinates of $\bar{O}$ ($\Delta \bar{t}, \Delta \bar{x}, \Delta \bar{y}, \Delta \bar{z}$) also satisfy $(\Delta \bar{x})^2 + (\Delta \bar{y})^2 + (\Delta \bar{z})^2 - (\Delta \bar{t})^2 = 0$. We shall define the *interval* between *any* two events (not necessarily on the same light beam's world line) that are separated by coordinate increments ($\Delta t, \Delta x, \Delta y, \Delta z$) to be[6]

◆ $$\Delta s^2 = -(\Delta t)^2 + (\Delta x)^2 + (\Delta y)^2 + (\Delta z)^2. \tag{1.1}$$

---

6   The student to whom this is new should probably regard the notation Δs^2 as a single symbol, not as the square of a quantity Δs. Since Δs^2 can be either positive or negative, it is not convenient to take its square root. Some authors do, however, call Δs^2 the 'squared interval', reserving the name 'interval' for $\Delta s = \sqrt{(\Delta s^2)}$. Note also that the notation Δs^2 *never* means $\Delta(s^2)$.

It follows that if $\Delta s^2 = 0$ for two events using their coordinates in $\mathcal{O}$, then $\Delta \bar{s}^2 = 0$ for the same two events using their coordinates in $\bar{\mathcal{O}}$. What does this imply about the relation between the coordinates of the two frames? To answer this question, we shall assume that the relation between the coordinates of $\mathcal{O}$ and $\bar{\mathcal{O}}$ is *linear* and that we choose their origins to coincide (i.e. that the events $\bar{t} = \bar{x} = \bar{y} = \bar{z} = 0$ and $t = x = y = z = 0$ are the same). Then in the expression for $\Delta \bar{s}^2$,

$$\Delta \bar{s}^2 = -(\Delta \bar{t})^2 + (\Delta \bar{x})^2 + (\Delta \bar{y})^2 + (\Delta \bar{z})^2,$$

the numbers $(\Delta \bar{t}, \Delta \bar{x}, \Delta \bar{y}, \Delta \bar{z})$ are linear combinations of their unbarred counterparts, which means that $\Delta \bar{s}^2$ is a *quadratic* function of the unbarred coordinate increments. We can therefore write

$$\Delta \bar{s}^2 = \sum_{\alpha=0}^{3} \sum_{\beta=0}^{3} M_{\alpha\beta}(\Delta x^{\alpha})(\Delta x^{\beta}) \tag{1.2}$$

for some numbers $\{M_{\alpha\beta}; \alpha, \beta = 0, \ldots, 3\}$, which may be functions of v, the relative velocity of the two frames. Note that we can suppose that $M_{\alpha\beta} = M_{\beta\alpha}$ for all α and β, since only the sum $M_{\alpha\beta} + M_{\beta\alpha}$ ever appears in Eq. (1.2) when $\alpha \neq \beta$. Now we again suppose that $\Delta s^2 = 0$, so that from Eq. (1.1) we have

$$\Delta t = \Delta r, \qquad \Delta r = [(\Delta x)^2 + (\Delta y)^2 + (\Delta z)^2]^{1/2}.$$

(We have supposed $\Delta t > 0$ for convenience.) Putting this into Eq. (1.2) gives

$$\Delta \bar{s}^2 = M_{00}(\Delta r)^2 + 2\left(\sum_{i=1}^{3} M_{0i}\Delta x^i\right)\Delta r$$

$$+ \sum_{i=1}^{3} \sum_{j=1}^{3} M_{ij}\Delta x^i \Delta x^j. \tag{1.3}$$

But we have already observed that $\Delta \bar{s}^2$ must vanish if Δs^2 does, and this must be true for *arbitrary* $\{\Delta x^i, i = 1, 2, 3\}$. It is easy to show (see Exer. 8, § 1.14) that this implies

$$M_{0i} = 0 \qquad i = 1, 2, 3 \tag{1.4a}$$

and

$$M_{ij} = -(M_{00})\delta_{ij} \qquad (i, j = 1, 2, 3), \tag{1.4b}$$

where δ_{ij} is the Kronecker delta, defined by

$$\delta_{ij} = \begin{cases} 1 & \text{if } i = j, \\ 0 & \text{if } i \neq j. \end{cases} \tag{1.4c}$$

From this and Eq. (1.2) we conclude that

$$\Delta \bar{s}^2 = M_{00}[(\Delta t)^2 - (\Delta x)^2 - (\Delta y)^2 - (\Delta z)^2].$$

a function

$) = -M_{00},$

'e proved the following theorem: *The universality of the speed
lies that the intervals Δs^2 and $\Delta \bar{s}^2$ between any two events as
computed by different observers satisfy the relation*

$$\Delta \bar{s}^2 = \phi(v)\Delta s^2. \tag{1.5}$$

We shall now show that, in fact, $\phi(v) = 1$, which is the statement that
the interval is independent of the observer. The proof of this has two
parts. The first part shows that $\phi(v)$ depends only on $|v|$. Consider a rod
which is oriented perpendicular to the velocity v of $\bar{O}$ relative to O.
Suppose the rod is at rest in O, lying on the y axis. In the spacetime
diagram of O (Fig. 1.6), the world lines of its ends are drawn and the

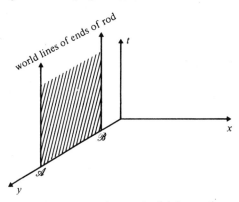

Fig. 1.6   A rod at rest in O, lying on the y-axis.

region between shaded. It is easy to see that the square of its length is
just the interval between the two events $\mathcal{A}$ and $\mathcal{B}$ that are simultaneous
in O (at $t = 0$) and occur at the ends of the rod. This is because, for these
events, $(\Delta x)_{\mathcal{A}\mathcal{B}} = (\Delta z)_{\mathcal{A}\mathcal{B}} = (\Delta t)_{\mathcal{A}\mathcal{B}} = 0$. Now comes the key point of the
first part of the proof: the events $\mathcal{A}$ and $\mathcal{B}$ are simultaneous as measured
by $\bar{O}$ as well. The reason is most easily seen by the construction shown
in Fig. 1.7, which is the same spacetime diagram as Fig. 1.6, but in which
the world line of a clock in $\bar{O}$ is drawn. This line is perpendicular to the
y axis and parallel to the $t-x$ plane, i.e. parallel to the $\bar{t}$ axis shown in
Fig. 1.5(a).

Suppose this clock emits light rays at event $\mathcal{P}$ which reach events $\mathcal{A}$
and $\mathcal{B}$. (Not every clock can do this, so we have chosen the one clock
in $\bar{O}$ which passes through the y axis at $t = 0$ and can send out such light
rays.) The light rays reflect from $\mathcal{A}$ and $\mathcal{B}$, and one can see from the

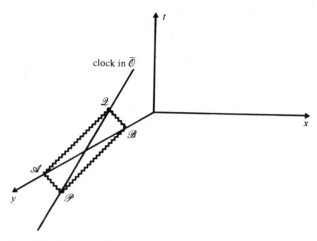

Fig. 1.7   A clock of $\bar{O}$'s frame, moving in the x-direction in O's frame.

geometry (if you can allow for the perspective in the diagram) that they arrive back at $\bar{O}$'s clock at the *same* event $\mathcal{2}$. Therefore, from $\bar{O}$'s point of view, the two events occur at the same time. (This is the *same* construction we used to determine the $\bar{x}$ axis.) But if $\mathcal{A}$ and $\mathcal{B}$ are simultaneous in $\bar{O}$, then the interval between them in $\bar{O}$ is also the square of their length in $\bar{O}$. The result is:

$$\text{(length of rod in } \bar{O})^2 = \phi(v)(\text{length of rod in } O)^2.$$

On the other hand, the length of the rod cannot depend on the *direction* of the velocity, because the rod is perpendicular to it and there are no preferred directions of motion (the principle of relativity). Hence the first part of the proof concludes that

$$\phi(v) = \phi(|v|). \tag{1.6}$$

The second step of the proof is easier. It uses the principle of relativity to show that $\phi(|v|) = 1$. Consider three frames, O, $\bar{O}$, and $\bar{\bar{O}}$. Frame $\bar{O}$ moves with speed v in, say, the x direction relative to O. Frame $\bar{\bar{O}}$ moves with speed v in the negative x direction relative to $\bar{O}$. It is clear that $\bar{\bar{O}}$ is in fact identical to O, but for the sake of clarity we shall keep separate notation for the moment. We have, from Eqs. (1.5) and (1.6),

$$\left.\begin{array}{l} \Delta\bar{\bar{s}}^2 = \phi(v)\Delta\bar{s}^2 \\ \Delta\bar{s}^2 = \phi(v)\Delta s^2 \end{array}\right\} \Rightarrow \Delta\bar{\bar{s}}^2 = [\phi(v)]^2\Delta s^2.$$

But since O and $\bar{\bar{O}}$ are identical, $\Delta\bar{\bar{s}}^2$ and Δs^2 are equal. It follows that

$$\phi(v) = \pm 1.$$

We must choose the plus sign, since in the first part of this proof the square of the length of a rod must be positive. We have therefore proved the fundamental theorem that *the interval between any two events is the same when calculated by any inertial observer*:

◆ $\Delta \bar{s}^2 = \Delta s^2.$ (1.7)

Notice that from the first part of this proof we can also conclude now that *the length of a rod oriented perpendicular to the relative velocity of two frames is the same when measured by either frame.* It is also worth reiterating that the construction in Fig. 1.7 above proved a related result, that *two events which are simultaneous in one frame are simultaneous in any frame moving in a direction perpendicular to their separation relative to the first frame.*

Because Δs^2 is a property only of the two events and not of the observer, it can be used to classify the relation between the events. If Δs^2 is positive (so that the spatial increments dominate Δt) the events are said to be *spacelike separated*. If Δs^2 is negative the events are said to be *timelike separated*. If Δs^2 is zero (so the events are on the same light path) the events are said to be *lightlike* or *null separated*.

The events that are lightlike separated from any particular event $\mathscr{A}$, lie on a cone whose apex is $\mathscr{A}$. This cone is illustrated in Fig. 1.8. This

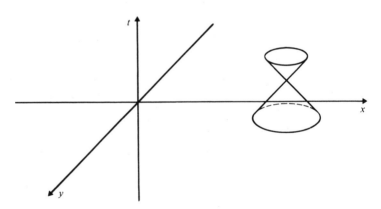

Fig. 1.8   The light cone of an event. The z-dimension is suppressed.

is called the *light cone of $\mathscr{A}$*. All events within the light cone are timelike separated from $\mathscr{A}$; all events outside it are spacelike separated. Therefore all events inside the cone can be reached from $\mathscr{A}$ on a world line which everywhere moves in a timelike direction. Since we will see later that nothing can move faster than light, all world lines of physical objects move in a timelike direction. Therefore events inside the light cone are reachable from $\mathscr{A}$ by a physical object, whereas those outside are not. For this reason the events inside the 'future' or 'forward' light cone are sometimes called the *absolute future* of the apex; those within the 'past' or 'backward' light cone are called the *absolute past*; and those outside are called the *absolute elsewhere*. The events on the cone are therefore the boundary of the absolute past and future. Thus, although 'time' and 'space' can in some sense be transformed into one another in SR, it is important to realize that we can still talk about 'future' and 'past' in an invariant manner. To Galileo and Newton the past was everything 'earlier' than now; all of spacetime was the union of the past and the future, whose boundary was 'now'. In SR, the past is only everything inside the past light cone, and spacetime has *three* invariant divisions: SR adds the notion of 'elsewhere'. What is more, although *all* observers agree on what constitutes the past, future and elsewhere of a given event (because the interval is invariant), each different event has a *different* past and future; no two events have identical pasts and futures, even though they can overlap. These ideas are illustrated in Fig. 1.9.

1.7   Invariant hyperbolae

We can now calibrate the axes of $\bar{\mathscr{O}}$'s coordinates in the spacetime diagram of $\mathscr{O}$, Fig. 1.5. We restrict ourselves to the t–x plane. Consider a curve with the equation

$$-t^2 + x^2 = a^2,$$

where a is a real constant. This is a hyperbola in the spacetime diagram of $\mathscr{O}$, and it passes through all events whose interval from the origin is a^2. By the invariance of the interval, these same events have interval a^2 from the origin in $\bar{\mathscr{O}}$, so they also lie on the curve $-\bar{t}^2 + \bar{x}^2 = a^2$. This is a hyperbola spacelike separated from the origin. Similarly, the events on the curve

$$-t^2 + x^2 = -b^2$$

all have timelike interval $-b^2$ from the origin, and also lie on the curve $-\bar{t}^2 + \bar{x}^2 = -b^2$. These hyperbolae are drawn in Fig. 1.10. They are all asymptotic to the lines with slope ± 1, which are of course the light paths through the origin. In a three-dimensional diagram (in which we add the y axis, as in Fig. 1.8), hyperbolae of revolution would be asymptotic to the light cone.

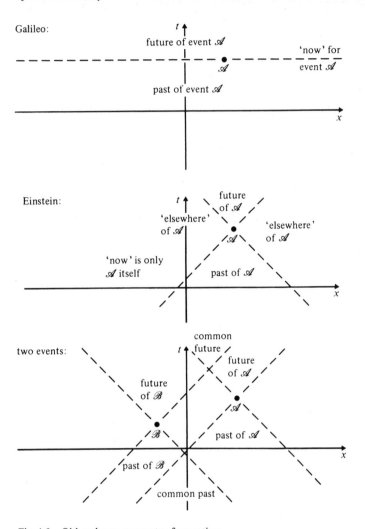

Fig. 1.9   Old and new concepts of spacetime.

We can now calibrate the axes of $\bar{O}$. In Fig. 1.11 are drawn the axes of O and $\bar{O}$, and an invariant hyperbola of timelike interval -1 from the origin. Event $\mathscr{A}$ is on the t axis, so has $x = 0$. Since the hyperbola has the equation

$$-t^2 + x^2 = -1,$$

it follows that event $\mathscr{A}$ has $t = 1$. Similarly, event $\mathscr{B}$ lies on the $\bar{t}$ axis, so has $\bar{x} = 0$. Since the hyperbola also has the equation

$$-\bar{t}^2 + \bar{x}^2 = -1,$$

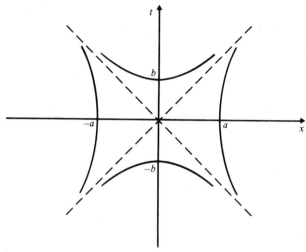

Fig. 1.10   Invariant hyperbolae, for $a > b$.

it follows that event $\mathscr{B}$ has $\bar{t} = 1$. We have therefore used the hyperbolae to calibrate the $\bar{t}$ axis. In the same way, the invariant hyperbola

$$-t^2 + x^2 = 4$$

shows that event $\mathscr{E}$ has coordinates $t = 0$, $x = 2$ and that event $\mathscr{F}$ has coordinates $\bar{t} = 0$, $\bar{x} = 2$. This kind of hyperbola calibrates the spatial axes of $\bar{O}$.

Notice that event $\mathscr{B}$ looks to be 'further' from the origin than $\mathscr{A}$. This again shows the inappropriateness of using geometrical intuition based

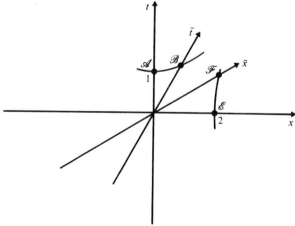

Fig. 1.11   Using the hyperbolae through events $\mathscr{A}$ and $\mathscr{E}$ to calibrate the $\bar{x}$ and $\bar{t}$ axes.

upon Euclidean geometry. Here the important physical quantity is the interval $-(\Delta t)^2 + (\Delta x)^2$, not the Euclidean distance $(\Delta t)^2 + (\Delta x)^2$. The student of relativity has to learn to use Δs^2 as the physical measure of 'distance' in spacetime, and he has to adapt his intuition accordingly. This is not, of course, in conflict with everyday experience. Everyday experience asserts that 'space' (e.g. the section of spacetime with $t = 0$) is Euclidean. For events that have $\Delta t = 0$ (simultaneous to observer $\mathcal{O}$), the interval is

$$\Delta s^2 = (\Delta x)^2 + (\Delta y)^2 + (\Delta z)^2.$$

This is just their Euclidean distance. The new feature of SR is that time can (and must) be brought into the computation of distance. It is not possible to define 'space' uniquely since different observers identify different sets of events to be simultaneous (Fig. 1.5). But there is still a distinction between space and time, since temporal increments enter Δs^2 with the opposite sign from spatial ones.

In order to use the hyperbolae to derive the effects of time dilation and Lorentz contraction, as we do in the next section, we must point out a simple but important property of the tangent to the hyperbolae.

In Fig. 1.12(a) we have drawn a hyperbola and its tangent at $x = 0$, which is obviously a line of simultaneity $t = \text{const}$. In Fig. 1.12(b) we have drawn the same curves from the point of view of observer $\bar{\mathcal{O}}$ who moves to the left relative to $\mathcal{O}$. The event $\mathcal{P}$ has been shifted to the right: it could be shifted anywhere on the hyperbola by choosing the Lorentz transformation properly. The lesson of Fig. 1.12(b) is that the tangent to a hyperbola at any event $\mathcal{P}$ is a line of simultaneity of the Lorentz frame whose time axis joins $\mathcal{P}$ to the origin. If this frame has velocity v, the tangent has slope $1/v$.

1.8   Particularly important results

Time dilation. From Fig. 1.11 and the calculation following it, we deduce that when a clock moving on the $\bar{t}$ axis reaches $\mathcal{B}$ it has a reading of $\bar{t} = 1$, but that event $\mathcal{B}$ has coordinate $t = 1/\sqrt{(1 - v^2)}$ in $\mathcal{O}$. So to $\mathcal{O}$ it appears to run slowly:

$$(\Delta t)_{\text{measured in } \mathcal{O}} = \frac{(\Delta \bar{t})_{\text{measured in } \bar{\mathcal{O}}}}{\sqrt{(1 - v^2)}}. \tag{1.8}$$

Notice that $\Delta \bar{t}$ is the time actually measured by a single clock which moves on a world line from the origin to $\mathcal{B}$, while Δt is the difference in the readings of two clocks at rest in $\mathcal{O}$, one on a world line through the origin and one on a world line through $\mathcal{B}$. We shall return to this

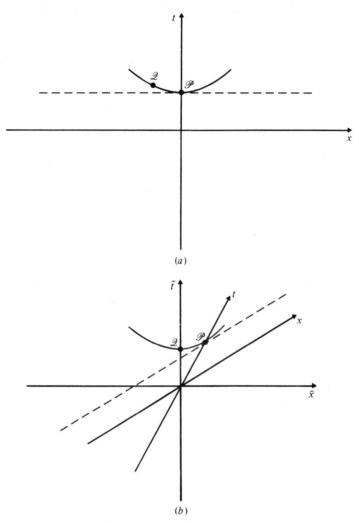

Fig. 1.12 (a) A line of simultaneity in O is tangent to the hyperbola at $\mathscr{P}$. (b) The same tangency as seen by $\bar{O}$.

observation later. For now, we define the *proper time* between events $\mathscr{B}$ and the origin to be the time ticked off by a clock which actually passes through both events. It is a directly measurable quantity, and it is closely related to the interval. Let the clock be at rest in frame $\bar{O}$, so that the proper time $\Delta\tau$ is the same as the coordinate time $\Delta\bar{t}$. Then, since the clock is at rest in $\bar{O}$, we have $\Delta\bar{x} = \Delta\bar{y} = \Delta\bar{z} = 0$, so

$$\Delta s^2 = -\Delta\bar{t}^2 = -\Delta\tau^2. \tag{1.9}$$

The proper time is just the square root of the negative of the interval. By expressing the interval in terms of O's coordinates we get

$$\Delta\tau = [(\Delta t)^2 - (\Delta x)^2 - (\Delta y)^2 - (\Delta z)^2]^{1/2}$$
$$= \Delta t\sqrt{(1 - v^2)}. \qquad (1.10)$$

This is the time dilation all over again.

Lorentz contraction. In Fig. 1.13 below we show the world path of a rod at rest in $\bar{O}$. Its length in $\bar{O}$ is the square root of $\Delta s^2_{\mathscr{A}\mathscr{C}}$, while its length

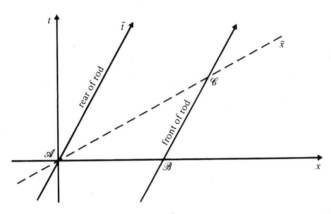

Fig. 1.13   The proper length of $\mathscr{A}\mathscr{C}$ is the length of the rod in its rest frame, while that of $\mathscr{A}\mathscr{B}$ is its length in O.

in O is the square root of $\Delta s^2_{\mathscr{A}\mathscr{B}}$. If event $\mathscr{C}$ has coordinates $\bar{t} = 0$, $\bar{x} = l$, then by the identical calculation from before it has x coordinate in O

$$x_{\mathscr{C}} = l/\sqrt{(1 - v^2)},$$

and since the $\bar{x}$ axis is the line $t = vx$, we have

$$t_{\mathscr{C}} = vl/\sqrt{(1 - v^2)}.$$

The line $\mathscr{B}\mathscr{C}$ has slope

$$\Delta x/\Delta t = v,$$

and so we have

$$\frac{x_{\mathscr{C}} - x_{\mathscr{B}}}{t_{\mathscr{C}} - t_{\mathscr{B}}} = v,$$

and we want to know $x_{\mathscr{B}}$ when $t_{\mathscr{B}} = 0$. Thus,

$$x_{\mathscr{B}} = x_{\mathscr{C}} - vt_{\mathscr{C}}$$
$$= \frac{l}{\sqrt{(1 - v^2)}} - \frac{v^2 l}{\sqrt{(1 - v^2)}} = l\sqrt{(1 - v^2)}. \qquad (1.11)$$

This is the Lorentz contraction.

Conventions. The interval Δs^2 is one of the most important mathematical concepts of SR but there is no universal agreement on its definition: many authors define $\Delta s^2 = (\Delta t)^2 - (\Delta x)^2 - (\Delta y)^2 - (\Delta z)^2$. This overall sign is a matter of convention (like the use of Latin and Greek indices we referred to earlier), since invariance of Δs^2 implies invariance of $-\Delta s^2$. The physical result of importance is just this invariance, which arises from the difference in sign between the $(\Delta t)^2$ and $[(\Delta x)^2 + (\Delta y)^2 + (\Delta z)^2]$ parts. As with other conventions, students should ensure they know which sign is being used: it affects all sorts of formulae, for example Eq. (1.9).

Failure of relativity? Newcomers to SR, and others who don't understand it well enough, often worry at this point that the theory is inconsistent. We began by assuming the principle of relativity, which asserts that all observers are equivalent. Now we have shown that if $\bar{\mathcal{O}}$ moves relative to $\mathcal{O}$, the clocks of $\bar{\mathcal{O}}$ will be measured by $\mathcal{O}$ to be running more slowly than those of $\mathcal{O}$. So isn't it therefore the case that $\bar{\mathcal{O}}$ will measure $\mathcal{O}$'s clocks to be running faster than his own? If so, this violates the principle of relativity, since we could as easily have begun with $\bar{\mathcal{O}}$ and deduced that $\mathcal{O}$'s clocks run more slowly than $\bar{\mathcal{O}}$'s.

This is what is known as a 'paradox', but like all 'paradoxes' in SR, this comes from not having reasoned correctly. We will now demonstrate, using spacetime diagrams, that $\bar{\mathcal{O}}$ measures $\mathcal{O}$'s clocks to be running more slowly. Clearly, one could simply draw the spacetime diagram from $\bar{\mathcal{O}}$'s point of view, and the result would follow. But it is more instructive to stay in $\mathcal{O}$'s spacetime diagram.

Different observers will agree on the outcome of certain kinds of experiments. For example, if A flips a coin, *every* observer will agree on the result. Similarly, if two clocks are right next to each other, all observers will agree which is reading an earlier time than the other. But the question of the *rate* at which clocks run can only be settled by comparing the same two clocks on two different occasions, and if the clocks are moving relative to one another then they can be next to each other on only one of these occasions. On the other occasion they must be compared over some distance, and different observers may draw different conclusions. The reason for this is that they actually perform different and inequivalent experiments. In the following analysis, we will see that each observer uses *two* of his own clocks and one of the other's. This asymmetry in the 'design' of the experiment gives the asymmetric result.

Let us analyze $\mathcal{O}$'s measurement first, in Fig. 1.14. This consists of comparing the reading on a single clock of $\bar{\mathcal{O}}$ (which travels from $\mathscr{A}$ to

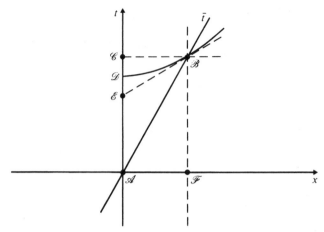

Fig. 1.14   The proper length of $\mathscr{AB}$ is the time ticked by a clock at rest in $\bar{O}$, while that of $\mathscr{AC}$ is the time it takes to do so as measured by O.

$\mathscr{B}$) with *two* clocks of his own: the first is the clock at the origin, which reads $\bar{O}$'s clock at event $\mathscr{A}$; and the second is the clock which is at $\mathscr{F}$ at $t = 0$ and coincides with $\bar{O}$'s clock at $\mathscr{B}$. This second clock of O moves parallel to the first one, on the vertical dashed line. What O says is that both clocks at $\mathscr{A}$ read $t = 0$, while at $\mathscr{B}$ the clock of $\bar{O}$ reads $\bar{t} = 1$, while that of O reads a later time, $t = (1 - v^2)^{-1/2}$. Clearly, $\bar{O}$ agrees with this, as he is just as capable of looking at clock dials as O is. But for O to claim that $\bar{O}$'s clock is running slowly, he must be sure that his own two clocks are synchronized, for otherwise there is no particular significance in observing that at $\mathscr{B}$ the clock of $\bar{O}$ lags behind that of O. Now, from O's point of view his clocks *are* synchronized, and the measurement and its conclusion are valid. Indeed, they are the only conclusions he can properly make.

But $\bar{O}$ need not accept them, because to him O's clocks are *not* synchronized. The dotted line through $\mathscr{B}$ is the locus of events that $\bar{O}$ regards as simultaneous to $\mathscr{B}$. Event $\mathscr{C}$ is on this line, and is the tick of O's first clock which $\bar{O}$ measures to be simultaneous with event $\mathscr{B}$. A simple calculation shows this to be at $t = (1 - v^2)^{1/2}$, earlier than O's other clock at $\mathscr{B}$, which is reading $(1 - v^2)^{-1/2}$. So $\bar{O}$ can reject O's measurement since the clocks involved aren't synchronized. Moreover, if $\bar{O}$ studies O's first clock, he concludes that it ticks from $t = 0$ to $t = (1 - v^2)^{1/2}$ (i.e. from $\mathscr{A}$ to $\mathscr{C}$) in the time it takes his own clock to tick from $\bar{t} = 0$ to $\bar{t} = 1$ (i.e. from $\mathscr{A}$ to $\mathscr{B}$). So he regards O's clocks as running more slowly than his own.

It follows that the principle of relativity is not contradicted: each observer measures the other's clock to be running slowly. The reason they seem to disagree is that they measure different things. Observer $\mathcal{O}$ compares the interval from $\mathcal{A}$ to $\mathcal{B}$ with that from $\mathcal{A}$ to $\mathcal{C}$. The other observer compares that from $\mathcal{A}$ to $\mathcal{B}$ with that from $\mathcal{A}$ to $\mathcal{E}$. All observers agree on the values of the intervals involved. What they disagree on is which pair to use in order to decide on the rate at which a clock is running. This disagreement arises directly from the fact that the observers do not agree on which events are simultaneous. And, to reiterate a point that needs to be understood, simultaneity (clock synchronization) is at the heart of clock comparisons: $\mathcal{O}$ uses *two* of his clocks to 'time' the rate of $\bar{\mathcal{O}}$'s one clock, whereas $\bar{\mathcal{O}}$ uses two of his own clocks to time one clock of $\mathcal{O}$.

Is this disagreement worrisome? It should not be, but it should make the student very cautious. The fact that different observers disagree on clock rates or simultaneity just means that such concepts are not invariant: they are coordinate dependent. It does *not* prevent any given observer from using such concepts consistently himself. For example, $\mathcal{O}$ can say that $\mathcal{A}$ and $\mathcal{F}$ are simultaneous, and he is correct in the sense that they have the same value of the coordinate t. For him this is a useful thing to know, as it helps locate the events in spacetime. Any single observer can make consistent observations using concepts which are valid for him but which may not transfer to other observers. All the so-called paradoxes of relativity involve, not the inconsistency of a single observer's deductions, but the inconsistency of assuming that certain concepts are independent of the observer when they are in fact very observer dependent.

Two more points should be made before we turn to the calculation of the Lorentz transformation. The first is that we have not had to define a 'clock', so our statements apply to any good timepiece: atomic clocks, wrist watches, circadian rhythm, or the half-life of the decay of an elementary particle. Truly, all time is 'slowed' by these effects. Put more properly, since time dilation is a consequence of the failure of simultaneity, it has nothing to do with the physical construction of the clock and it is certainly not noticeable to an observer who looks only at his *own* clocks. Observer $\bar{\mathcal{O}}$ sees all his clocks running at the same rate as each other and as his psychological awareness of time, so *all* these processes run more slowly as measured by $\mathcal{O}$. This leads to the twin 'paradox', which we discuss later.

The second point is that these effects are *not* optical illusions, since our observers exercise as much care as possible in performing their experiments. Beginning students often convince themselves that the problem arises in the finite transmission speed of signals, but this is incorrect. Observers define 'now' as described in § 1.5 for observer $\bar{O}$, and this is the most reasonable way to do it. The problem is that when two different observers each define 'now' in the most reasonable way, they don't agree. This is an inescapable consequence of the universality of the speed of light.

1.9   The Lorentz transformation

We shall now make our reasoning less dependent on geometrical logic by studying the algebra of SR: the Lorentz transformation, which expresses the coordinates of $\bar{O}$ in terms of those of O. Without losing generality, we orient our axes so that $\bar{O}$ moves with speed v on the positive x axis relative to O. We know that lengths perpendicular to the x axis are the same when measured by O or $\bar{O}$. The most general linear transformation we need consider, then, is

$$\bar{t} = \alpha t + \beta x \qquad \bar{y} = y,$$
$$\bar{x} = \gamma t + \sigma x \qquad \bar{z} = z,$$

where α, β, γ, and σ depend only on v.

From our construction in § 1.5 (Fig. 1.4) it is clear that the $\bar{t}$ and $\bar{x}$ axes have the equations:

$$\bar{t} \text{ axis } (\bar{x} = 0): vt - x = 0,$$
$$\bar{x} \text{ axis } (\bar{t} = 0): vx - t = 0.$$

The equations of the axes imply, respectively:

$$\gamma/\sigma = -v, \beta/\alpha = -v,$$

which gives the transformation

$$\bar{t} = \alpha(t - vx),$$
$$\bar{x} = \sigma(x - vt).$$

Fig. 1.4 gives us one other bit of information: events $(\bar{t} = 0, \bar{x} = a)$ and $(\bar{t} = a, \bar{x} = 0)$ are connected by a light ray. This can easily be shown to imply that $\alpha = \sigma$. Therefore we have, just from the geometry:

$$\bar{t} = \alpha(t - vx),$$
$$\bar{x} = \alpha(x - vt).$$

Now we use the invariance of the interval:

$$-(\Delta\bar{t})^2 + (\Delta\bar{x})^2 = -(\Delta t)^2 + (\Delta x)^2.$$

This gives, after some straightforward algebra,

$$\alpha = \pm 1/\sqrt{(1 - v^2)}.$$

We must select the + sign so that when $v = 0$ we get an identity rather than an inversion of the coordinates. The complete Lorentz transformation is, therefore,

♦ $$\bar{t} = \frac{t}{\sqrt{(1 - v^2)}} - \frac{vx}{\sqrt{(1 - v^2)}},$$

♦ $$\bar{x} = \frac{-vt}{\sqrt{(1 - v^2)}} + \frac{x}{\sqrt{(1 - v^2)}},$$

♦ $$\bar{y} = y,$$ $\qquad\qquad$ (1.12)

♦ $$\bar{z} = z.$$

This is called a *boost* of velocity v in the x direction.

This gives the simplest form of the relation between the coordinates of $\bar{O}$ and O. For this form to apply, the spatial coordinates must be oriented in a particular way: $\bar{O}$ must move with speed v in the positive x direction as seen by O, and the axes of $\bar{O}$ must be parallel to the corresponding ones in O. Spatial rotations of the axes relative to one another produce more complicated sets of equations than Eq. (1.12), but we will be able to get away with Eq. (1.12).

1.10 The velocity-composition law

The Lorentz transformation contains all the information one needs to derive the standard formulae, such as those for time dilation and Lorentz contraction. As an example of its use we will generalize the Galilean law of addition of velocities (§ 1.1).

Suppose a particle has speed W in the $\bar{x}$ direction of $\bar{O}$, i.e. $\Delta\bar{x}/\Delta\bar{t} = W$. In another frame O its velocity will be $W' = \Delta x/\Delta t$, and we can deduce Δx and Δt from the Lorentz transformation. If $\bar{O}$ moves with velocity v with respect to O, then Eq. (1.12) implies $\Delta x = (\Delta\bar{x} + v\Delta\bar{t})/(1 - v^2)^{1/2}$ and $\Delta t = (\Delta\bar{t} + v\Delta\bar{x})/(1 - v^2)^{1/2}$. Then we have

$$W' = \frac{\Delta x}{\Delta t} = \frac{(\Delta\bar{x} + v\Delta\bar{t})/(1 - v^2)^{1/2}}{(\Delta\bar{t} + v\Delta\bar{x})/(1 - v^2)^{1/2}}$$

$$= \frac{\Delta\bar{x}/\Delta\bar{t} + v}{1 + v\Delta\bar{x}/\Delta\bar{t}} = \frac{W + v}{1 + Wv}. \qquad (1.13)$$

This is the Einstein law of composition of velocities. The important point is that $|W'|$ never exceeds 1 if $|W|$ and $|v|$ are both smaller than 1. To see this, set $W' = 1$. Then Eq. (1.13) implies

$$(1 - v)(1 - W) = 0,$$

i.e. that either v or W must also equal 1. Therefore, two 'subluminal' velocities produce another subluminal one. Moreover, if $W = 1$ then

$W' = 1$ independently of v: this is the universality of the speed of light. What is more, if $|W| \ll 1$ and $|v| \ll 1$, then, to first order, Eq. (1.13) gives

$$W' = W + v.$$

This is the Galilean law of velocity addition, which we know to be valid for small velocities. This was true for our previous formulae in § 1.8: the relativistic 'corrections' to the Galilean expressions are of order v^2, and so are negligible for small v.

1.11 Paradoxes and physical intuition

Elementary introductions to SR often try to illustrate the physical differences between Galilean relativity and SR by posing certain problems called 'paradoxes'. The commonest ones include the 'twin paradox', the 'pole-in-the-barn paradox', and the 'space-war paradox'. The idea is to pose these problems in language that makes predictions of SR seem inconsistent or paradoxical, and then to resolve them by showing that a careful application of the fundamental principles of SR leads to no inconsistencies at all: the paradoxes are apparent, not real, and result invariably from mixing Galilean concepts with modern ones. Unfortunately, the careless student (or the attentive student of a careless teacher) often comes away with the idea that SR does in fact lead to paradoxes. This is pure nonsense. Students should realize that all 'paradoxes' are really mathematically ill-posed problems, that SR is a perfectly consistent picture of spacetime which has been experimentally verified countless times in situations where gravitational effects can be neglected, and that SR forms the framework in which every modern physicist must construct his theories. (For the student who really wants to study a paradox in depth, the appendix to this chapter discusses the twin 'paradox'.)

Psychologically, the reason that newcomers to SR have trouble and perhaps give 'paradoxes' more weight than they deserve is that we have so little direct experience with velocities comparable to that of light (see Fig. 1.15). The only remedy is to solve problems in SR and to study carefully its 'counter-intuitive' predictions. One of the best methods for developing a modern intuition is to be completely familiar with the geometrical picture of SR: Minkowski space, the effect of Lorentz transformations on axes, and the 'picture' of such things as time dilation and Lorentz contraction. This geometrical picture should be in the back of your mind as we go on from here to study vector and tensor calculus; we shall bring it to the front again when we study GR.

'Its top speed is 186 mph – that's 1/3 600 000 the speed of light.'

Fig. 1.15 The speed of light is rather far from our usual experience!

1.12 Bibliography

There are many good introductions to SR, but a very readable one which has guided our own treatment and is far more detailed is Taylor & Wheeler (1966). More traditional introductions include Durrell (1960), French (1968), Sears & Brehme (1968), and Smith (1965). For treatments that take a more thoughtful look at the fundamentals of the theory, consult Arzeliès (1966), Bohm (1965), Borel (1960), Dixon (1978), Geroch (1978), Jammer (1969), or Williams (1968). Paradoxes are discussed in some detail by Arzeliès (1966), Marder (1971) and Terletskii (1968). For a scientific biography of Einstein, see Pais (1982).

Our interest in SR in this text is primarily because it is a simple special case of GR in which it is possible to develop the mathematics we shall later need. But SR is itself the underpinning of all the other fundamental theories of physics, such as electromagnetism and quantum theory, and as such it rewards much more study than we shall give it. See the advanced discussions in Synge (1965), Schrödinger (1950), Møller (1972), or Robertson & Noonan (1968).

The original papers on SR may be found in Kilmister (1970).

1.13   Appendix: A full discussion of the twin 'paradox'

The problem

Diana leaves her twin Artemis behind on Earth and travels in her rocket for 2.2×10^8 s (≈ 7 yr) of *her* time at $24/25 = 0.96$ the speed of light. She then instantaneously reverses her direction (fearlessly braving those *g*s) and returns to Earth in the same manner. Who is older at the reunion of the twins? A spacetime diagram can be very helpful.

Brief solution. Refer to Fig. 1.16 below. Diana travels out on line $\mathcal{PB}$. In her frame, Artemis' event $\mathcal{A}$ is simultaneous with event $\mathcal{B}$, so Artemis is indeed ageing slowly. But as soon as Diana turns around she changes inertial reference frames: now she regards $\mathcal{B}$ as simultaneous with Artemis' event $\mathcal{C}$! Effectively, Diana sees Artemis age incredibly quickly for a moment. This one spurt more than makes up for the slowness Diana observed all along. Numerically, Artemis ages 50 years for Diana's 14.

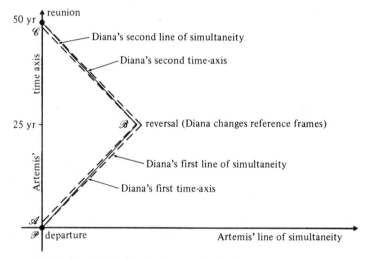

Fig. 1.16   The idealized twin 'paradox' in the spacetime diagram of the stay-at-home twin.

Fuller discussion. For readers who are unsatisfied with the statement 'Diana sees Artemis age incredibly quickly for a moment', or who wonder what physics lies underneath such a statement, we will discuss this in more detail, bearing in mind that the statement 'Diana sees' really means 'Diana observes', using the rods, clocks, and data bank that every good relativistic observer has.

Diana might make her measurements in the following way. Blasting off from Earth, she leaps on to an inertial frame called $\bar{O}$ rushing away from the Earth at $v = 0.96$. As soon as she gets settled in this new frame she orders all clocks synchronized with hers, which read $\bar{t} = 0$ upon leaving Earth. She further places a graduate student on every one of her clocks and orders each of them who rides a clock that passes Earth to note the time on Earth's clock at the event of passage. After traveling seven years by her own watch, she leaps off inertial frame $\bar{O}$ and grabs hold of another one $\bar{\bar{O}}$ that is flying *toward* Earth at $v = 0.96$ (measured in Earth's frame, of course). When she settles into this frame she again distributes her graduate students on the clocks and orders all clocks to be synchronized with hers, which read $\bar{\bar{t}} = 7$ yr at the changeover. (All clocks were already synchronized with each other – she just adjusts only their zero of time.) She further orders that every graduate student who passes Earth from $\bar{\bar{t}} = 7$ yr until she gets there herself should record the time of passage and the reading of Earth's clocks at that event.

Diana finally arrives home after ageing 14 years. Knowing a little about time dilation, she expects Artemis to have aged much less, but to her surprise Artemis is a wizened old prune, a full 50 years older! Diana keeps her surprise to herself and runs over to the computer room to check out the data. She reads the dispatches from the graduate students riding the clocks of the outgoing frame. Sure enough, Artemis seems to have aged very slowly by their reports. At Diana's time $\bar{t} = 7$ yr, the graduate student passing Earth recorded that Earth's clocks read only slightly less than two years of elapsed time. But then Diana checks the information from her graduate students riding the clocks of the ingoing frame. She finds that at her time $\bar{\bar{t}} = 7$ yr, the graduate student reported a reading of Earth's clocks at more than *48 years* of elapsed time! How could one student see Earth to be at $t = 2$ yr, and another student, *at the same time*, see it at $t = 48$ yr? Diana leaves the computer room muttering about the declining standards of undergraduate education today.

We know the mistake Diana made, however. Her two messengers did *not* pass Earth at the same time. Their clocks read the same amount, but they encountered Earth at the very different events $\mathcal{A}$ and $\mathcal{C}$. Diana should have asked the first frame students to continue recording information until they saw the second frame's $\bar{\bar{t}} = 7$ yr student pass Earth. What does it matter, after all, that they would have sent her dispatches dated $\bar{t} = 171$ yr? Time is only a coordinate. One must be sure to catch *all* the events.

What Diana really did was use a bad coordinate system. By demanding information only before $\bar{t} = 7$ yr in the outgoing frame and only after $\bar{\bar{t}} = 7$ yr in the ingoing frame, she left the whole interior of the triangle $\mathcal{ABC}$ out of her coordinate patches (Fig. 1.17(a)). Small wonder that a lot happened that she did not discover! Had she allowed the first frame's students to gather data until $\bar{t} = 171$ yr, she could have covered the interior of that triangle.

One can devise an analogy with rotations in the plane (Fig. 1.17(b)). Consider trying to measure the length of the curve $ABCD$, but being forced to rotate coordinates in the middle of the measurement, say after you have measured from A to B in the x–y system. If you then rotate to $\bar{x}$–$\bar{y}$, you must resume the measuring

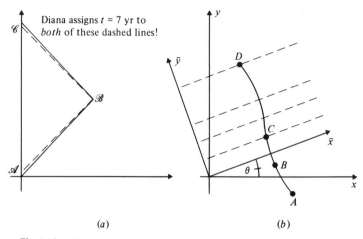

(a)   (b)

Fig. 1.17   Diana's change of frame is analogous to a rotation of coordinates in Euclidean geometry.

at B again, which might be at a coordinate $\bar{y} = -5$, whereas originally B had coordinate $y = 2$. If you were to measure the curve's length starting at whatever point had $\bar{y} = 2$ (same $\bar{y}$ as the y value you ended at in the other frame), you would begin at C and get much too short a length for the curve.

Now, nobody would make that error in measurements in a plane. But lots of people would if they were confronted by the twin paradox. This comes from our refusal to see time as simply a coordinate. We are used to thinking of a universal time, the same everywhere to everyone regardless of their motion. But it is not the same to everyone, and one must treat it as a coordinate, and make sure that one's coordinates cover all of spacetime.

Coordinates that do not cover all of spacetime have caused a lot of problems in GR. When we study gravitational collapse and black holes we will see that the usual coordinates for the spacetime outside the black hole do not reach inside the black hole. For this reason, a particle falling into a black hole takes infinite coordinate time to go a finite distance. This is purely the fault of the coordinates: the particle falls in a finite proper time, into a region not covered by the 'outside' coordinates. A coordinate system that covers both inside and outside satisfactorily was not discovered until the mid-1950s.

1.14   Exercises

1   Convert the following to units in which $c = 1$, expressing everything in terms of m and kg:

(a)   Worked example: 10 J. In SI units, $10\,\text{J} = 10\,\text{kg}\,\text{m}^2\,\text{s}^{-2}$. Since $c = 1$, we have $1\,\text{s} = 3 \times 10^8\,\text{m}$, and so $1\,\text{s}^{-2} = (9 \times 10^{16})^{-1}\,\text{m}^{-2}$. Therefore we get

$10 \text{ J} = 10 \text{ kg m}^2 (9 \times 10^{16})^{-1} \text{ m}^{-2} = 1.1 \times 10^{-16} \text{ kg}$. Alternatively, treat c as a conversion factor:

$$1 = 3 \times 10^8 \text{ m s}^{-1},$$
$$1 = (3 \times 10^8)^{-1} \text{ m}^{-1} \text{ s},$$
$$10 \text{ J} = 10 \text{ kg m}^2 \text{ s}^{-2} = 10 \text{ kg m}^2 \text{ s}^{-2} \times (1)^2$$
$$= 10 \text{ kg m}^2 \text{ s}^{-2} \times (3 \times 10^8)^{-2} \text{ s}^2 \text{ m}^{-2}$$
$$= 1.1 \times 10^{-16} \text{ kg}.$$

One is allowed to multiply or divide by as many factors of c as are necessary to cancel out the seconds.

(b) The power output of 100 W.
(c) Planck's constant, $\hbar = 1.05 \times 10^{-34} \text{ J s}$.
(d) Velocity of a car, $v = 30 \text{ m s}^{-1}$.
(e) Momentum of a car, $3 \times 10^4 \text{ kg m s}^{-1}$.
(f) Pressure of one atmosphere $= 10^5 \text{ N m}^{-2}$.
(g) Density of water, 10^3 kg m^{-3}.
(h) Luminosity flux $10^6 \text{ J s}^{-1} \text{ cm}^{-2}$.

2 Convert the following from natural units ($c = 1$) to SI units:
(a) A velocity $v = 10^{-2}$.
(b) Pressure $10^{19} \text{ kg m}^{-3}$.
(c) Time $t = 10^{18} \text{ m}$.
(d) Energy density $u = 1 \text{ kg m}^{-3}$.
(e) Acceleration 10 m^{-1}.

3 Draw the t and x axes of the spacetime coordinates of an observer $\mathcal{O}$ and then draw:
(a) The world line of $\mathcal{O}$'s clock at $x = 1 \text{ m}$.
(b) The world line of a particle moving with velocity $dx/dt = 0.1$, and which is at $x = 0.5 \text{ m}$ when $t = 0$.
(c) The $\bar{t}$ and $\bar{x}$ axes of an observer $\bar{\mathcal{O}}$ who moves with velocity $v = 0.5$ in the positive x direction relative to $\mathcal{O}$ and whose origin ($\bar{x} = \bar{t} = 0$) coincides with that of $\mathcal{O}$.
(d) The locus of events whose interval Δs^2 from the origin is -1 m^2.
(e) The locus of events whose interval Δs^2 from the origin is $+1 \text{ m}^2$.
(f) The calibration ticks at one meter intervals along the $\bar{x}$ and $\bar{t}$ axes.
(g) The locus of events whose interval Δs^2 from the origin is 0.
(h) The locus of events, all of which occur at the time $t = 2 \text{ m}$ (simultaneous as seen by $\mathcal{O}$).
(i) The locus of events, all of which occur at the time $\bar{t} = 2 \text{ m}$ (simultaneous as seen by $\bar{\mathcal{O}}$).
(j) The event which occurs at $\bar{t} = 0$ and $\bar{x} = 0.5 \text{ m}$.
(k) The locus of events $\bar{x} = 1 \text{ m}$.
(l) The world line of a photon which is emitted from the event $t = -1 \text{ m}$, $x = 0$, travels in the negative x direction, is reflected when it encounters a mirror located at $\bar{x} = -1 \text{ m}$, and is absorbed when it encounters a detector located at $x = 0.75 \text{ m}$.

4 Write out all the terms of the following sums, substituting the coordinate names (t, x, y, z) for (x^0, x^1, x^2, x^3):

(a) $\sum_{\alpha=0}^{3} V_\alpha \Delta x^\alpha$, where $\{V_\alpha, \ \alpha = 0, \ldots, 3\}$ is a collection of four arbitrary numbers.

(b) $\sum_{i=1}^{3} (\Delta x^i)^2$.

5(a) Use the spacetime diagram of an observer $\mathcal{O}$ to describe the following experiment performed by $\mathcal{O}$. Two bursts of particles of speed $v = 0.5$ are emitted from $x = 0$ at $t = -2$ m, one traveling in the positive x direction and the other in the negative x direction. These encounter detectors located at $x = \pm 2$ m. After a delay of 0.5 m of time, the detectors send signals back to $x = 0$ at speed $v = 0.75$.

(b) The signals arrive back at $x = 0$ at the same event. (Make sure your spacetime diagram shows this!) From this the experimenter concludes that the particle detectors did indeed send out their signals simultaneously, since he knows they are equal distances from $x = 0$. Explain why this conclusion is valid.

(c) A second observer $\bar{\mathcal{O}}$ moves with speed $v = 0.75$ in the *negative x* direction relative to $\mathcal{O}$. Draw the spacetime diagram of $\bar{\mathcal{O}}$ and in it depict the experiment performed by $\mathcal{O}$. Does $\bar{\mathcal{O}}$ conclude that particle detectors sent out their signals simultaneously? If not, which signal was sent first?

(d) Compute the interval Δs^2 between the events at which the detectors emitted their signals, using both the coordinates of $\mathcal{O}$ and those of $\bar{\mathcal{O}}$.

6 Show that Eq. (1.2) contains only $M_{\alpha\beta} + M_{\beta\alpha}$ when $\alpha \neq \beta$, not $M_{\alpha\beta}$ and $M_{\beta\alpha}$ independently. Argue that this enables us to set $M_{\alpha\beta} = M_{\beta\alpha}$ without loss of generality.

7 In the discussion leading up to Eq. (1.2), assume that the coordinates of $\bar{\mathcal{O}}$ are given as the following linear combinations of those of $\mathcal{O}$:

$\bar{t} = \alpha t + \beta x$,

$\bar{x} = \mu t + \nu x$,

$\bar{y} = ay$,

$\bar{z} = bz$,

where α, β, μ, ν, a, and b may be functions of the velocity v of $\bar{\mathcal{O}}$ relative to $\mathcal{O}$, but they do not depend on the coordinates. Find the numbers $\{M_{\alpha\beta}, \alpha, b = 0, \ldots, 3\}$ of Eq. (1.2) in terms of α, β, μ, ν, a, and b.

8(a) Derive Eq. (1.3) from Eq. (1.2), for general $\{M_{\alpha\beta}, \alpha, \beta = 0, \ldots, 3\}$.

(b) Since $\Delta \bar{s}^2 = 0$ in Eq. (1.3) for any $\{\Delta x^i\}$, replace Δx^i by $-\Delta x^i$ in Eq. (1.3) and subtract the resulting equation from Eq. (1.3) to establish that $M_{0i} = 0$ for $i = 1, 2, 3$.

(c) Use Eq. (1.3) with $\Delta \bar{s}^2 = 0$ to establish Eq. (1.4b). (Hint: Δx, Δy, and Δz are arbitrary.)

9 Explain why the line $\mathscr{P}2$ in Fig. 1.7 is drawn in the manner described in the text.

10 For the pairs of events whose coordinates (t, x, y, z) in some frame are given below, classify their separations as timelike, spacelike, or null.
- (a) $(0, 0, 0, 0)$ and $(-1, 1, 0, 0)$,
- (b) $(1, 1, -1, 0)$ and $(-1, 1, 0, 2)$,
- (c) $(6, 0, 1, 0)$ and $(5, 0, 1, 0)$,
- (d) $(-1, 1, -1, 1)$ and $(4, 1, -1, 6)$.

11 Show that the hyperbolae $-t^2 + x^2 = a^2$ and $-t^2 + x^2 = -b^2$ are asymptotic to the lines $t = \pm x$, regardless of a and b.

12(a) Use the fact that the tangent to the hyperbola $\mathcal{DB}$ in Fig. 1.14 is the line of simultaneity for $\bar{O}$ to show that the time interval $\mathcal{AE}$ is shorter than the time recorded on $\bar{O}$'s clock as it moved from $\mathcal{A}$ to $\mathcal{B}$.
- (b) Calculate that
$$(\Delta s^2)_{\mathcal{AE}} = (1 - v^2)(\Delta s^2)_{\mathcal{AB}}.$$
- (c) Use (b) to show that $\bar{O}$ regards $\mathcal{O}$'s clocks to be running slowly, at just the 'right' rate.

13 The half-life of the elementary particle called the pi meson (or pion) is 2.5×10^{-8} s when the pion is at rest relative to the observer measuring its decay time. Show, by the principle of relativity, that pions moving at speed $v = 0.999$ must have a half-life of 5.6×10^{-7} s, as measured by an observer at rest.

14 Suppose that the velocity v of $\bar{O}$ relative to $\mathcal{O}$ is small, $|v| \ll 1$. Show that the time dilation, Lorentz contraction, and velocity-addition formulae can be approximated by, respectively:
- (a) $\Delta t \approx (1 + \tfrac{1}{2}v^2)\Delta \bar{t}$.
- (b) $\Delta x \approx (1 - \tfrac{1}{2}v^2)\Delta \bar{x}$.
- (c) $w' \approx w + v - wv (w + v)$ (with $|w| \ll 1$ as well).
- (d) What are the relative errors in these approximations when $|v| = w = 0.1$?

15 Suppose that the velocity v of $\bar{O}$ relative to $\mathcal{O}$ is nearly that of light, $|v| = 1 - \varepsilon, \ 0 < \varepsilon \ll 1$.
Show that the same formulae of Exer. 14 become
- (a) $\Delta t \approx \Delta \bar{t}/\sqrt{(2\varepsilon)}$,
- (b) $\Delta x \approx \Delta \bar{x}\sqrt{(2\varepsilon)}$,
- (c) $w' \approx 1 - \varepsilon(1 - w)/(1 + w)$.
- (d) What are the relative errors on these approximations when $\varepsilon = 0.1$ and $w = 0.9$?

16 Use the Lorentz transformation, Eq. (1.12), to derive (a) the time dilation, and (b) the Lorentz contraction formulae. Do this by identifying the pairs of events whose separations (in time or space) are to be compared, and then using the Lorentz transformation to accomplish the algebra that the invariant hyperbolae had been used for in the text.

17   A lightweight pole 20 m long lies on the ground next to a barn 15 m
long. An Olympic athlete picks up the pole, carries it far away, and
runs with it toward the end of the barn at a speed 0.8c. His friend
remains at rest, standing by the door of the barn. Attempt all parts of
this question, even if you can't answer some.

(a)   How long does the friend measure the pole to be, as it approaches the
barn?

(b)   The barn door is initially open, and immediately after the runner and
pole are entirely inside the barn, the friend shuts the door. How long
after the door is shut does the front of the pole hit the other end of the
barn, as measured by the friend? Compute the interval between the
events of shutting the door and hitting the wall. Is it spacelike, timelike,
or null?

(c)   In the reference frame of the runner, what is the length of the barn and
the pole?

(d)   Does the runner believe that the pole is entirely inside the barn when
its front hits the end of the barn? Can you explain why?

(e)   After the collision, the pole and runner come to rest relative to the barn.
From the friend's point of view, the 20 m pole is now inside a 15 m
barn, since the barn door was shut before the pole stopped. How is this
possible? Alternatively, from the runner's point of view, the collision
should have stopped the pole *before* the door closed, so the door could
not be closed at all. Was or was not the door closed with the pole inside?

(f)   Draw a spacetime diagram from the friend's point of view and use it
to illustrate and justify all your conclusions.

18(a)   The Einstein velocity-addition law, Eq. (1.13), has a simpler form if we
introduce the concept of the *velocity parameter u*, defined by the equation
$v = \tanh u$.

Notice that for $-\infty < u < \infty$, the velocity is confined to the acceptable
limits $-1 < v < 1$. Show that if

$v = \tanh u$

and

$w = \tanh U$,

then Eq. (1.13) implies

$w' = \tanh (u + U)$.

This means that velocity parameters add linearly.

(b)   Use this to solve the following problem. A star measures a second star
to be moving away at speed $v = 0.9c$. The second star measures a third
to be receding in the same direction at $0.9c$. Similarly, the third measures
a fourth, and so on, up to some large number N of stars. What is the
velocity of the Nth star relative to the first? Give an exact answer and
an approximation useful for large N.

19(a) Using the velocity parameter introduced in Exer. 1.18, show that the Lorentz transformation equations, Eq. (1.12), can be put in the form

$$\bar{t} = t \cosh u - x \sinh u, \qquad \bar{y} = y,$$
$$\bar{x} = -t \sinh u + x \cosh u, \qquad \bar{z} = z.$$

(b) Use the identity $\cosh^2 u - \sinh^2 u = 1$ to demonstrate the invariance of the interval from these equations.

(c) Draw as many parallels as you can between the geometry of spacetime and ordinary two-dimensional Euclidean geometry, where the coordinate transformation analogous to the Lorentz transformation is

$$\bar{x} = x \cos \theta + y \sin \theta$$
$$\bar{y} = -x \sin \theta + y \cos \theta.$$

What is the analogue of the interval? Of the invariant hyperbolae?

20 Write the Lorentz transformation equations in matrix form.

21(a) Show that if two events are timelike separated, there is a Lorentz frame in which they occur at the same point, i.e. at the same spatial coordinate values.

(b) Similarly, show that if two events are spacelike separated, there is a Lorentz frame in which they are simultaneous.

2
Vector analysis
in special relativity

2.1 Definition of a vector

For the moment we will use the notion of a vector that one carries over from Euclidean geometry, that a vector is something whose components transform like the coordinates do under a coordinate transformation. Later on we shall define vectors in a more satisfactory manner.

The typical vector is the displacement vector, which points from one event to another and has components equal to the coordinate differences:

◆ $$\Delta \vec{x} \underset{\mathcal{O}}{\rightarrow} (\Delta t, \Delta x, \Delta y, \Delta z). \tag{2.1}$$

In this line we have introduced several new notations: an *arrow* over a symbol denotes a vector (so that $\vec{x}$ is a vector having nothing particular to do with the coordinate x); the arrow after $\Delta \vec{x}$ means 'has components' and the $\mathcal{O}$ underneath it means 'in the frame $\mathcal{O}$'; the components will always be in the order t, x, y, z (equivalently, indices in the order 0, 1, 2, 3). The notation $\rightarrow_{\mathcal{O}}$ is used in order to emphasize the distinction between the vector and its components. The vector $\Delta \vec{x}$ is an arrow between two events, while the collection of components is a set of four coordinate-dependent numbers. We shall always emphasize the notion of a vector (and, later, any tensor) as a *geometrical object*: something which can be defined and (sometimes) visualized without referring to a specific coordinate system. Another important notation is

$$\Delta \vec{x} \underset{\mathcal{O}}{\rightarrow} \{\Delta x^\alpha\}, \qquad (2.2)$$

where by $\{\Delta x^\alpha\}$ we mean all of Δx^0, Δx^1, Δx^2, Δx^3. If we ask for this vector's components in another coordinate system, say the frame $\bar{\mathcal{O}}$, we write

$$\Delta \vec{x} \underset{\bar{\mathcal{O}}}{\rightarrow} \{\Delta x^{\bar{\alpha}}\}.$$

That is, we put a bar over the *index* to denote the new coordinates. The vector $\Delta \vec{x}$ is the *same*, and no new notation is needed for it when the frame is changed. Only the components of it change.[1] What *are* the new components $\Delta x^{\bar{\alpha}}$? We get them from the Lorentz transformation:

$$\Delta x^{\bar{0}} = \frac{\Delta x^0}{\sqrt{(1 - v^2)}} - \frac{v \Delta x^1}{\sqrt{(1 - v^2)}}, \quad \text{etc.}$$

Since this is a linear transformation, it can be written

$$\Delta x^{\bar{0}} = \sum_{\beta=0}^{3} \Lambda^{\bar{0}}{}_\beta \Delta x^\beta,$$

where $\{\Lambda^{\bar{0}}{}_\beta\}$ are four numbers, one for each value of β. In this case

$$\Lambda^{\bar{0}}{}_0 = 1/\sqrt{(1 - v^2)}, \qquad \Lambda^{\bar{0}}{}_1 = -v/\sqrt{(1 - v^2)},$$
$$\Lambda^{\bar{0}}{}_2 = \Lambda^{\bar{0}}{}_3 = 0.$$

A similar equation holds for $\Delta x^{\bar{1}}$, and so in general we write

$$\Delta x^{\bar{\alpha}} = \sum_{\beta=0}^{3} \Lambda^{\bar{\alpha}}{}_\beta \Delta x^\beta, \quad \text{for arbitrary } \bar{\alpha}. \qquad (2.3)$$

Now $\{\Lambda^{\bar{\alpha}}{}_\beta\}$ is a collection of 16 numbers, which constitutes the Lorentz transformation matrix. The reason we have written one index up and the other down will become clear when we study differential geometry. For now, it enables us to introduce the final bit of notation, the *Einstein summation convention*: Whenever an expression contains one index as a superscript and the *same* index as a subscript, a summation is implied over all values that index can take. That is,

$$A_\alpha B^\alpha \text{ and } T^\gamma E_{\gamma\alpha}$$

are shorthand for the summations

$$\sum_{\alpha=0}^{3} A_\alpha B^\alpha \text{ and } \sum_{\gamma=0}^{3} T^\gamma E_{\gamma\alpha},$$

---

1   This is what some books on linear algebra call a 'passive' transformation: the coordinates change, but the vector does not.

while

$$A_\alpha B^\beta, \quad T^\gamma E_{\beta\alpha}, \quad \text{and} \quad A_\beta A_\beta$$

do *not* represent sums on any index. The Lorentz transformation, Eq. (2.3), can now be abbreviated to

◆ $$\Delta x^{\bar\alpha} = \Lambda^{\bar\alpha}{}_\beta \Delta x^\beta,$$ (2.4)

saving some messy writing.

Notice that Eq. (2.4) is identically equal to

$$\Delta x^{\bar\alpha} = \Lambda^{\bar\alpha}{}_\gamma \Delta x^\gamma.$$

Since the repeated index (β in one case, γ in the other) merely denotes a summation from 0 to 3, it doesn't matter what letter is used. Such a summed index is called a *dummy index*, and relabeling a dummy index (as we have done, replacing β by γ) is often a useful tool in tensor algebra. There is only one thing one should *not* replace the dummy index β with: a Latin index. The reason is that Latin indices can (by our convention) only take the values 1, 2, 3, whereas β must be able to equal zero as well. Thus, the expressions

$$\Lambda^{\bar\alpha}{}_\beta \Delta x^\beta \quad \text{and} \quad \Lambda^{\bar\alpha}{}_i \Delta x^i$$

are not the same; in fact we have

◆ $$\Lambda^{\bar\alpha}{}_\beta \Delta x^\beta = \Lambda^{\bar\alpha}{}_0 \Delta x^0 + \Lambda^{\bar\alpha}{}_i \Delta x^i.$$ (2.5)

Eq. (2.4) is really four different equations, one for each value that $\bar\alpha$ can assume. An index like $\bar\alpha$, on which no sum is performed, is called a *free index*. Whenever an equation is written down with one or more free indices, it is valid if and only if it is true for *all* possible values the free indices can assume. As with a dummy index, the name given to a free index is largely arbitrary. Thus, Eq. (2.4) can be rewritten as

$$\Delta x^{\bar\gamma} = \Lambda^{\bar\gamma}{}_\beta \Delta x^\beta.$$

This is equivalent to Eq. (2.4) because $\bar\gamma$ can assume the same four values that $\bar\alpha$ could assume. If a free index is renamed, it must be renamed everywhere. For example, the following modification of Eq. (2.4),

$$\Delta x^{\bar\gamma} = \Lambda^{\bar\alpha}{}_\beta \Delta x^\beta,$$

makes no sense and should never be written. The difference between these last two expressions is that the first guarantees that, whatever value $\bar\gamma$ assumes, both $\Delta x^{\bar\gamma}$ on the left and $\Lambda^{\bar\gamma}{}_\beta$ on the right will have the *same* free index. The second expression does not link the indices in this way, so it is not equivalent to Eq. (2.4).

The *general vector*[2] is defined by a collection of numbers (its components in some frame, say $\mathcal{O}$)

$$\vec{A} \underset{\mathcal{O}}{\rightarrow} (A^0, A^1, A^2, A^3) = \{A^\alpha\}, \tag{2.6}$$

and by the rule that its components in a frame $\bar{\mathcal{O}}$ are

$$A^{\bar{\alpha}} = \Lambda^{\bar{\alpha}}{}_\beta A^\beta. \tag{2.7}$$

That is, its components transform the same way the coordinates do. Remember that a vector can be defined by giving four numbers (e.g. (10^8, -10^{-16}, 5.8368, π)) in some frame; then its components in all other frames are uniquely determined. Vectors in spacetime obey the usual rules: if $\vec{A}$ and $\vec{B}$ are vectors and μ is a number, then $\vec{A} + \vec{B}$ and $\mu\vec{A}$ are also vectors, with components

$$\left.\begin{aligned} \vec{A} + \vec{B} &\underset{\mathcal{O}}{\rightarrow} (A^0 + B^0, A^1 + B^1, A^2 + B^2, A^3 + B^3), \\ \mu\vec{A} &\underset{\mathcal{O}}{\rightarrow} (\mu A^0, \mu A^1, \mu A^2, \mu A^3). \end{aligned}\right\} \tag{2.8}$$

Thus, vectors add by the usual parallelogram rule. Notice that one can give any four numbers to make a vector, except that if the numbers are not dimensionless they must all have the same dimensions, since under a transformation they will be added together.

2.2   Vector algebra

Basis vectors. In any frame $\mathcal{O}$ there are four special vectors, defined by giving their components:

$$\left.\begin{aligned} \vec{e}_0 &\underset{\mathcal{O}}{\rightarrow} (1, 0, 0, 0), \\ \vec{e}_1 &\underset{\mathcal{O}}{\rightarrow} (0, 1, 0, 0), \\ \vec{e}_2 &\underset{\mathcal{O}}{\rightarrow} (0, 0, 1, 0), \\ \vec{e}_3 &\underset{\mathcal{O}}{\rightarrow} (0, 0, 0, 1). \end{aligned}\right\} \tag{2.9}$$

These definitions define the basis vectors of the frame $\mathcal{O}$. Similarly, $\bar{\mathcal{O}}$

---

2   Such a vector, with four components, is sometimes called a *four-vector* to distinguish it from the three-component vectors one is used to in elementary physics, which we shall call *three-vectors*. Unless we say otherwise, a 'vector' is always a four-vector. We denote four-vectors by arrows, e.g. $\vec{A}$, and three-vectors by boldface, e.g. **A**.

has basis vectors

$$\vec{e}_{\bar{0}} \underset{\bar{O}}{\to} (1, 0, 0, 0), \text{ etc.}$$

Generally, $\vec{e}_{\bar{0}} \neq \vec{e}_0$, since they are defined in different frames. The reader should verify that the definition of the basis vectors is equivalent to

$$(\vec{e}_\alpha)^\beta = \delta_\alpha{}^\beta, \tag{2.10}$$

that is, the β component of $\vec{e}_\alpha$ is the Kronocker delta: 1 if $\beta = \alpha$ and 0 if $\beta \neq \alpha$.

Any vector can be expressed in terms of the basis vectors. If

$$\vec{A} \underset{O}{\to} (A^0, A^1, A^2, A^3),$$

then

$$\vec{A} = A^0 \vec{e}_0 + A^1 \vec{e}_1 + A^2 \vec{e}_2 + A^3 \vec{e}_3,$$

◆ $$\vec{A} = A^\alpha \vec{e}_\alpha. \tag{2.11}$$

In the last line we use the summation convention (remember always to write the index on $\vec{e}$ as a *subscript* in order to employ the convention in this manner). The meaning of Eq. (2.11) is that $\vec{A}$ is the linear sum of four vectors $A^0 \vec{e}_0$, $A^1 \vec{e}_1$, etc.

Transformation of basis vectors. The discussion leading up to Eq. (2.11) could have been applied to any frame, so it is equally true in $\bar{O}$:

$$\vec{A} = A^{\bar{\alpha}} \vec{e}_{\bar{\alpha}}.$$

This says that $\vec{A}$ is also the sum of the four vectors $A^{\bar{0}} \vec{e}_{\bar{0}}$, $A^{\bar{1}} \vec{e}_{\bar{1}}$, etc. These are not the same four vectors as in Eq. (2.11), since they are parallel to the basis vectors of $\bar{O}$ and not of O, but they add up to the same vector $\vec{A}$. It is important to understand that the expressions $A^\alpha \vec{e}_\alpha$ and $A^{\bar{\alpha}} \vec{e}_{\bar{\alpha}}$ are not obtained from one another merely by relabeling dummy indices. Barred and unbarred indices cannot be interchanged, since they have different meanings. Thus, $\{A^{\bar{\alpha}}\}$ is a different set of numbers from $\{A^\alpha\}$, just as the set of vectors $\{\vec{e}_{\bar{\alpha}}\}$ is different from $\{\vec{e}_\alpha\}$. But, by definition, the two sums are the same:

$$A^\alpha \vec{e}_\alpha = A^{\bar{\alpha}} \vec{e}_{\bar{\alpha}}, \tag{2.12}$$

and this has an important consequence: from it we deduce the transformation law for the basis vectors, i.e. the relation between $\{\vec{e}_\alpha\}$ and $\{\vec{e}_{\bar{\alpha}}\}$. Using Eq. (2.7) for $A^{\bar{\alpha}}$, we write Eq. (2.12) as

$$\Lambda^{\bar{\alpha}}{}_\beta A^\beta \vec{e}_{\bar{\alpha}} = A^\alpha \vec{e}_\alpha.$$

On the left we have *two* sums. Since they are finite sums their order doesn't matter. Since the numbers $\Lambda^{\bar{\alpha}}{}_\beta$ and A^β *are* just numbers, their order doesn't matter, and we can write

$$A^\beta \Lambda^{\bar{\alpha}}{}_\beta \vec{e}_{\bar{\alpha}} = A^\alpha \vec{e}_\alpha.$$

Now we use the fact that β and $\bar{\alpha}$ are dummy indices: we change β to α and $\bar{\alpha}$ to $\bar{\beta}$,

$$A^\alpha \Lambda^{\bar{\beta}}{}_\alpha \vec{e}_{\bar{\beta}} = A^\alpha \vec{e}_\alpha.$$

This equation must be true for *all* sets $\{A^\alpha\}$, since $\vec{A}$ is an arbitrary vector. Writing it as

$$A^\alpha (\Lambda^{\bar{\beta}}{}_\alpha \vec{e}_{\bar{\beta}} - \vec{e}_\alpha) = 0$$

we deduce

$$\Lambda^{\bar{\beta}}{}_\alpha \vec{e}_{\bar{\beta}} - \vec{e}_\alpha = 0 \text{ for every value of } \alpha,$$

or

◆ $$\vec{e}_\alpha = \Lambda^{\bar{\beta}}{}_\alpha \vec{e}_{\bar{\beta}}. \tag{2.13}$$

This gives the law by which basis vectors change. It is *not* a component transformation: it gives the basis $\{\vec{e}_\alpha\}$ of $\mathcal{O}$ as a linear sum over the basis $\{\vec{e}_{\bar{\alpha}}\}$ of $\bar{\mathcal{O}}$. Comparing this to the law for components, Eq. (2.7),

$$A^{\bar{\beta}} = \Lambda^{\bar{\beta}}{}_\alpha A^\alpha,$$

we see that it is different indeed.

The above discussion introduced many new techniques, so study it carefully. Notice that the omission of the summation signs keeps things neat. Notice also that a step of key importance was relabeling the dummy indices: this allowed us to isolate the arbitrary A^α from the rest of the things in the equation.

An example. Let $\bar{\mathcal{O}}$ move with velocity v in the x direction relative to $\mathcal{O}$. Then the matrix $[\Lambda^{\bar{\beta}}{}_\alpha]$ is

$$[\Lambda^{\bar{\beta}}{}_\alpha] = \begin{bmatrix} \gamma & -v\gamma & 0 & 0 \\ -v\gamma & \gamma & 0 & 0 \\ 0 & 0 & 1 & 0 \\ 0 & 0 & 0 & 1 \end{bmatrix},$$

where we use the standard notation

$$\gamma \equiv 1/\sqrt{(1 - v^2)}.$$

Then, if $\vec{A} \underset{\mathcal{O}}{\rightarrow} (5, 0, 0, 2)$, we find its components in $\bar{\mathcal{O}}$ by

$$\begin{aligned} A^{\bar{0}} &= \Lambda^{\bar{0}}{}_0 A^0 + \Lambda^{\bar{0}}{}_1 A^1 + \cdots \\ &= \gamma \cdot 5 + (-v\gamma) \cdot 0 + 0 \cdot 0 + 0 \cdot 2 \\ &= 5\gamma. \end{aligned}$$

Similarly,

$$A^{\bar{1}} = -5v\gamma,$$
$$A^{\bar{2}} = 0,$$
$$A^{\bar{3}} = 2.$$

Therefore $\vec{A} \underset{\bar{O}}{\rightarrow} (5\gamma, -5v\gamma, 0, 2)$.

The basis vectors are expressible as

$$\vec{e}_\alpha = \Lambda^{\bar{\beta}}{}_\alpha \vec{e}_{\bar{\beta}}$$

or

$$\vec{e}_0 = \Lambda^{\bar{0}}{}_0 \vec{e}_{\bar{0}} + \Lambda^{\bar{1}}{}_0 \vec{e}_{\bar{1}} + \cdots$$
$$= \gamma \vec{e}_{\bar{0}} - v\gamma \vec{e}_{\bar{1}}.$$

Similarly,

$$\vec{e}_1 = -v\gamma \vec{e}_{\bar{0}} + \gamma \vec{e}_{\bar{1}},$$
$$\vec{e}_2 = \vec{e}_{\bar{2}},$$
$$\vec{e}_3 = \vec{e}_{\bar{3}}.$$

This gives O's basis in terms of $\bar{O}$'s, so let us draw the picture (Fig. 2.1) in $\bar{O}$'s frame: This transformation is of course exactly what is needed to keep the basis vectors pointing along the axes of their respective frames. Compare this with Fig. 1.5(b).

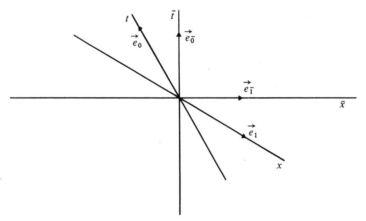

Fig. 2.1   Basis vectors of O and $\bar{O}$ as drawn by $\bar{O}$.

Inverse transformations. The only thing the Lorentz transformation $\Lambda^{\bar{\beta}}{}_\alpha$ depends on is the relative velocity of the two frames. Let us for the moment show this explicitly by writing

$$\Lambda^{\bar{\beta}}{}_\alpha = \Lambda^{\bar{\beta}}{}_\alpha(v).$$

Then

$$\vec{e}_\alpha = \Lambda^{\bar\beta}{}_\alpha(\boldsymbol{v})\vec{e}_{\bar\beta}. \tag{2.14}$$

If the basis of $\mathcal{O}$ is obtained from that of $\bar{\mathcal{O}}$ by the transformation with velocity $\boldsymbol{v}$, then the reverse must be true if we use $-\boldsymbol{v}$. Thus we must have

$$\vec{e}_{\bar\mu} = \Lambda^\nu{}_{\bar\mu}(-\boldsymbol{v})\vec{e}_\nu. \tag{2.15}$$

In this equation I have used $\bar\mu$ and ν as indices to avoid confusion with the previous formula. The bars still refer, of course, to the frame $\bar{\mathcal{O}}$. The matrix $[\Lambda^\nu{}_{\bar\mu}]$ is exactly the matrix $[\Lambda^{\bar\beta}{}_\alpha]$ except with $\boldsymbol{v}$ changed to $-\boldsymbol{v}$. The bars on the indices only serve to indicate the *names* of the observers involved: they affect the entries in the matrix $[\Lambda]$ only in that the matrix is *always* constructed using the velocity of the upper-index frame relative to the lower-index frame. This is made explicit in Eqs. (2.14) and (2.15). Since $\boldsymbol{v}$ is the velocity of $\bar{\mathcal{O}}$ (the upper-index frame in Eq. (2.14)) relative to $\mathcal{O}$, then $-\boldsymbol{v}$ is the velocity of $\mathcal{O}$ (the upper-index frame in Eq. (2.15)) relative to $\bar{\mathcal{O}}$. Exer. 11, § 2.9, will help you understand this point.

We can rewrite the last expression as

$$\vec{e}_{\bar\beta} = \Lambda^\nu{}_{\bar\beta}(-\boldsymbol{v})\vec{e}_\nu.$$

Here we have just changed $\bar\mu$ to $\bar\beta$. This doesn't change anything: it is still the same four equations, one for each value of $\bar\beta$. In this form we can put it into the expression for $\vec{e}_\alpha$, Eq. (2.14):

$$\vec{e}_\alpha = \Lambda^{\bar\beta}{}_\alpha(\boldsymbol{v})\vec{e}_{\bar\beta} = \Lambda^{\bar\beta}{}_\alpha(\boldsymbol{v})\Lambda^\nu{}_{\bar\beta}(-\boldsymbol{v})\vec{e}_\nu. \tag{2.16}$$

In this equation only the basis of $\mathcal{O}$ appears. It must therefore be an identity *for all* $\boldsymbol{v}$. On the right there are two sums, one on $\bar\beta$ and one on ν. If we imagine performing the $\bar\beta$ sum first, then the right is a sum over the basis $\{\vec{e}_\nu\}$ in which each basis vector $\vec{e}_\nu$ has coefficient

$$\sum_{\bar\beta} \Lambda^{\bar\beta}{}_\alpha(\boldsymbol{v})\Lambda^\nu{}_{\bar\beta}(-\boldsymbol{v}). \tag{2.17}$$

Imagine evaluating Eq. (2.16) for some fixed value of the index α. If the right side of Eq. (2.16) is equal to the left, the coefficient of $\vec{e}_\alpha$ on the *right* must be 1 and all other coefficients must vanish. The mathematical way of saying this is

$$\Lambda^{\bar\beta}{}_\alpha(\boldsymbol{v})\Lambda^\nu{}_{\bar\beta}(-\boldsymbol{v}) = \delta^\nu{}_\alpha,$$

where $\delta^\nu{}_\alpha$ is the Kronecker delta again. This would imply

$$\vec{e}_\alpha = \delta^\nu{}_\alpha \vec{e}_\nu,$$

which is an identity.

Let us change the order of multiplication above and write down the key formula

$$\blacklozenge \qquad \Lambda^\nu{}_{\bar\beta}(-\boldsymbol{v})\Lambda^{\bar\beta}{}_\alpha(\boldsymbol{v}) = \delta^\nu{}_\alpha. \tag{2.18}$$

This expresses the fact that the matrix $[\Lambda^{\nu}{}_{\bar{\beta}}(-\boldsymbol{v})]$ is the *inverse* of $[\Lambda^{\bar{\beta}}{}_{\alpha}(\boldsymbol{v})]$, because the sum on $\bar{\beta}$ is exactly the operation one performs when one multiplies two matrices. The matrix $(\delta^{\nu}{}_{\alpha})$ is, of course, the identity matrix.

The expression for the change of a vector's components,

$$A^{\bar{\beta}} = \Lambda^{\bar{\beta}}{}_{\alpha}(\boldsymbol{v})A^{\alpha},$$

also has its inverse. Let us multiply both sides by $\Lambda^{\nu}{}_{\bar{\beta}}(-\boldsymbol{v})$ and sum on $\bar{\beta}$. We get

$$\Lambda^{\nu}{}_{\bar{\beta}}(-\boldsymbol{v})A^{\bar{\beta}} = \Lambda^{\nu}{}_{\bar{\beta}}(-\boldsymbol{v})\Lambda^{\bar{\beta}}{}_{\alpha}(\boldsymbol{v})A^{\alpha}$$
$$= \delta^{\nu}{}_{\alpha}A^{\alpha}$$
$$= A^{\nu}.$$

This says that the components of $\vec{A}$ in $\mathcal{O}$ are obtained from those in $\bar{\mathcal{O}}$ by the transformation with $-\boldsymbol{v}$, which is, of course, correct.

The operations we have performed should be familiar to you in concept from vector algebra in Euclidean space. The new element we have introduced here is the index notation, which will be a permanent and powerful tool in the rest of the course. Make sure that you understand the geometrical meaning of all our results as well as their algebraic justification.

2.3   The four velocity

A particularly important vector is the four-velocity of a world line. In the three-geometry of Galileo, the velocity was a vector tangent to a particle's path. In our four-geometry we define the four-velocity $\vec{U}$ to be a vector tangent to the world line of the particle, and of such a length that it stretches one unit of time in that particle's frame. For a uniformly moving particle, let us look at this definition in the inertial frame in which it is at rest. Then the four-velocity points parallel to the time axis and is one unit of time long. That is, it is identical with $\vec{e}_0$ of that frame. Thus we could also use as our definition of the four-velocity of a uniformly moving particle that it is the vector $\vec{e}_0$ in its inertial rest frame. The word 'velocity' is justified by the fact that the spatial components of $\vec{U}$ are closely related to the particle's ordinary velocity $\boldsymbol{v}$, which is called the three-velocity. This will be demonstrated in the example below, Eq. (2.21).

An *accelerated particle* has no inertial frame in which it is always at rest. However, there *is* an inertial frame which momentarily has the same velocity as the particle, but which a moment later is of course no longer comoving with it. This frame is the *momentarily comoving reference frame* (MCRF), and is an important concept. (Actually, there are an infinity of

MCRFs for a given accelerated particle at a given event; they all have the same velocity, but their spatial axes are obtained from one another by rotations. This ambiguity will usually not be important.) The four-velocity of an accelerated particle is *defined* as the $\vec{e}_0$ basis vector of its MCRF at that event. This vector is tangent to the (curved) world line of the particle. In Fig. 2.2 the particle at event $\mathcal{A}$ has MCRF $\bar{O}$, whose basis vectors are shown. The vector $\vec{e}_{\bar{0}}$ is identical to $\vec{U}$ there.

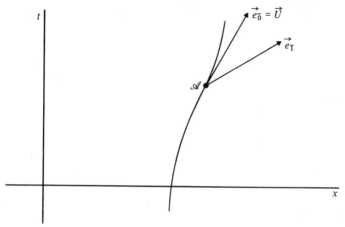

Fig. 2.2   The four-velocity and MCRF basis vectors of the world line at $\mathcal{A}$.

2.4   The four-momentum

The four-momentum $\vec{p}$ is defined as

$$\vec{p} = m\vec{U}, \tag{2.19}$$

where m is the *rest mass* of the particle, which is its mass as measured in its rest frame. In some frame O it has components conventionally denoted by

$$\vec{p} \underset{O}{\rightarrow} (E, p^1, p^2, p^3). \tag{2.20}$$

We call p^0 the *energy* E of the particle in the frame O. The other components are its spatial momentum p^i.

An example. A particle of rest mass m moves with velocity v in the x direction of frame O. What are the components of the four-velocity and four-momentum? Its rest frame $\bar{O}$ has time basis vector $\vec{e}_{\bar{0}}$, so, by definition of $\vec{p}$ and $\vec{U}$, we have

$$\vec{U} = \vec{e}_{\bar{0}}, \qquad\qquad \vec{p} = m\vec{U},$$
$$U^\alpha = \Lambda^\alpha{}_{\bar{\beta}}(\vec{e}_{\bar{0}})^{\bar{\beta}} = \Lambda^\alpha{}_{\bar{0}}, \qquad p^\alpha = m\Lambda^\alpha{}_{\bar{0}}. \tag{2.21}$$

Therefore we have

$$U^0 = (1 - v^2)^{-1/2}, \qquad p^0 = m(1 - v^2)^{-1/2}$$
$$U^1 = v(1 - v^2)^{-1/2}, \qquad p^1 = mv(1 - v^2)^{-1/2},$$
$$U^2 = 0, \qquad\qquad\qquad p^2 = 0,$$
$$U^3 = 0, \qquad\qquad\qquad p^3 = 0.$$

For small v the spatial components of $\vec{U}$ are $(v, 0, 0)$, which justifies calling it the four-velocity, while the spatial components of $\vec{p}$ are $(mv, 0, 0)$, justifying its name. For small v the energy is

$$E \equiv p^0 = m(1 - v^2)^{-1/2} \simeq m + \tfrac{1}{2}mv^2.$$

This is the rest-mass energy plus the Galilean kinetic energy.

Conservation of four-momentum. The interactions of particles in Galilean physics are governed by the laws of conservation of energy and of momentum. Since the components of $\vec{p}$ reduce in the nonrelativistic limit to the familiar Galilean energy and momentum, it is natural to postulate that the correct relativistic law is that the four-vector $\vec{p}$ is conserved. That is, if several particles interact, then

$$\vec{p} \equiv \sum_{\substack{\text{all} \\ \text{particles} \\ (i)}} \vec{p}_{(i)}, \tag{2.22}$$

where $\vec{p}_{(i)}$ is the ith particle's momentum, is the same before and after each interaction.

This law has the status of an extra *postulate*, since it is only one of many whose nonrelativistic limit is correct. However, like the two fundamental postulates of SR, this one is amply verified by experiment. Not the least of its new predictions is that the energy conservation law must include rest mass: rest mass can be decreased and the difference turned into kinetic energy and hence into heat. This happens every day in nuclear power stations.

There is an important point glossed over in the above statement of the conservation of four-momentum. What is meant by 'before' and 'after' a collision? Suppose there are two collisions, involving different particles, which occur at spacelike separated events, as below. When adding up the total four-momentum, should one take them as they are on the line of constant time t or on the line of constant $\bar{t}$? As measured by $\mathcal{O}$, event $\mathcal{A}$ in Fig. 2.3 occurs before $t = 0$ and $\mathcal{B}$ after, so the total momentum at $t = 0$ is the sum of the momenta after $\mathcal{A}$ and before $\mathcal{B}$. On the other hand, to $\bar{\mathcal{O}}$ they both occur before $\bar{t} = 0$ and so the total momentum at $\bar{t} = 0$ is the sum of the momenta after $\mathcal{A}$ and after $\mathcal{B}$. There

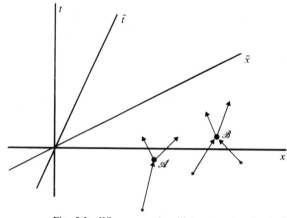

Fig. 2.3 When several collisions are involved, the individual 4-momentum vectors contributing to the total 4-momentum at any particular time may depend upon the frame, but the total 4-momentum is the same 4-vector in all frames; its components transform from frame to frame by the Lorentz transformation.

is even a frame in which $\mathscr{B}$ is *later* than $\mathscr{A}$ and the adding-up may be the reverse of $\mathcal{O}$'s. There is really no problem here, though. Since each collision conserves momentum, the sum of the momenta before $\mathscr{A}$ is the same as that after $\mathscr{A}$, and likewise for $\mathscr{B}$. So *every* inertial observer will get the same total four-momentum vector $\vec{p}$. (Its components will still be different in different frames, but it will be the same vector.) This is an important point: *any* observer can define his line of constant time (this is actually a three-space of constant time, which is called a *hypersurface* of constant time in the four-dimensional spacetime), at that time add up all the momenta, and get the same vector as any other observer does. It is important to understand this, because such conservation laws will appear again.

Center of momentum (CM) frame. This is defined as the inertial frame where

$$\sum_i \vec{p}_{(i)} \xrightarrow[CM]{} (E_{\text{TOTAL}}, 0, 0, 0). \tag{2.23}$$

As with MCRFs, any other frame at rest relative to a CM frame is also a CM frame.

2.5 Scalar product

Magnitude of a vector. By analogy with the interval we define

$$\vec{A}^2 = -(A^0)^2 + (A^1)^2 + (A^2)^2 + (A^3)^2 \tag{2.24}$$

to be the *magnitude* of the vector $\vec{A}$. Because we *defined* the components to transform under a Lorentz transformation in the same manner as $(\Delta t, \Delta x, \Delta y, \Delta z)$, we are *guaranteed* that

$$-(A^0)^2 + (A^1)^2 + (A^2)^2 + (A^3)^2 = -(A^{\bar{0}})^2 + (A^{\bar{1}})^2 + (A^{\bar{2}})^2 + (A^{\bar{3}})^2.$$

(2.25)

The magnitude so defined is a frame-independent number, i.e. a scalar under Lorentz transformations.

This magnitude doesn't have to be positive, of course. As with intervals we adopt the following names: if $\vec{A}^2$ is positive, $\vec{A}$ is a *spacelike* vector; if zero, a *null* vector; and if negative, a *timelike* vector. Thus, spatially pointing vectors have positive magnitude, as is usual in Euclidean space. It is particularly important to understand that a null vector is *not* a zero vector. That is, a null vector has $\vec{A}^2 = 0$, but not all A^α vanish; a zero vector is defined as one, all of whose components vanish. Only in a space where $\vec{A}^2$ is positive-definite does $\vec{A}^2 = 0$ require $A^\alpha = 0$ for all α.

Scalar product of two vectors. We define

◆ $$\vec{A} \cdot \vec{B} = -A^0 B^0 + A^1 B^1 + A^2 B^2 + A^3 B^3$$ (2.26)

in some frame $\mathcal{O}$. We now prove that this is the same number in all other frames. We note first that $\vec{A} \cdot \vec{A}$ is just $\vec{A}^2$, which we know is invariant. Therefore $(\vec{A} + \vec{B}) \cdot (\vec{A} + \vec{B})$, which is the magnitude of $\vec{A} + \vec{B}$, is also invariant. But from Eqs. (2.24) and (2.26) it follows that

$$(\vec{A} + \vec{B}) \cdot (\vec{A} + \vec{B}) = \vec{A}^2 + \vec{B}^2 + 2\vec{A} \cdot \vec{B}.$$

Since the left-hand side is the same in all frames and the first two terms on the right also are, then the last term on the right must be as well. This proves the frame invariance of the scalar product.

Two vectors $\vec{A}$ and $\vec{B}$ are said to be *orthogonal* if $\vec{A} \cdot \vec{B} = 0$. The minus sign in the definition of the scalar product means that two vectors orthogonal to one another are not necessarily at right angles in the spacetime diagram (see examples below). An extreme example is the null vector, which is orthogonal to *itself*! Such a phenomenon is not encountered in spaces where the scalar product is positive-definite.

Example. The basis vectors of a frame $\mathcal{O}$ satisfy:

$$\vec{e}_0 \cdot \vec{e}_0 = -1,$$
$$\vec{e}_1 \cdot \vec{e}_1 = \vec{e}_2 \cdot \vec{e}_2 = \vec{e}_3 \cdot \vec{e}_3 = +1,$$
$$\vec{e}_\alpha \cdot \vec{e}_\beta = 0 \quad \text{if } \alpha \neq \beta.$$

They thus make up a tetrad of mutually orthogonal vectors: an *orthonormal* tetrad, which means *ortho*gonal and *normal*ized to unit magnitude.

(A timelike vector has 'unit magnitude' if its magnitude is -1.) The relations above can be summarized as

◆　　　$\vec{e}_\alpha \cdot \vec{e}_\beta = \eta_{\alpha\beta}$,　　　　　　　　　　　　　(2.27)

where $\eta_{\alpha\beta}$ is similar to a Kronecker delta in that it is zero when $\alpha \neq \beta$, but it differs in that $\eta_{00} = -1$, while $\eta_{11} = \eta_{22} = \eta_{33} = +1$. We will see later that $\eta_{\alpha\beta}$ is in fact of central importance: it is the metric tensor. But for now we treat it as a generalized Kronecker delta.

Example. The basis vectors of $\bar{\mathcal{O}}$ also satisfy

　　　$\vec{e}_{\bar\alpha} \cdot \vec{e}_{\bar\beta} = \eta_{\bar\alpha\bar\beta}$,

so that, in particular, $\vec{e}_{\bar0} \cdot \vec{e}_{\bar1} = 0$. Look at this in the spacetime diagram of $\mathcal{O}$, Fig. 2.4: The two vectors certainly are not perpendicular in the

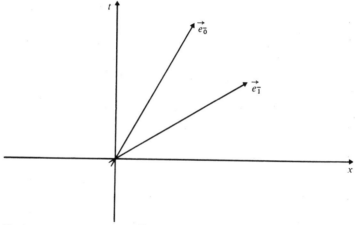

Fig. 2.4　The basis vectors of $\bar{\mathcal{O}}$ are not 'perpendicular' (in the Euclidean sense) when drawn in $\mathcal{O}$, but they *are* orthogonal with respect to the dot product of Minkowski spacetime.

picture. Nevertheless, their scalar product is zero. The rule is that two vectors are orthogonal if they make equal angles with the 45° line representing the path of a light ray. Thus, a vector tangent to the light ray is orthogonal to itself. This is just another way in which SR cannot be 'visualized' in terms of notions we have developed in Euclidean space.

Example. The four-velocity $\vec{U}$ of a particle is just the time basis vector of its MCRF, so from Eq. (2.27) we have

　　　$\vec{U} \cdot \vec{U} = -1$.　　　　　　　　　　　　(2.28)

2.6   Applications

Four-velocity and acceleration as derivatives. Suppose a particle makes an infinitesimal displacement $d\vec{x}$, whose components in O are (dt, dx, dy, dz). The magnitude of this displacement is, by Eq. (2.24), just $-dt^2 + dx^2 + dy^2 + dz^2$. Comparing this with Eq. (1.1), we see that this is just the interval, ds^2:

$$ds^2 = d\vec{x} \cdot d\vec{x}. \tag{2.29}$$

Since the world line is timelike, this is negative. This led us (Eq. (1.9)) to define the proper time by

$$d\tau^2 = -d\vec{x} \cdot d\vec{x}. \tag{2.30}$$

Now consider the vector $d\vec{x}/d\tau$, where $d\tau$ is the square root of Eq. (2.30) (Fig. 2.5). This vector is tangent to the world line since it is a multiple of $d\vec{x}$. Its magnitude is

$$\frac{d\vec{x}}{d\tau} \cdot \frac{d\vec{x}}{d\tau} = \frac{d\vec{x} \cdot d\vec{x}}{(d\tau)^2} = -1.$$

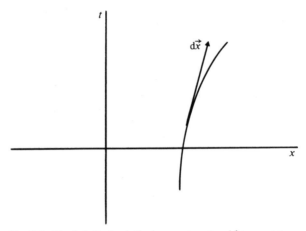

Fig. 2.5   The infinitesimal displacement vector $d\vec{x}$ tangent to a world line.

It is therefore a timelike vector of unit magnitude tangent to the world line. In an MCRF,

$$d\vec{x} \xrightarrow[\substack{\text{MCRF} \\ d\tau = dt}]{} (dt, 0, 0, 0),$$

so that

$$\frac{d\vec{x}}{d\tau} \xrightarrow[\text{MCRF}]{} (1, 0, 0, 0)$$

or

$$\frac{d\vec{x}}{d\tau} = (\vec{e}_0)_{\text{MRCF}}.$$

This was the *definition* of the four-velocity. So we have the useful expression

◆ $\qquad \vec{U} = d\vec{x}/d\tau.$ (2.31)

Moreover, let us examine

$$\frac{d\vec{U}}{d\tau} = \frac{d^2\vec{x}}{d\tau^2},$$

which is some sort of four-acceleration. First we differentiate Eq. (2.24) and use Eq. (2.26):

$$\frac{d}{d\tau}(\vec{U}\cdot\vec{U}) = 2\vec{U}\cdot\frac{d\vec{U}}{d\tau}.$$

But since $\vec{U}\cdot\vec{U} = -1$ is a constant we have

$$\vec{U}\cdot\frac{d\vec{U}}{d\tau} = 0.$$

Since, in the MCRF, $\vec{U}$ has only a 0 component, this orthogonality means that

$$\frac{d\vec{U}}{d\tau} \xrightarrow[\text{MCRF}]{} (0, a^1, a^2, a^3).$$

This vector is defined as the *acceleration* four-vector $\vec{a}$:

◆ $\qquad \vec{a} = \frac{d\vec{U}}{d\tau}, \qquad \vec{U}\cdot\vec{a} = 0.$ (2.32)

Exer. 19, § 2.9, justifies the name 'acceleration'.

Energy and momentum. Consider a particle whose momentum is $\vec{p}$. Then

$$\vec{p}\cdot\vec{p} = m^2\vec{U}\cdot\vec{U} = -m^2. \qquad (2.33)$$

But

$$\vec{p}\cdot\vec{p} = -E^2 + (p^1)^2 + (p^2)^2 + (p^3)^2.$$

Therefore

$$E^2 = m^2 + \sum_{i=1}^{3} (p^i)^2. \qquad (2.34)$$

This is the familiar expression for the total energy of a particle.

Suppose an observer $\bar{\mathcal{O}}$ moves with four-velocity $\vec{U}_{\text{obs}}$ not necessarily equal to the particle's four-velocity. Then

$$\vec{p}\cdot\vec{U}_{\text{obs}} = \vec{p}\cdot\vec{e}_{\bar{0}},$$

where $\vec{e}_{\bar{0}}$ is the basis vector of the frame of the oserver. In that frame the four-momentum has components

$$\vec{p} \underset{\sigma}{\to} (\bar{E}, p^{\bar{1}}, p^{\bar{2}}, p^{\bar{3}}).$$

Therefore we obtain, from Eq. (2.26),

$$-\vec{p} \cdot \vec{U}_{\text{obs}} = \bar{E}. \tag{2.35}$$

This is an important equation. It says that the energy of the particle relative to the observer, $\bar{E}$, can be computed by anyone in any frame by taking the scalar product $\vec{p} \cdot \vec{U}_{\text{obs}}$. This is called a 'frame-invariant' expression for the energy relative to the observer. It is almost always helpful in calculations to use such expressions.

2.7 Photons

No four-velocity. Photons move on null lines, so, for a photon path,

$$d\vec{x} \cdot d\vec{x} = 0. \tag{2.36}$$

Therefore $d\tau$ is *zero* and Eq. (2.31) shows that the *four-velocity cannot be defined.* Another way of saying the same thing is to note that there is no frame in which light is at rest (the second postulate of SR), so there is no MCRF for a photon. Thus, no $\vec{e}_0$ in any frame will be tangent to a photon's world line.

Note carefully that it is still possible to find vectors tangent to a photon's path (which, being a straight line, has the same tangent everywhere): $d\vec{x}$ is one. The problem is finding a tangent of *unit magnitude*, since they all have vanishing magnitude.

Four-momentum. The four-momentum of a particle is *not* a unit vector. Instead, it is a vector whose components in some frame give the particle energy and momentum relative to that frame. If a photon carries energy E in some frame, then in that frame $p^0 = E$. If it moves in the x direction, then $p^y = p^z = 0$, and in order for the four-momentum to be parallel to its world line (hence be null) we must have $p^x = E$. This ensures that

$$\vec{p} \cdot \vec{p} = -E^2 + E^2 = 0. \tag{2.37}$$

So we conclude that photons have *spatial momentum* equal to their energy.

We know from quantum mechanics that a photon has energy

$$E = h\nu, \tag{2.38}$$

where ν is its frequency and h is Planck's constant, $h = 6.6256 \times 10^{-34}$ J s^{-1}.

This relation and the Lorentz transformation of the four-momentum immediately give us the Doppler-shift formula for photons. Suppose, for instance, that in frame $\mathcal{O}$ a photon has frequency ν and moves in the x direction. Then, in $\bar{\mathcal{O}}$, which has velocity v in the x direction relative to $\mathcal{O}$, the photon's energy is

$$\bar{E} = E/\sqrt{(1-v^2)} - p^x v/\sqrt{(1-v^2)}$$
$$= h\nu/\sqrt{(1-v^2)} - h\nu v/\sqrt{(1-v^2)}.$$

Setting this equal to $h\bar{\nu}$ gives $\bar{\nu}$, the frequency in $\bar{\mathcal{O}}$:

$$\bar{\nu}/\nu = (1-v)/\sqrt{(1-v^2)} = \sqrt{[(1-v)/(1+v)]}. \tag{2.39}$$

This is generalized in Exer. 25, § 2.9.

Zero rest-mass particles. The rest mass of a photon must be zero, since

$$m^2 = -\vec{p} \cdot \vec{p} = 0. \tag{2.40}$$

Any particle whose four-momentum is null must have rest mass zero, and conversely. The only two known zero rest-mass particles are the photon and the neutrino. (Sometimes the 'graviton' is added to this list, since gravitational waves also travel at the speed of light, as we shall see later. But 'photon' and 'graviton' are concepts that come from quantum mechanics, and there is as yet no satisfactory quantized theory of gravity, so that 'graviton' is not really a well-defined notion yet.) The idea that only particles with zero rest mass can travel at the speed of light is reinforced by the fact that no particle of finite rest mass can be accelerated to the speed of light, since then its energy would be infinite. Put another way, a particle traveling at the speed of light (in, say, the x direction) has $p^1/p^0 = 1$, while a particle of rest mass m moving in the x direction has, from the equation $\vec{p} \cdot \vec{p} = -m^2$, $p^1/p^0 = [1 - m^2/(p^0)^2]^{1/2}$, which is always less than one, no matter how much energy the particle is given. Although it may seem to get close to the speed of light, there is an important distinction: the particle with $m \neq 0$ always has an MCRF, a Lorentz frame in which it is at rest, namely that whose velocity v is p^1/p^0 relative to the old frame. A photon has *no* rest frame.

2.8 Bibliography

We have only scratched the surface of relativistic kinematics and particle dynamics. These are particularly important in particle physics, which in turn provides the most stringent tests of SR. See Muirhead (1973) or Hagedorn (1963).

2.9 Exercises

1 Given the numbers $\{A^0 = 5, A^1 = 0, A^2 = -1, A^3 = -6\}$, $\{B_0 = 0, B_1 = -2, B_2 = 4, B_3 = 0\}$, $\{C_{00} = 1, C_{01} = 0, C_{02} = 2, C_{03} = 3, C_{30} = -1, C_{10} = 5, C_{11} = -2, C_{12} = -2, C_{13} = 0, C_{21} = 5, C_{22} = 2, C_{23} = -2, C_{20} = 4, C_{31} = -1, C_{32} = -3, C_{33} = 0\}$, find:
(a) $A^\alpha B_\alpha$; (b) $A^\alpha C_{\alpha\beta}$ for all β; (c) $A^\gamma C_{\gamma\sigma}$ for all σ; (d) $A^\nu C_{\mu\nu}$ for all μ; (e) $A^\alpha B_\beta$ for all α,β; (f) $A^i B_i$; (g) $A^j B_k$ for all j, k.

2 Identify the free and dummy indices in the following equations and change them into equivalent expressions with different indices. How many different equations does each expression represent?
(a) $A^\alpha B_\alpha = 5$; (b) $A^{\bar{\mu}} = \Lambda^{\bar{\mu}}{}_\nu A^\nu$; (c) $T^{\alpha\mu\lambda} A_\mu C_\lambda{}^\gamma = D^{\gamma\alpha}$; (d) $R_{\mu\nu} - \frac{1}{2} g_{\mu\nu} R = G_{\mu\nu}$.

3 Prove Eq. (2.5).

4 Given the vectors $\vec{A} \to_\mathcal{O} (5, -1, 0, 1)$ and $\vec{B} \to_\mathcal{O} (-2, 1, 1, -6)$, find the components in $\mathcal{O}$ of (a) $-6\vec{A}$; (b) $3\vec{A} + \vec{B}$; (c) $-6\vec{A} + 3\vec{B}$.

5 A collection of vectors $\{\vec{a}, \vec{b}, \vec{c}, \vec{d}\}$ is said to be linearly independent if no linear combination of them is zero except the trivial one, $0\vec{a} + 0\vec{b} + 0\vec{c} + 0\vec{d} = 0$.

(a) Show that the basis vectors in Eq. (1.9) are linearly independent.

(b) Is the following set linearly independent?
$\{\vec{a}, \vec{b}, \vec{c}, 5\vec{a} + 3\vec{b} - 2\vec{c}\}$.

6 In the t–x spacetime diagram of $\mathcal{O}$, draw the basis vectors $\vec{e}_0$ and $\vec{e}_1$. Draw the corresponding basis vectors of $\bar{\mathcal{O}}$, who moves with speed 0.6 in the positive x direction relative to $\mathcal{O}$. Draw the corresponding basis vectors of $\bar{\bar{\mathcal{O}}}$, who moves with speed 0.6 in the positive x direction relative to $\bar{\mathcal{O}}$.

7(a) Verify Eq. (2.10) for all α, β.
(b) Prove Eq. (2.11) from Eq. (2.9).

8(a) Prove that the zero vector $(0, 0, 0, 0)$ has these same components in *all* reference frames.
(b) Use (a) to prove that if two vectors have equal components in one frame, they have equal components in all frames.

9 Prove, by writing out all the terms, that
$$\sum_{\bar{\alpha}=0}^{3} \left(\sum_{\beta=0}^{3} \Lambda^{\bar{\alpha}}{}_\beta A^\beta \vec{e}_{\bar{\alpha}} \right) = \sum_{\beta=0}^{3} \left(\sum_{\bar{\alpha}=0}^{3} \Lambda^{\bar{\alpha}}{}_\beta A^\beta \vec{e}_{\bar{\alpha}} \right).$$

Since the order of summation doesn't matter, we are justified in using the Einstein summation convention to write simply $\Lambda^{\bar{\alpha}}{}_{\beta}A^{\beta}\vec{e}_{\bar{\alpha}}$, which doesn't specify the order of summation.

10 Prove Eq. (2.13) from the equation $A^{\alpha}(\Lambda^{\bar{\beta}}{}_{\alpha}\vec{e}_{\bar{\beta}} - \vec{e}_{\alpha}) = 0$ by making specific choices for the components of the arbitrary vector $\vec{A}$.

11 Let $\Lambda^{\bar{\alpha}}{}_{\beta}$ be the matrix of the Lorentz transformation from $\mathcal{O}$ to $\bar{\mathcal{O}}$, given in Eq. (1.12). Let $\vec{A}$ be an arbitrary vector with components (A^0, A^1, A^2, A^3) in frame $\mathcal{O}$.

(a) Write down the matrix of $\Lambda^{\nu}{}_{\bar{\mu}}(-v)$.
(b) Find $A^{\bar{\alpha}}$ for all $\bar{\alpha}$.
(c) Verify Eq. (2.18) by performing the indicated sum for all values of ν and α.
(d) Write down the Lorentz transformation matrix from $\bar{\mathcal{O}}$ to $\mathcal{O}$, justifying each entry.
(e) Use (d) to find A^{β} from $A^{\bar{\alpha}}$. How is this related to Eq. (2.18)?
(f) Verify, in the same manner as (c), that
$$\Lambda^{\nu}{}_{\bar{\beta}}(v)\Lambda^{\bar{\alpha}}{}_{\nu}(-v) = \delta^{\bar{\alpha}}{}_{\bar{\beta}}.$$
(g) Establish that
$$\vec{e}_{\alpha} = \delta^{\nu}{}_{\alpha}\vec{e}_{\nu}$$
and
$$A^{\beta} = \delta^{\beta}{}_{\mu}A^{\mu}.$$

12 Given $\vec{A} \to_{\mathcal{O}}(0, -2, 3, 5)$, find
(a) the components of $\vec{A}$ in $\bar{\mathcal{O}}$, which moves at speed 0.8 relative to $\mathcal{O}$ in the positive x direction;
(b) the components of $\vec{A}$ in $\bar{\bar{\mathcal{O}}}$, which moves at speed 0.6 relative to $\bar{\mathcal{O}}$ in the positive x direction;
(c) the magnitude of $\vec{A}$ from its components in $\mathcal{O}$;
(d) the magnitude of $\vec{A}$ from its components in $\bar{\bar{\mathcal{O}}}$.

13 Let $\bar{\mathcal{O}}$ move with velocity v relative to $\mathcal{O}$, and let $\bar{\bar{\mathcal{O}}}$ move with velocity v' relative to $\bar{\mathcal{O}}$.
(a) Show that the Lorentz transformation from $\mathcal{O}$ to $\bar{\bar{\mathcal{O}}}$ is
$$\Lambda^{\bar{\bar{\alpha}}}{}_{\mu} = \Lambda^{\bar{\bar{\alpha}}}{}_{\bar{\gamma}}(v')\Lambda^{\bar{\gamma}}{}_{\mu}(v). \tag{2.41}$$
(b) Show that Eq. (2.41) is just the matrix product of the matrices of the individual Lorentz transformations.
(c) Let $v = 0.6\vec{e}_x$, $v' = 0.8\vec{e}_{\bar{y}}$. Find $\Lambda^{\bar{\bar{\alpha}}}{}_{\mu}$ for all μ and $\bar{\bar{\alpha}}$.
(d) Verify that the transformation found in (c) is indeed a Lorentz transformation by showing explicitly that $\Delta \bar{\bar{s}}^2 = \Delta s^2$ for any $(\Delta t, \Delta x, \Delta y, \Delta z)$.
(e) Compute
$$\Lambda^{\bar{\bar{\alpha}}}{}_{\bar{\gamma}}(v)\Lambda^{\bar{\gamma}}{}_{\mu}(v')$$
for v and v', as given in (c), and show that the result does not equal that of (c). Interpret this physically.

14   The following matrix gives a Lorentz transformation from $\mathcal{O}$ to $\bar{\mathcal{O}}$:

$$\begin{pmatrix} 1.25 & 0 & 0 & .75 \\ 0 & 1 & 0 & 0 \\ 0 & 0 & 1 & 0 \\ .75 & 0 & 0 & 1.25 \end{pmatrix}.$$

(a)   What is the velocity (speed and direction) of $\bar{\mathcal{O}}$ relative to $\mathcal{O}$?

(b)   What is the inverse matrix to the given one?

(c)   Find the components in $\mathcal{O}$ of a vector $\vec{A} \underset{\bar{\mathcal{O}}}{\rightarrow} (1, 2, 0, 0)$.

15(a)   Compute the four-velocity components in $\mathcal{O}$ of a particle whose speed in $\mathcal{O}$ is v in the positive x direction, by using the Lorentz transformation from the rest frame of the particle.

(b)   Generalize this result to find the four-velocity components when the particle has arbitrary velocity v, with $|v| < 1$.

(c)   Use your result in (b) to express v in terms of the components $\{U^\alpha\}$.

(d)   Find the three-velocity v of a particle whose four-velocity components are $(2, 1, 1, 1)$.

16   Derive the Einstein velocity-addition formula by performing a Lorentz transformation with velocity v on the four-velocity of a particle whose speed in the original frame was W.

17(a)   Prove that any timelike vector $\vec{U}$ for which $U^0 > 0$ and $\vec{U} \cdot \vec{U} = -1$ is the four-velocity of *some* world line.

(b)   Use this to prove that for any timelike vector $\vec{V}$ there is a Lorentz frame in which $\vec{V}$ has zero spatial components.

18(a)   Show that the sum of any two orthogonal spacelike vectors is spacelike.

(b)   Show that a timelike vector and a null vector cannot be orthogonal.

19   A body is said to be *uniformly accelerated* if its acceleration four-vector $\vec{a}$ has constant spatial direction and magnitude, say $\vec{a} \cdot \vec{a} = \alpha^2 \geqslant 0$.

(a)   Show that this implies that $\vec{a}$ always has the same components in the body's MCRF, and that these components are what one would call 'acceleration' in Galilean terms. (This would be the physical situation for a rocket whose engine always gave the same acceleration.)

(b)   Suppose a body is uniformly accelerated with $\alpha = 1\,g = 10\,\mathrm{m\,s^{-2}}$, the acceleration of the earth's gravity. If the body starts from rest, find its speed after time t. (Be sure to use the correct units.) How far has it traveled in this time? How long does it take to reach $v = 0.999$?

(c)   Find the elapsed *proper* time for the body in (b), as a function of t. (Integrate $\mathrm{d}\tau$ along its world line.) How much proper time has elapsed by the time its speed is $v = 0.999$? How much would a person accelerated as in (b) age on a trip from Earth to the centre of our Galaxy, a distance of about $2 \times 10^{20}\,\mathrm{m}$?

20 The world line of a particle is described by the equations

$$x(t) = at + b \sin \omega t, \qquad y(t) = b \cos \omega t,$$
$$z(t) = 0, \qquad |b\omega| < 1,$$

in some inertial frame. Describe the motion and compute the components of the particle's four-velocity and four-acceleration.

21 The world line of a particle is described by the parametric equations in some Lorentz frame

$$t(\lambda) = a \sinh\left(\frac{\lambda}{a}\right), \qquad x(\lambda) = a \cosh\left(\frac{\lambda}{a}\right),$$

where λ is the parameter and a is a constant. Describe the motion and compute the particle's four-velocity and acceleration components. Show that λ is proper time along the world line and that the acceleration is uniform. Interpret a.

22(a) Find the energy, rest mass, and three-velocity v of a particle whose four-momentum has the components $(4, 1, 1, 0)$ *kg*.

(b) The collision of two particles of four-momenta

$$\vec{p}_1 \underset{\mathcal{O}}{\to} (3, -1, 0, 0)kg, \qquad \vec{p}_2 \underset{\mathcal{O}}{\to} (2, 1, 1, 0) \ kg$$

results in the destruction of the two particles and the production of three new ones, two of which have four-momenta

$$\vec{p}_3 \underset{\mathcal{O}}{\to} (1, 1, 0, 0)kg, \qquad \vec{p}_4 \underset{\mathcal{O}}{\to} (1, -\tfrac{1}{2}, 0, 0).$$

Find the four-momentum, energy, rest mass, and three-velocity of the third particle produced. Find the CM frame's three-velocity.

23 A particle of rest mass m has three-velocity v. Find its energy correct to terms of order $|v|^4$. At what speed $|v|$ does the absolute value of $0(|v|^4)$ term equal $\tfrac{1}{2}$ of the kinetic-energy term $\tfrac{1}{2}m|v|^2$?

24 Prove that conservation of four-momentum forbids a reaction in which an electron and position annihilate and produce a single photon (γ-ray). Prove that the production of two photons is not forbidden.

25(a) Let frame $\bar{\mathcal{O}}$ move with speed v in the x-direction relative to $\mathcal{O}$. Let a photon have frequency ν in $\mathcal{O}$ and move at an angle θ with respect to $\mathcal{O}$'s x axis. Show that its frequency in $\bar{\mathcal{O}}$ is

$$\bar{\nu}/\nu = (1 - v \cos \theta)/\sqrt{(1 - v^2)}. \tag{2.42}$$

(b) Even when the motion of the photon is perpendicular to the x axis ($\theta = \pi/2$) there is a frequency shift. This is called the *transverse Doppler shift*, and arises because of the time dilation. At what angle θ does the photon have to move so that there is *no* doppler shift between $\mathcal{O}$ and $\bar{\mathcal{O}}$?

(c) Use Eqs. (2.35) and (2.38) to calculate Eq. (2.42).

26 Calculate the energy that is required to accelerate a particle of rest mass $m \neq 0$ from speed v to speed $v + \delta v$ ($\delta v \ll v$), to first order in δv. Show that it would take an infinite amount of energy to accelerate the particle to the speed of light.

27 Two identical bodies of mass 10 kg are at rest at the same temperature. One of them is heated by the addition of 100 J of heat. Both are then subjected to the same force. Which accelerates faster, and by how much?

28 Let $\vec{A} \to {}_O(5, 1, -1, 0)$, $\vec{B} \to {}_O(-2, 3, 1, 6)$, $\vec{C} \to {}_O(2, -2, 0, 0)$. Let $\bar{O}$ be a frame moving at speed $v = 0.6$ in the positive x direction relative to O, with its spatial axes oriented parallel to O's.
(a) Find the components of $\vec{A}$, $\vec{B}$, and $\vec{C}$ in $\bar{O}$.
(b) Form the dot products $\vec{A} \cdot \vec{B}$, $\vec{B} \cdot \vec{C}$, $\vec{A} \cdot \vec{C}$ and $\vec{C} \cdot \vec{C}$ using the components in $\bar{O}$. Verify the frame independence of these numbers.
(c) Classify $\vec{A}$, $\vec{B}$, and $\vec{C}$ as timelike, spacelike, or null.

29 Prove, using the component expressions, Eqs. (2.24) and (2.26), that
$$\frac{d}{d\tau}(\vec{U} \cdot \vec{U}) = 2\vec{U} \cdot \frac{d\vec{U}}{d\tau}.$$

30 The four-velocity of a rocket ship is $\vec{U} \to {}_O(2, 1, 1, 1)$. It encounters a high-velocity cosmic ray whose momentum is $\vec{P} \to {}_O(300, 299, 0, 0) \times 10^{-27}$ kg. Compute the energy of the cosmic ray as measured by the rocket ship's passengers, using each of the two following methods.
(a) Find the Lorentz transformations from O to the MCRF of the rocket ship, and use it to transform the components of $\vec{P}$.
(b) Use Eq. (2.35).
(c) Which method is quicker? Why?

31 A photon of frequency ν is reflected without change of frequency from a mirror, with an angle of incidence θ. Calculate the momentum transferred to the mirror. What momentum would be transferred if the photon were absorbed rather than reflected?

32 Let a particle of charge e and rest mass m, initially at rest in the laboratory, scatter a photon of initial frequency ν_i. This is called *Compton scattering*. Suppose the scattered photon comes off at an angle θ from the incident direction. Use conservation of four-momentum to deduce that the photon's final frequency ν_f is given by
$$\frac{1}{\nu_f} = \frac{1}{\nu_i} + h\left(\frac{1 - \cos\theta}{m}\right). \tag{2.43}$$

33 Space is filled with cosmic rays (high-energy protons) and the cosmic microwave background radiation. These can Compton scatter off one another. Suppose a photon of energy $h\nu = 2 \times 10^{-4}$ eV scatters off a

proton of energy $10^9 m_P = 10^{18}$ eV, energies measured in the Sun's rest frame. Use Eq. (2.43) in the proton's initial rest frame to calculate the maximum final energy the photon can have in the solar rest frame after the scattering. What energy range is this (X-ray, visible, *etc.*)?

34 Show that, if $\vec{A}$, $\vec{B}$, and $\vec{C}$ are any vectors and α and β any real numbers,

$$(\alpha\vec{A})\cdot\vec{B} = \alpha(\vec{A}\cdot\vec{B}),$$
$$\vec{A}\cdot(\beta\vec{B}) = \beta(\vec{A}\cdot\vec{B}),$$
$$\vec{A}\cdot(\vec{B}+\vec{C}) = \vec{A}\cdot\vec{B} + \vec{A}\cdot\vec{C},$$
$$(\vec{A}+\vec{B})\cdot\vec{C} = \vec{A}\cdot\vec{C} + \vec{B}\cdot\vec{C}.$$

35 Show that the vectors $\{\vec{e}_{\bar{\beta}}\}$ obtained from $\{\vec{e}_\alpha\}$ by Eq. (2.15) satisfy
$\vec{e}_{\bar{\alpha}}\cdot\vec{e}_{\bar{\beta}} = \eta_{\bar{\alpha}\bar{\beta}}$ for all $\bar{\alpha}$, $\bar{\beta}$.

3
Tensor analysis in special relativity

3.1 The metric tensor

Consider the representation of two vectors $\vec{A}$ and $\vec{B}$ on the basis $\{\vec{e}_\alpha\}$ of some frame $\mathcal{O}$:

$$\vec{A} = A^\alpha \vec{e}_\alpha, \qquad \vec{B} = B^\beta \vec{e}_\beta.$$

Their scalar product is

$$\vec{A} \cdot \vec{B} = (A^\alpha \vec{e}_\alpha) \cdot (B^\beta \vec{e}_\beta).$$

(Note the importance of using *different* indices α and β to distinguish the first summation from the second.) Following Exer. 34, § 2.9, we can rewrite this as

$$\vec{A} \cdot \vec{B} = A^\alpha B^\beta (\vec{e}_\alpha \cdot \vec{e}_\beta),$$

which, by Eq. (2.28), is

$$\blacklozenge \qquad \vec{A} \cdot \vec{B} = A^\alpha B^\beta \eta_{\alpha\beta}. \qquad (3.1)$$

This is a *frame-invariant* way of writing

$$-A^0 B^0 + A^1 B^1 + A^2 B^2 + A^3 B^3.$$

The numbers $\eta_{\alpha\beta}$ are called 'components of the metric tensor'. We will justify this name later. Right now we observe that they essentially give a 'rule' for associating with two vectors $\vec{A}$ and $\vec{B}$ a single *number*, which we call their scalar product. The rule is that the number is the double sum $A^\alpha B^\beta \eta_{\alpha\beta}$. Such a rule is at the heart of the meaning of 'tensor', as we now discuss.

3.2 Definition of tensors

We make the following definition of a tensor:

◆ A tensor of type $\binom{0}{N}$ is a function of N vectors into the real numbers, which is linear in each of its N arguments.

Let us see what this definition means. For the moment, we will just accept the notation $\binom{0}{N}$; its justification will come later in this chapter. The rule for the scalar product, Eq. (3.1), satisfies our definition of a $\binom{0}{2}$ tensor. It is a rule which takes two vectors, $\vec{A}$ and $\vec{B}$, and produces a single real number $\vec{A} \cdot \vec{B}$. To say that it is linear in its arguments means what is proved in Exer. 34, § 2.9. Linearity on the first argument means

$$(\alpha \vec{A}) \cdot \vec{B} = \alpha(\vec{A} \cdot \vec{B}),$$

and

$$(\vec{A} + \vec{B}) \cdot \vec{C} = \vec{A} \cdot \vec{C} + \vec{B} \cdot \vec{C},$$

$$(3.2)$$

while linearity on the second argument means

$$\vec{A} \cdot (\beta \vec{B}) = \beta(\vec{A} \cdot \vec{B}),$$
$$\vec{A} \cdot (\vec{B} + \vec{C}) = \vec{A} \cdot \vec{B} + \vec{A} \cdot \vec{C}.$$

This definition of linearity is of central importance for tensor algebra, and the student should study it carefully.

To give concreteness to this notion of the dot product being a tensor, we introduce a name and notation for it. We let **g** be the *metric tensor* and write, by definition,

$$\mathbf{g}(\vec{A}, \vec{B}) \equiv \vec{A} \cdot \vec{B}. \tag{3.3}$$

Then we regard **g**(,) as a function which can take two arguments, and which is linear in that

$$\mathbf{g}(\alpha \vec{A} + \beta \vec{B}, \vec{C}) = \alpha \mathbf{g}(\vec{A}, \vec{C}) + \beta \mathbf{g}(\vec{B}, \vec{C}), \tag{3.4}$$

and similarly for the second argument. The value of **g** on two arguments, denoted by $\mathbf{g}(\vec{A}, \vec{B})$, is their dot product, a real number.

Notice that the definition of a tensor does not mention components of the vectors. A tensor must be a rule which gives the same real number independently of the reference frame in which the vectors' components are calculated. We showed in the previous chapter that Eq. (3.1) satisfies this requirement. This enables us to regard a tensor as a function of the vectors themselves rather than of their components, and this can some-times be helpful conceptually.

Notice that an ordinary function of position, $f(t, x, y, z)$, is a real-valued function of no vectors at all. It is therefore classified as a $\binom{0}{0}$ tensor.

Aside on the usage of the term 'function'. The most familiar notion of a function is expressed in the equation

$$y = f(x),$$

where y and x are real numbers. But this can be written more precisely as: f is a 'rule' (called a mapping) which associates a real number (symbolically called y, above) with another real number, which is the argument of f (symbolically called x, above). The function itself is *not* $f(x)$, since $f(x)$ is y, which is a real number called the 'value' of the function. The function itself is f, which we can write as $f(\ \ )$ in order to show that it has one argument. In algebra this seems like hair-splitting since we unconsciously think of x and y as two things at once: they are, on the one hand, specific real numbers and, on the other hand, *names* for general and arbitrary real numbers. In tensor calculus we will make this distinction explicit: $\vec{A}$ and $\vec{B}$ are *specific* vectors, $\vec{A} \cdot \vec{B}$ is a specific real number, and **g** is the name of the function that associates $\vec{A} \cdot \vec{B}$ with $\vec{A}$ and $\vec{B}$.

Components of a tensor. Just like a vector, a tensor has components. They are defined as

> The components in a frame $\mathcal{O}$ of a tensor of type $\binom{0}{N}$ are the values of the function when its arguments are the basis vectors $\{\vec{e}_\alpha\}$ of the frame $\mathcal{O}$.

Thus we have the notion of components as frame-dependent numbers (frame-dependent because the basis refers to a specific frame). For the metric tensor this gives the components as

◆ $$\mathbf{g}(\vec{e}_\alpha, \vec{e}_\beta) = \vec{e}_\alpha \cdot \vec{e}_\beta = \eta_{\alpha\beta}. \tag{3.5}$$

So the matrix $\eta_{\alpha\beta}$ that we introduced before is to be thought of as an array of the components of **g** on the basis. We will have many more examples of this later. First we study a particularly important class of tensors.

3.3   The $\binom{0}{1}$ tensors: one-forms

A tensor of the type $\binom{0}{1}$ is called a covector, a covariant vector, or a one-form. Often these names are used interchangeably, even in a single text-book or reference.

General properties. Let an arbitrary one-form be called $\tilde{p}$ (we adopt the notation that $\sim$ above a symbol denotes a one-form, just as $\rightarrow$ above denotes a vector). Then $\tilde{p}$, supplied with one vector argument, gives a real number: $\tilde{p}(\vec{A})$ is a real number. Suppose $\tilde{q}$ is another one-form. Then we can define

$$\left.\begin{array}{l} \tilde{s} = \tilde{p} + \tilde{q}, \\ \tilde{r} = \alpha\tilde{p}, \end{array}\right\} \tag{3.6a}$$

to be the one-forms whose values for an argument $\vec{A}$ are

$$\left.\begin{array}{l} \tilde{s}(\vec{A}) = \tilde{p}(\vec{A}) + \tilde{q}(\vec{A}), \\ \tilde{r}(\vec{A}) = \alpha\tilde{p}(\vec{A}). \end{array}\right\} \tag{3.6b}$$

With these rules, the set of all one-forms satisfies the axioms for a vector space, which accounts for their other names. This space is called the 'dual vector space' to distinguish it from the space of all vectors like $\vec{A}$. When discussing vectors we relied heavily on components and their transformations. Let us look at those of $\tilde{p}$. The components of $\tilde{p}$ are called p_α:

$$p_\alpha \equiv \tilde{p}(\vec{e}_\alpha). \tag{3.7}$$

Any component with a single lower index is, by convention, the component of a one-form; an upper index denotes the component of a vector. In terms of components, $\tilde{p}(\vec{A})$ is

$$\begin{aligned} \tilde{p}(\vec{A}) &= \tilde{p}(A^\alpha\vec{e}_\alpha) \\ &= A^\alpha\tilde{p}(\vec{e}_\alpha), \\ \tilde{p}(\vec{A}) &= A^\alpha p_\alpha. \end{aligned} \tag{3.8}$$

The second step follows from the linearity which is the heart of the definition we gave of a tensor. So the real number $\tilde{p}(\vec{A})$ is easily found to be the sum $A^0 p_0 + A^1 p_1 + A^2 p_2 + A^3 p_3$. Notice that *all* terms have plus signs: this operation is called *contraction* of $\vec{A}$ and $\tilde{p}$, and is *more* fundamental in tensor analysis than the scalar product because it can be performed between any one-form and vector without reference to other tensors. We have seen that two vectors cannot make a scalar (their dot product) without the help of a third tensor, the metric.

The components of $\tilde{p}$ on a basis $\{\vec{e}_{\bar{\beta}}\}$ are

$$\begin{aligned} p_{\bar{\beta}} &\equiv \tilde{p}(\vec{e}_{\bar{\beta}}) = \tilde{p}(\Lambda^\alpha{}_{\bar{\beta}}\vec{e}_\alpha) \\ &= \Lambda^\alpha{}_{\bar{\beta}}\tilde{p}(\vec{e}_\alpha) = \Lambda^\alpha{}_{\bar{\beta}}p_\alpha. \end{aligned} \tag{3.9}$$

Comparing this with

$$\vec{e}_{\bar{\beta}} = \Lambda^\alpha{}_{\bar{\beta}}\vec{e}_\alpha,$$

we see that components of one-forms transform in exactly the same manner as basis vectors and in the opposite manner to components of vectors. By 'opposite', we mean using the inverse transformation. This use of the inverse guarantees that $A^\alpha p_\alpha$ is frame independent for any vector $\vec{A}$ and one-form $\tilde{p}$. This is such an important observation that we shall prove it explicitly:

$$A^{\bar{\alpha}} p_{\bar{\alpha}} = (\Lambda^{\bar{\alpha}}{}_{\beta} A^{\beta})(\Lambda^{\mu}{}_{\bar{\alpha}} p_{\mu}), \tag{3.10a}$$

$$= \Lambda^{\mu}{}_{\bar{\alpha}} \Lambda^{\bar{\alpha}}{}_{\beta} A^{\beta} p_{\mu}, \tag{3.10b}$$

$$= \delta^{\mu}{}_{\beta} A^{\beta} p_{\mu}, \tag{3.10c}$$

$$= A^{\beta} p_{\beta}. \tag{3.10d}$$

(This is the same way in which the vector $A^{\alpha} \vec{e}_{\alpha}$ is kept frame independent.) This inverse transformation gives rise to the word 'dual' in 'dual vector space'. The property of transforming *with* basis vectors gives rise to the *co* in 'covariant vector' and its shorter form 'covector'. Since components of ordinary vectors transform oppositely to basis vectors (in order to keep $A^{\beta} \vec{e}_{\beta}$ frame independent), they are often called 'contravariant' vectors. Most of these names are old-fashioned; 'vectors' and 'dual vectors' or 'one-forms' are the modern names. The reason that 'co' and 'contra' have been abandoned is that they mix up two very different things: the transformation of a basis is the expression of *new* vectors in terms of *old* ones; the transformation of components is the expression of the *same* object in terms of the new basis. It is important for the student to be sure of these distinctions before proceeding further.

Basis one-forms. Since the set of all one-forms is a vector space, one can use any set of four linearly independent one-forms as a basis. (As with any vector space, one-forms are said to be linearly independent if no nontrivial linear combination equals the zero one-form. The zero one-form is the one whose value on any vector is zero.) However, in the previous section we have already used the basis vectors $\{\vec{e}_{\alpha}\}$ to define the components of a one-form. This suggests that we should be able to use the basis vectors to define an associated one-form basis $\{\tilde{\omega}^{\alpha}, \alpha = 0, \ldots, 3\}$, which we shall call the basis *dual* to $\{\vec{e}_{\alpha}\}$, upon which a one-form has the components defined above. That is, we want a set $\{\tilde{\omega}^{\alpha}\}$ such that

$$\tilde{p} = p_{\alpha} \tilde{\omega}^{\alpha}. \tag{3.11}$$

(Notice that using a raised index on $\tilde{\omega}^{\alpha}$ permits the summation convention to operate.) The $\{\tilde{\omega}^{\alpha}\}$ are *four distinct* one-forms, just as the $\{\vec{e}_{\alpha}\}$ are four distinct vectors. This equation must imply Eq. (3.8) for any vector $\vec{A}$ and one-form $\tilde{p}$:

$$\tilde{p}(\vec{A}) = p_{\alpha} A^{\alpha}.$$

But from Eq. (3.11) we get

$$\tilde{p}(\vec{A}) = p_\alpha \tilde{\omega}^\alpha(\vec{A})$$
$$= p_\alpha \tilde{\omega}^\alpha(A^\beta \vec{e}_\beta)$$
$$= p_\alpha A^\beta \tilde{\omega}^\alpha(\vec{e}_\beta).$$

(Notice the use of β as an index in the second line, in order to distinguish its summation from the one on α.) Now, this final line can only equal $p_\alpha A^\alpha$ for all A^β and p_α if

$$\blacklozenge \qquad \tilde{\omega}^\alpha(\vec{e}_\beta) = \delta^\alpha{}_\beta. \qquad\qquad (3.12)$$

Comparing with Eq. (3.7), we see that this equation gives the βth component of the αth basis one-form. It therefore *defines* the αth basis one-form. We can write out these components as

$$\tilde{\omega}^0 \underset{\mathcal{O}}{\rightarrow} (1, 0, 0, 0),$$

$$\tilde{\omega}^1 \underset{\mathcal{O}}{\rightarrow} (0, 1, 0, 0),$$

$$\tilde{\omega}^2 \underset{\mathcal{O}}{\rightarrow} (0, 0, 1, 0),$$

$$\tilde{\omega}^3 \underset{\mathcal{O}}{\rightarrow} (0, 0, 0, 1).$$

It is important to understand two points here. One is that Eq. (3.12) defines the basis $\{\tilde{\omega}^\alpha\}$ in terms of the basis $\{\vec{e}_\beta\}$. The vector basis induces a unique and convenient one-form basis. This is not the only possible one-form basis, but it is so useful to have the relationship, Eq. (3.12), between the bases that we will always use it. The relationship, Eq. (3.12), is between the two bases, not between individual pairs, such as $\tilde{\omega}^0$ and $\vec{e}_0$. That is, if we change $\vec{e}_0$, while leaving $\vec{e}_1$, $\vec{e}_2$, and $\vec{e}_3$ unchanged, then in general this induces changes not only in $\tilde{\omega}^0$ but also in $\tilde{\omega}^1$, $\tilde{\omega}^2$, and $\tilde{\omega}^3$. The second point to understand is that, although we can describe both vectors and one-forms by giving a set of four components, their geometrical significance is very different. The student should not lose sight of the fact that the components tell only part of the story. The basis contains the rest of the information. That is, a set of numbers (0, 2, −1, 5) alone does not define anything; to make it into something, one must say whether these are components on a vector basis or a one-form basis and, indeed, which of the infinite number of possible bases is being used.

It remains to determine how $\{\tilde{\omega}^\alpha\}$ transforms under a change of basis. That is, each frame has its own unique set $\{\tilde{\omega}^\alpha\}$; how are those of two frames related? The derivation here is analogous to that for the basis vectors. It leads to the only equation one can write down with the indices in their correct positions:

$$\tilde{\omega}^{\bar{\alpha}} = \Lambda^{\bar{\alpha}}{}_{\beta}\tilde{\omega}^{\beta}. \tag{3.13}$$

This is the same as for components of a vector, and opposite that for components of a one-form.

Picture of a one-form. For vectors we usually imagine an arrow, if we need a picture. It is helpful to have an image of a one-form as well. First of all, it is not an arrow. Its picture must reflect the fact that it maps vectors into real numbers. A vector itself does not automatically map another vector into a real number. To do this it needs a metric tensor to define the scalar product. With a different metric, the *same* two vectors will produce a *different* scalar product. So two vectors by themselves don't give a number. We need a picture of a one-form which doesn't depend on any other tensors having been defined. The one generally used by mathematicians is shown in Fig. 3.1. The one-form consists of

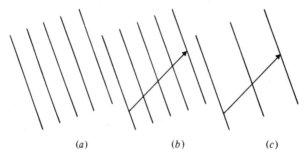

(a)   (b)   (c)

Fig. 3.1   (a) The picture of one-form complementary to that of a vector as an arrow. (b) The value of a one-form on a given vector is the number of surfaces the arrow pierces. (c) The value of a smaller one-form on the same vector is a smaller number of surfaces. The larger the one-form, the more 'intense' the slicing of space in its picture.

a series of surfaces. The 'magnitude' of it is given by the spacing between the surfaces: the larger the spacing the *smaller* the magnitude. In this picture, the number produced when a one-form acts on a vector is the number of surfaces that the arrow of the vector pierces. So the closer their spacing, the larger the number (compare (b) and (c) in Fig. 3.1). In a four-dimensional space, the surfaces are three-dimensional. The one-form doesn't define a unique direction, since it is not a vector. Rather, it defines a way of 'slicing' the space. In order to justify this picture we shall look at a particular one-form, the gradient.

Derivative of a function is a one-form. Consider a scalar field $\phi(\vec{x})$ defined at every event $\vec{x}$. The world line of some particle (or person) encounters a value of ϕ at each event on it (see Fig. 3.2), and this value changes from event to event. If we label (parametrize) each point on the curve by the value of proper time τ along it (i.e. the reading of a clock moving

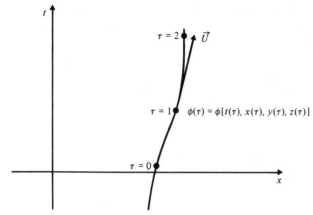

Fig. 3.2 A world line parametrized by proper time τ, and the values $\phi(\tau)$ of the scalar field $\phi(t, x, y, z)$ along it.

on the line), then we can express the coordinates of events on the curve as functions of τ:

$$[t = t(\tau), x = x(\tau), y = y(\tau), z = z(\tau)].$$

The four-velocity has components

$$\vec{U} \rightarrow \left(\frac{dt}{d\tau}, \frac{dx}{d\tau}, \ldots \right).$$

Since ϕ is a function of t, x, y and z, it is implicitly a function of τ on the curve:

$$\phi(\tau) = \phi[t(\tau), x(\tau), y(\tau), z(\tau)],$$

and its rate of change on the curve is

$$\frac{d\phi}{d\tau} = \frac{\partial\phi}{\partial t}\frac{dt}{d\tau} + \frac{\partial\phi}{\partial x}\frac{dx}{d\tau} + \frac{\partial\phi}{\partial y}\frac{dy}{d\tau} + \frac{\partial\phi}{\partial z}\frac{dz}{d\tau}$$

$$= \frac{\partial\phi}{\partial t}U^t + \frac{\partial\phi}{\partial x}U^x + \frac{\partial\phi}{\partial y}U^y + \frac{\partial\phi}{\partial z}U^z. \tag{3.14}$$

It is clear from this that in the last equation we have devised a means of producing from the vector $\vec{U}$ the number $d\phi/d\tau$ that represents the rate of change of ϕ on a curve on which $\vec{U}$ is the tangent. This number $d\phi/d\tau$ is clearly a linear function of $\vec{U}$, so we have defined a one-form.

By comparison with Eq. (3.8) we see that this one-form has components
$(\partial\phi/\partial t, \partial\phi/\partial x, \partial\phi/\partial y, \partial\phi/\partial z)$. This one-form is called the *gradient* of ϕ,
denoted by $\tilde{\mathbf{d}}\phi$:

$$\tilde{\mathbf{d}}\phi \underset{O}{\rightarrow} \left(\frac{\partial\phi}{\partial t}, \frac{\partial\phi}{\partial x}, \frac{\partial\phi}{\partial y}, \frac{\partial\phi}{\partial z}\right). \tag{3.15}$$

It is clear that the gradient fits our definition of a one-form. We will see
later how it comes about that the gradient is usually introduced in
three-dimensional vector calculus as a vector.

The gradient enables us to justify our picture of a one-form. In Fig.
3.3 we have drawn part of a topographical map, showing contours of
equal elevation. If h is the elevation, then the gradient $\tilde{\mathbf{d}}h$ is clearly
largest in an area like A, where the lines are closest together, and smallest

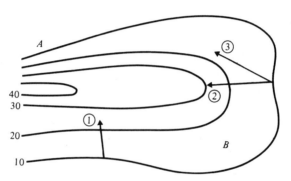

Fig. 3.3   A topographical map illustrates the gradient one-form (local contours
of constant elevation). The change of height along any trip (arrow) is the number
of contours crossed by the arrow.

near B, where the lines are spaced far apart. Moreover, suppose one
wanted to know how much elevation a walk between two points would
involve. On would lay out on the map a line (vector $\Delta\vec{x}$) between the
points. Then the number of contours the line crossed would give the
change in elevation. For example, line 1 crosses $1\frac{1}{2}$ contours, while 2
crosses 2 contours. Line 3 starts near 2 but goes in a different direction,
winding up only $\frac{1}{2}$ contour higher. But these numbers are just Δh, which
is the contraction of $\tilde{\mathbf{d}}h$ with $\Delta\vec{x}$: $\Delta h = \sum_i (\partial h/\partial x^i)\Delta x^i$ or the *value* of $\tilde{\mathbf{d}}h$
on $\Delta\vec{x}$ (see Eq. (3.8)).

Therefore, a one-form is represented by a series of surfaces (Fig. 3.4),
and its contraction with a vector $\vec{V}$ is the number of surfaces $\vec{V}$ crosses.
The closer the surfaces, the larger $\tilde{\omega}$. Properly, just as a vector is straight,

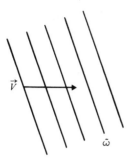

Fig. 3.4   The value $\tilde{\omega}(V)$ is 2.5.

the one-form's surfaces are straight and parallel. This is because we deal with one-forms at a point, not over an extended region: 'tangent' one-forms, in the same sense as tangent vectors.

These pictures show why one in general *cannot* call a gradient a vector. One would like to identify the *vector* gradient as that vector pointing 'up' the slope, i.e. in such a way that it crosses the greatest number of contours per unit length. The key phrase is 'per unit length'. If there is a metric, a measure of distance in the space, then a vector *can* be associated with a gradient. But the metric must intervene here in order to produce a vector. Geometrically, on its own, the gradient is a one-form.

Let us be sure that Eq. (3.15) is a consistent definition. How do the components transform? For a one-form we must have

$$(\tilde{\mathrm{d}}\phi)_{\bar{\alpha}} = \Lambda^{\beta}{}_{\bar{\alpha}}(\tilde{\mathrm{d}}\phi)_{\beta}. \tag{3.16}$$

But we know how to transform partial derivatives:

$$\frac{\partial\phi}{\partial x^{\bar{\alpha}}} = \frac{\partial\phi}{\partial x^{\beta}}\frac{\partial x^{\beta}}{\partial x^{\bar{\alpha}}},$$

which means

$$(\tilde{\mathrm{d}}\phi)_{\bar{\alpha}} = \frac{\partial x^{\beta}}{\partial x^{\bar{\alpha}}}(\tilde{\mathrm{d}}\phi)_{\beta}. \tag{3.17}$$

Are Eqs. (3.16) and (3.17) consistent? The answer, of course, is yes. The reason: since

$$x^{\beta} = \Lambda^{\beta}{}_{\bar{\alpha}}x^{\bar{\alpha}},$$

and since $\Lambda^{\beta}{}_{\bar{\alpha}}$ are just constants, then

♦ $$\qquad \partial x^{\beta}/\partial x^{\bar{\alpha}} = \Lambda^{\beta}{}_{\bar{\alpha}}. \tag{3.18}$$

This identity is fundamental. Components of the gradient transform according to the *inverse* of the components of vectors. So the gradient is the 'archetypal' one-form.

Notation for derivatives. From now on we shall employ the usual sub-scripted notation to indicate derivatives:

$$\frac{\partial \phi}{\partial x} \equiv \phi_{,x}$$

and, more generally,

$$\frac{\partial \phi}{\partial x^\alpha} \equiv \phi_{,\alpha}. \tag{3.19}$$

Note that the index α appears as a superscript in the denominator of the left-hand side of Eq. (3.19) and as a subscript on the right-hand side. As we have seen, this placement of indices is consistent with the transformation properties of the expression.

In particular, we have

$$x^\alpha_{,\beta} \equiv \delta^\alpha_{\ \beta},$$

which we can compare with Eq. (3.12) to conclude that

$$\tilde{d}x^\alpha \equiv \tilde{\omega}^\alpha. \tag{3.20}$$

This is a useful result, that the basis one-form is just $\tilde{d}x^\alpha$. We can use it to write, for any function f,

$$\tilde{d}f = \frac{\partial f}{\partial x^\alpha} \tilde{d}x^\alpha.$$

This looks very much like the physicist's 'sloppy-calculus' way of writing differentials or infinitesimals. The notation $\tilde{d}$ has been chosen partly to suggest this comparison, but this choice makes it doubly important for the student to avoid confusion on this point. The object $\tilde{d}f$ is a tensor, not a small increment in f; it *can* have a small ('infinitesimal') value if it is contracted with a small vector.

Normal one-forms. Like the gradient, the concept of a normal vector – a vector orthogonal to a surface – is one which is more naturally replaced by that of a normal one-form. For a normal vector to be defined we need to have a scalar product: the normal vector must be orthogonal to all vectors tangent to the surface. This can be defined only by using the metric tensor. But a normal one-form can be defined without reference to the metric. A one-form is said to be normal to a surface if its value is zero on every vector tangent to the surface. If the surface is closed and divides spacetime into an 'inside' and 'outside', a normal is said to be an *outward* normal one-form if it is a normal one-form and its value on vectors which point outwards from the surface is positive. In Exer. 13, § 3.10, we prove that $\tilde{d}f$ is normal to surfaces of constant f.

3.4 The $\binom{0}{2}$ tensors

These are tensors that have two vector arguments. We have encountered the metric tensor already, but the simplest of this type is the product of two one-forms, formed according to the following rule: if $\tilde{p}$ and $\tilde{q}$ are one-forms, then $\tilde{p} \otimes \tilde{q}$ is the $\binom{0}{2}$ tensor which, when supplied with vectors $\vec{A}$ and $\vec{B}$ as arguments, produces the number $\tilde{p}(\vec{A}) \cdot \tilde{q}(\vec{B})$, i.e. just the product of the numbers produced by the $\binom{0}{1}$ tensors. The symbol $\otimes$ is called an 'outer product sign' and is a formal notation to show how the $\binom{0}{2}$ tensor is formed from the one-forms. Notice that $\otimes$ is *not* commutative: $\tilde{p} \otimes \tilde{q}$ and $\tilde{q} \otimes \tilde{p}$ are *different* tensors. The first gives the value $\tilde{p}(\vec{A}) \, \tilde{q}(\vec{B})$, the second the value $\tilde{q}(\vec{A})\tilde{p}(\vec{B})$.

Components. The most general $\binom{0}{2}$ tensor is not a simple outer product, but it can always be represented as a sum of such tensors. To see this we must first consider the components of an arbitrary $\binom{0}{2}$ tensor $\mathbf{f}$:

$$f_{\alpha\beta} \equiv \mathbf{f}(\vec{e}_\alpha, \vec{e}_\beta). \tag{3.21}$$

Since each index can have four values, there are 16 components, and they can be thought of as being arrayed in a matrix. The value of $\mathbf{f}$ on arbitrary vectors is

$$\begin{aligned}
\mathbf{f}(\vec{A}, \vec{B}) &= \mathbf{f}(A^\alpha \vec{e}_\alpha, B^\beta \vec{e}_\beta) \\
&= A^\alpha B^\beta \mathbf{f}(\vec{e}_\alpha, \vec{e}_\beta) \\
&= A^\alpha B^\beta f_{\alpha\beta}.
\end{aligned} \tag{3.22}$$

(Again notice that two different dummy indices are used to keep the different summations distinct.) Can we form a basis for these tensors? That is, can we define a set of 16 $\binom{0}{2}$ tensors $\tilde{\omega}^{\alpha\beta}$ such that, analogous to Eq. (3.11),

$$\mathbf{f} = f_{\alpha\beta} \tilde{\omega}^{\alpha\beta} ? \tag{3.23}$$

For this to be the case we would have to have

$$f_{\mu\nu} = \mathbf{f}(\vec{e}_\mu, \vec{e}_\nu) = f_{\alpha\beta} \tilde{\omega}^{\alpha\beta}(\vec{e}_\mu, \vec{e}_\nu)$$

and this would imply, as before, that

$$\tilde{\omega}^{\alpha\beta}(\vec{e}_\mu, \vec{e}_\nu) = \delta^\alpha{}_\mu \delta^\beta{}_\nu. \tag{3.24}$$

But $\delta^\alpha{}_\mu$ is (by Eq. (3.12)), the value of $\tilde{\omega}^\alpha$ on $\vec{e}_\mu$, and analogously for $\delta^\beta{}_\nu$. Therefore, $\tilde{\omega}^{\alpha\beta}$ is a tensor whose value is just the product of the values of two basis one-forms, and we therefore conclude

$$\tilde{\omega}^{\alpha\beta} = \tilde{\omega}^\alpha \otimes \tilde{\omega}^\beta. \tag{3.25}$$

So the tensors $\tilde{\omega}^\alpha \otimes \tilde{\omega}^\beta$ are a basis for all $\binom{0}{2}$ tensors, and we can write

$$\blacklozenge \qquad \mathbf{f} = f_{\alpha\beta} \tilde{\omega}^\alpha \otimes \tilde{\omega}^\beta. \tag{3.26}$$

This is one way in which a general $\binom{0}{2}$ tensor is a sum over simple outer-product tensors.

Symmetries. A $\binom{0}{2}$ tensor takes two arguments, and their order is important, as we have seen. The behavior of the value of a tensor under an interchange of its arguments is an important property of it. A tensor **f** is called *symmetric* if

$$\mathbf{f}(\vec{A}, \vec{B}) = \mathbf{f}(\vec{B}, \vec{A}) \quad \forall \vec{A}, \vec{B}. \tag{3.27}$$

Setting $\vec{A} = \vec{e}_\alpha$ and $\vec{B} = \vec{e}_\beta$, this implies of its components that

$$f_{\alpha\beta} = f_{\beta\alpha}. \tag{3.28}$$

This is the same as the condition that the matrix array of the elements is symmetric. An arbitrary $\binom{0}{2}$ tensor **h** can define a new symmetric $\mathbf{h}_{(s)}$ by the rule

$$\mathbf{h}_{(s)}(\vec{A}, \vec{B}) = \tfrac{1}{2}\mathbf{h}(\vec{A}, \vec{B}) + \tfrac{1}{2}\mathbf{h}(\vec{B}, \vec{A}). \tag{3.29}$$

Make sure you understand that $\mathbf{h}_{(s)}$ satisfies Eq. (3.27) above. For the components this implies

$$h_{(s)\alpha\beta} = \tfrac{1}{2}(h_{\alpha\beta} + h_{\beta\alpha}). \tag{3.30}$$

This is such an important mathematical property that a special notation is used for it:

◆ $$h_{(\alpha\beta)} \equiv \tfrac{1}{2}(h_{\alpha\beta} + h_{\beta\alpha}). \tag{3.31}$$

Therefore the numbers $h_{(\alpha\beta)}$ are the components of the symmetric tensor formed from **h**.

Similarly, a tensor **f** is called *antisymmetric* if

$$\mathbf{f}(\vec{A}, \vec{B}) = -\mathbf{f}(\vec{B}, \vec{A}), \quad \forall \vec{A}, \vec{B}, \tag{3.32}$$

$$f_{\alpha\beta} = -f_{\beta\alpha}. \tag{3.33}$$

An antisymmetric $\binom{0}{2}$ tensor can always be formed as

$$\mathbf{h}_{(A)}(\vec{A}, \vec{B}) = \tfrac{1}{2}\mathbf{h}(\vec{A}, \vec{B}) - \tfrac{1}{2}\mathbf{h}(\vec{B}, \vec{A}),$$

$$h_{(A)\alpha\beta} = \tfrac{1}{2}(h_{\alpha\beta} - h_{\beta\alpha}).$$

The notation here is to use square brackets on the indices:

◆ $$h_{[\alpha\beta]} = \tfrac{1}{2}(h_{\alpha\beta} - h_{\beta\alpha}). \tag{3.34}$$

Notice that

$$h_{\alpha\beta} = \tfrac{1}{2}(h_{\alpha\beta} + h_{\beta\alpha}) + \tfrac{1}{2}(h_{\alpha\beta} - h_{\beta\alpha})$$

$$= h_{(\alpha\beta)} + h_{[\alpha\beta]}. \tag{3.35}$$

So any $\binom{0}{2}$ tensor can be split *uniquely* into its symmetric and antisymmetric parts.

The metric tensor **g** is symmetric, as can be deduced from Eq. (2.26):

$$\mathbf{g}(\vec{A}, \vec{B}) = \mathbf{g}(\vec{B}, \vec{A}). \tag{3.36}$$

3.5 Metric as a mapping of vectors into one-forms

We now introduce what we shall later see is the fundamental role of the metric in differential geometry, to act as a mapping between vectors and one-forms. To see how this works, consider $\mathbf{g}$ and a single vector $\vec{V}$. Since $\mathbf{g}$ requires two vectorial arguments, the expression $\mathbf{g}(\vec{V},\ )$ still lacks one: when another one is supplied, it becomes a number. Therefore, $\mathbf{g}(\vec{V},\ )$ considered as a function of vectors (which are to fill in the empty 'slot' in it), is a linear function of vectors producing real numbers: a one-form. We call it $\tilde{V}$:

$$\mathbf{g}(\vec{V},\ ) \equiv \tilde{V}(\ ), \tag{3.37}$$

where blanks inside parentheses are a way of indicating that a vector argument is to be supplied. Then $\tilde{V}$ is the one-form whose value on a vector $\vec{A}$ is $\vec{V} \cdot \vec{A}$:

$$\tilde{V}(\vec{A}) \equiv \mathbf{g}(\vec{V}, \vec{A}) = \vec{V} \cdot \vec{A}. \tag{3.38}$$

Note that since $\mathbf{g}$ is symmetric, we also can write

$$\mathbf{g}(\ , \vec{V}) \equiv \tilde{V}(\ ).$$

What are the components of $\tilde{V}$? They are

$$\begin{aligned}
V_\alpha \equiv \tilde{V}(\vec{e}_\alpha) &= \vec{V} \cdot \vec{e}_\alpha = \vec{e}_\alpha \cdot \vec{V} \\
&= \vec{e}_\alpha \cdot (V^\beta \vec{e}_\beta) \\
&= (\vec{e}_\alpha \cdot \vec{e}_\beta) V^\beta
\end{aligned}$$

◆ $$V_\alpha = \eta_{\alpha\beta} V^\beta. \tag{3.39}$$

It is important to notice here that we distinguish the components V^α of $\vec{V}$ from the components V_β of $\tilde{V}$ *only* by the position of the index: on a vector it is up; on a one-form, down. Then, from Eq. (3.39), we have as a special case

$$\begin{aligned}
V_0 = V^\beta \eta_{\beta 0} &= V^0 \eta_{00} + V^1 \eta_{10} + \cdots \\
&= V^0(-1) + 0 + 0 + 0 \\
&= -V^0, \tag{3.40}
\end{aligned}$$

$$\begin{aligned}
V_1 = V^\beta \eta_{\beta 1} &= V^0 \eta_{01} + V^1 \eta_{11} + \cdots \\
&= +V^1, \tag{3.41}
\end{aligned}$$

etc. This may be summarized as:

$$\text{if } \vec{V} \to (a, b, c, d),$$
$$\text{then } \tilde{V} \to (-a, b, c, d). \tag{3.42}$$

The components of $\tilde{V}$ are obtained from those of $\vec{V}$ by changing the sign of the time component. (Since this depended upon the components $\eta_{\alpha\beta}$, in situations we encounter later, where the metric has more complicated components, this rule of correspondence between $\tilde{V}$ and $\vec{V}$ will also be more complicated.)

The inverse: going from $\tilde{A}$ to $\vec{A}$. Does the metric also provide a way of finding a vector $\vec{A}$ that is related to a given one-form $\tilde{A}$? The answer is yes. Consider Eq. (3.39). It says that $\{V_\alpha\}$ is obtained by multiplying $\{V^\beta\}$ by a matrix $(\eta_{\alpha\beta})$. If this matrix has an inverse, then one could use it to obtain $\{V^\beta\}$ from $\{V_\alpha\}$. This inverse exists if and only if $(\eta_{\alpha\beta})$ has non-vanishing determinant. But since $(\eta_{\alpha\beta})$ is a diagonal matrix with entries $(-1, 1, 1, 1)$, its determinant is simply -1. An inverse does exist, and we call its components $\eta^{\alpha\beta}$. Then, given $\{A_\beta\}$ we can find $\{A^\alpha\}$:

♦ $$A^\alpha \equiv \eta^{\alpha\beta} A_\beta. \tag{3.43}$$

The use of the inverse guarantees that the two sets of components satisfy Eq. (3.39):

$$A_\beta = \eta_{\beta\alpha} A^\alpha.$$

So the mapping provided by **g** between vectors and one-forms is one-to-one and invertible.

In particular, with $\tilde{d}\phi$ we can associate a vector $\vec{d}\phi$, which is the one usually associated with the gradient. One can see that this vector is orthogonal to surfaces of constant ϕ as follows: its inner product with a vector in a surface of constant ϕ is, by this mapping, identical with the value of the one-form $\tilde{d}\phi$ on that vector. This, in turn, must be zero since $\tilde{d}\phi(\vec{V})$ is the rate of change of ϕ along $\vec{V}$, which in this case is zero since $\vec{V}$ is taken to be in a surface of constant ϕ.

It is important to know what $\{\eta^{\alpha\beta}\}$ is. You can easily verify that

$$\eta^{00} = -1, \qquad \eta^{0i} = 0, \qquad \eta^{ij} = \delta^{ij}, \tag{3.44}$$

so that $(\eta^{\alpha\beta})$ is *identical* to $(\eta_{\alpha\beta})$. Thus, to go from a one-form to a vector, simply change the sign of the time component.

Why distinguish one-forms from vectors? In Euclidean space, in Cartesian coordinates the metric is just $\{\delta_{ij}\}$, so the components of one-forms and vectors are the same. Therefore no distinction is ever made in elementary vector algebra. But in SR the components differ (by that one change in sign). Therefore, whereas the gradient has components

$$\tilde{d}\phi \rightarrow \left(\frac{\partial \phi}{\partial t}, \frac{\partial \phi}{\partial x}, \ldots \right),$$

the associated vector normal to surfaces of constant ϕ has components

$$\vec{d}\phi \rightarrow \left(-\frac{\partial \phi}{\partial t}, \frac{\partial \phi}{\partial x}, \ldots \right). \tag{3.45}$$

Had we simply tried to *define* the 'vector gradient' of a function as the vector with these components, without first discussing one-forms, the reader would have been justified in being more than a little sceptical. The nonEuclidean metric of SR forces us to be aware of the basic distinction between one-forms and vectors: it can't be swept under the rug.

As we remarked earlier, vectors and one-forms are dual to one another. Such dual spaces are important and are found elsewhere in mathematical physics. The simplest example is the space of column vectors

$$\begin{pmatrix} a \\ b \\ \vdots \end{pmatrix},$$

whose dual space is the space of row vectors $(a \ b \ \cdots)$. Notice that the product

$$(a \quad b \quad \cdots) \begin{pmatrix} p \\ q \\ \vdots \end{pmatrix} = ap + bq + \cdots \tag{3.46}$$

is a real number, so that a row vector can be considered to be a one-form on column vectors. The operation of finding an element of one space from one of the other is called the 'adjoint' and is 1–1 and invertible. A less trivial example arises in quantum mechanics. A wave-function (probability amplitude that is a solution to Schrödinger's equation) is a complex scalar field $\psi(\vec{x})$, and is drawn from the *Hilbert space* of all such functions. This Hilbert space is a vector space, since its elements (functions) satisfy the axioms of a vector space. What is the dual space of one-forms? The crucial hint is that the inner product of any two functions $\phi(\vec{x})$ and $\psi(\vec{x})$ is *not* $\int \phi(\vec{x})\psi(\vec{x})\,d^3x$ but, rather, is $\int \phi^*(\vec{x})\psi(\vec{x})\,d^3x$, the asterisk denoting complex conjugation. The function $\phi^*(\vec{x})$ acts like a one-form whose value on $\psi(\vec{x})$ is its integral with it (analogous to the sum in Eq. (3.8)). The operation of complex conjugation acts like our metric tensor, transforming a vector $\phi(\vec{x})$ (in the Hilbert space) into a one-form $\phi^*(\vec{x})$. The fact that $\phi^*(\vec{x})$ is also a function in the Hilbert space is, at this level, a distraction. (It is equivalent to saying that members of the set $(1, -1, 0, 0)$ can be components of either a vector or a one-form.) The important point is that in the integral $\int \phi^*(\vec{x})\psi(\vec{x})\,d^3x$, the function $\phi^*(\vec{x})$ is acting as a one-form, producing a real number from the vector $\psi(\vec{x})$. This dualism is most clearly brought out in the Dirac 'bra' and 'ket' notation. Elements of the space of all states of the system are called $|\ \rangle$ (with identifying labels written inside), while the elements of the dual (adjoint with complex conjugate) space are called $\langle\ |$. Two 'vectors' $|1\rangle$ and $|2\rangle$ don't form a number, but a vector and a dual vector $|1\rangle$ and $\langle 2|$ do: $\langle 2|1\rangle$ is the name of this number.

In such ways the concept of a dual vector space arises very frequently in advanced mathematical physics.

Magnitudes and scalar products of one-forms. A one-form $\tilde{p}$ is defined to have the same magnitude as its associated vector $\vec{p}$. Thus we write

$$\tilde{p}^2 = \vec{p}^2 = \eta_{\alpha\beta} p^\alpha p^\beta. \qquad (3.47)$$

This would seem to involve finding $\{p^\alpha\}$ from $\{p_\alpha\}$ before using Eq. (3.47), but we can easily get around this. We use Eq. (3.43) for both p^α and p^β in Eq. (3.47):

$$\tilde{p}^2 = \eta_{\alpha\beta}(\eta^{\alpha\mu} p_\mu)(\eta^{\beta\nu} p_\nu). \qquad (3.48)$$

(Notice that each independent summation uses a different dummy index.) But since $\eta_{\alpha\beta}$ and $\eta^{\beta\nu}$ are inverse matrices to each other, the sum on β collapses:

$$\eta_{\alpha\beta}\eta^{\beta\nu} = \delta^\nu{}_\alpha. \qquad (3.49)$$

Using this in Eq. (3.48) gives

◆ $$\tilde{p}^2 = \eta^{\alpha\mu} p_\mu p_\alpha. \qquad (3.50)$$

Thus, the inverse metric tensor can be used directly to find the magnitude of $\tilde{p}$ from its components. We can use Eq. (3.44) to write this explicitly as

$$\tilde{p}^2 = -(p_0)^2 + (p_1)^2 + (p_2)^2 + (p_3)^2. \qquad (3.51)$$

This is the same rule, in fact, as Eq. (2.24) for vectors. By its definition, this is frame invariant. One-forms are timelike, spacelike, or null, as their associated vectors are.

As with vectors, we can now define an inner product of one-forms. This is

$$\tilde{p} \cdot \tilde{q} \equiv \tfrac{1}{2}[(\tilde{p} + \tilde{q})^2 - \tilde{p}^2 - \tilde{q}^2]. \qquad (3.52)$$

Its expression in terms of components is, not surprisingly,

$$\tilde{p} \cdot \tilde{q} = -p_0 q_0 + p_1 q_1 + p_2 q_2 + p_3 q_3. \qquad (3.53)$$

Normal vectors and unit normal one-forms. A vector is said to be normal to a surface if its associated one-form is a normal one-form. Eq. (3.38) shows that this definition is equivalent to the usual one that the vector be orthogonal to all tangent vectors. A normal vector or one-form is said to be a *unit normal* if its magnitude is ±1. (We can't demand that it be $+1$, since timelike vectors will have negative magnitudes. All we can do is to multiply the vector or form by an overall factor to scale its magnitude to ±1.) Note that null normals cannot be unit normals.

A three-dimensional surface is said to be timelike, spacelike, or null according to which of these classes its normal falls into. (Exer. 12, § 3.10, proves that this definition is self-consistent.) In Exer. 21, § 3.10, we explore the following curious properties normal vectors have on account of our metric. An outward normal vector is the vector associated with an outward normal one-form, as defined earlier. This ensures that its scalar product with any vector which points outwards is positive. If the surface is spacelike, the outward normal vector points outwards. If the surface is timelike, however, the outward normal vector points *inwards*. And if the surface is null, the outward vector is *tangent* to the surface! These peculiarities simply reinforce the view that it is more natural to regard the normal as a one-form, where the metric doesn't enter the definition.

3.6 Finally: $\binom{M}{N}$ tensors

Vector as a function of one-forms. The dualism discussed above is in fact complete. Although we defined one-forms as functions of vectors, we can now see that vectors can perfectly well be regarded as linear functions that map one-forms into real numbers. Given a vector $\vec{V}$, once we supply a one-form we get a real number:

$$\vec{V}(\tilde{p}) \equiv \tilde{p}(\vec{V}) \equiv p_\alpha V^\alpha \equiv \langle \tilde{p}, \vec{V} \rangle. \tag{3.54}$$

In this way we dethrone vectors from their special position as things 'acted on' by tensors, and regard them as tensors themselves, specifically as linear functions of single one-forms into real numbers. The last notation on Eq. (3.54) is new, and emphasizes the equal status of the two objects.

$\binom{M}{0}$ *Tensors.* Generalizing this, we define:

> An $\binom{M}{0}$ tensor is a linear function of M one-forms into the real numbers.

All our previous discussions of $\binom{0}{N}$ tensors apply here. A simple $\binom{2}{0}$ tensor is $\vec{V} \otimes \vec{W}$, which, when supplied with two arguments $\tilde{p}$ and $\tilde{q}$, gives the number $\vec{V}(\tilde{p})\vec{W}(\tilde{q}) \equiv \tilde{p}(\vec{V})\tilde{q}(\vec{W}) = V^\alpha p_\alpha W^\beta q_\beta$. So $\vec{V} \otimes \vec{W}$ has components $V^\alpha W^\beta$. A basis for $\binom{2}{0}$ tensors is $\vec{e}_\alpha \otimes \vec{e}_\beta$. The components of an $\binom{M}{0}$ tensor are its values when the basis one-form $\tilde{\omega}^\alpha$ are its arguments. Notice that $\binom{M}{0}$ tensors have components all of whose indices are superscripts.

$\binom{M}{N}$ *tensors.* The final generalization is:

◆ > An $\binom{M}{N}$ tensor is a linear function of M one-forms *and* N vectors into the real numbers.

For instance, if **R** is a $\binom{1}{1}$ tensor then it requires a one-form $\tilde{p}$ and a vector $\vec{A}$ to give a number $\mathbf{R}(\tilde{p}; \vec{A})$. It has components $\mathbf{R}(\tilde{\omega}^{\alpha}; \vec{e}_{\beta}) \equiv R^{\alpha}{}_{\beta}$. In general, the components of a $\binom{M}{N}$ tensor will have M indices up and N down. In a new frame,

$$
\begin{aligned}
R^{\bar{\alpha}}{}_{\bar{\beta}} &= \mathbf{R}(\tilde{\omega}^{\bar{\alpha}}; \vec{e}_{\bar{\beta}}) \\
&= \mathbf{R}(\Lambda^{\bar{\alpha}}{}_{\mu}\tilde{\omega}^{\mu}; \Lambda^{\nu}{}_{\bar{\beta}}\vec{e}_{\nu}) \\
&= \Lambda^{\bar{\alpha}}{}_{\mu}\Lambda^{\nu}{}_{\bar{\beta}}R^{\mu}{}_{\nu}.
\end{aligned}
\tag{3.55}
$$

So the transformation of components is simple: each index transforms by bringing in a Λ whose indices are arranged in the only way permitted by the summation convention. Some old names that are still in current use are: upper indices are called 'contravariant' (because they transform *contrary* to basis vectors) and lower ones 'covariant'. An $\binom{M}{N}$ tensor is said to be 'M-times contravariant and N-times covariant'.

Circular reasoning? At this point the student might worry that all of tensor algebra has become circular: one-forms were defined in terms of vectors, but now we have defined vectors in terms of one-forms. This 'duality' is at the heart of the theory, but is not circularity. It means we can do as physicists do, which is to identify the vectors with displacements $\Delta \vec{x}$ and things like it (such as $\vec{p}$ and $\vec{v}$) and then generate all $\binom{M}{N}$ tensors by the rules of tensor algebra; these tensors inherit a physical meaning from the original meaning we gave vectors. But we could equally well have associated one-forms with some physical objects (gradients, for example) and recovered the whole algebra from that starting point. The power of the mathematics is that it doesn't need (or want) to say *what* the original vectors or one-forms are. It simply gives rules for manipulating them. The association of, say, $\vec{p}$ with a vector is at the interface between physics and mathematics: it is how we make a mathematical model of the physical world. A geometer does the same. He adds to the notion of these abstract tensor spaces the idea of what a vector in a curved space is. The modern geometer's idea of a vector is something we shall learn about when we come to curved spaces. For now we will get some practice with tensors in physical situations, where we stick with our (admittedly imprecise) notion of vectors 'like' $\Delta \vec{x}$.

3.7 Index 'raising' and 'lowering'

In the same way that the metric maps a vector $\vec{V}$ into a one-form $\tilde{V}$, it maps an $\binom{N}{M}$ tensor into an $\binom{N-1}{M+1}$ tensor. Similarly, the inverse maps an $\binom{N}{M}$ tensor into an $\binom{N+1}{M-1}$ tensor. Normally these are given the same name, and are distinguished only by the positions of their indices. Suppose $T^{\alpha\beta}{}_\gamma$ are the components of a $\binom{2}{1}$ tensor. Then

$$T^\alpha{}_{\beta\gamma} \equiv \eta_{\beta\mu} T^{\alpha\mu}{}_\gamma \tag{3.56}$$

are the components of a $\binom{1}{2}$ tensor (obtained by mapping the *second* one-form argument of $T^{\alpha\beta}{}_\gamma$ into a vector), and

$$T_\alpha{}^\beta{}_\gamma \equiv \eta_{\alpha\mu} T^{\mu\beta}{}_\gamma \tag{3.57}$$

are the components of another (inequivalent) $\binom{1}{2}$ tensor (mapping on the *first* index), while

$$T^{\alpha\beta\gamma} \equiv \eta^{\gamma\mu} T^{\alpha\beta}{}_\mu \tag{3.58}$$

are the components of a $\binom{3}{0}$ tensor. These operations are, naturally enough, called index 'raising' and 'lowering'. Whenever we speak of raising or lowering an index we mean this map generated by the metric. The rule in SR is simple: when raising or lowering a '0' index, the sign of the component changes; when raising or lowering a '1' or '2' or '3' index (in general, an 'i' index) the component is unchanged.

Mixed components of metric. The numbers $\{\eta_{\alpha\beta}\}$ are the components of the metric, and $\{\eta^{\alpha\beta}\}$ those of its inverse. Suppose we raise an index of $\eta_{\alpha\beta}$ using the inverse. Then we get the 'mixed' components of the metric,

$$\eta^\alpha{}_\beta \equiv \eta^{\alpha\mu} \eta_{\mu\beta}. \tag{3.59}$$

But on the right we have just the matrix product of two matrices that are the inverse of each other (readers who aren't sure of this should verify the following equation by direct calculation), so it is the unit identity matrix. Since one index is up and one down, it is the Kronecker delta, written as

$$\blacklozenge \qquad \eta^\alpha{}_\beta \equiv \delta^\alpha{}_\beta. \tag{3.60}$$

By raising the other index we merely obtain an identity, $\eta^{\alpha\beta} = \eta^{\alpha\beta}$. So we can regard $\eta^{\alpha\beta}$ as the components of the $\binom{2}{0}$ tensor which is mapped from the $\binom{0}{2}$ tensor $\mathbf{g}$ by $\mathbf{g}^{-1}$. So, for $\mathbf{g}$, its 'contravariant' components equal the elements of the matrix inverse to the matrix of its 'covariant' components. It is the only tensor for which this is true.

Metric and nonmetric vector algebras. It is of some interest to ask why the metric is the one that generates the correspondence between one-forms and vectors. Why not some other $\binom{0}{2}$ tensor that has an inverse? We'll explore that idea in stages.

First, why a correspondence at all? Suppose we had a 'nonmetric' vector algebra, complete with all the dual spaces and $\binom{M}{N}$ tensors. Why make a correspondence between one-forms and vectors? The answer is that sometimes one does and sometimes one doesn't. Without one, the inner product of two vectors is undefined, since numbers are produced only when one-forms act on vectors and vice-versa. In physics, scalar products are useful, so one needs a metric. But there are *some* vector spaces in mathematical physics where metrics are not important. An example is phase space of classical and quantum mechanics.

Second, why the metric and not another tensor? If a metric weren't defined but another symmetric tensor did the mapping, a mathematician would just call the other tensor the metric. That is, he would define it as the one generating a mapping. To a mathematician, the metric is an added bit of *structure* in the vector algebra. Different spaces in mathematics can have different metric structures. A *Riemannian* space is characterized by a metric that gives positive-definite magnitudes of vectors. One like ours, with indefinite sign, is called *pseudo-Riemannian.* One can even define a 'metric' that is *antisymmetric*: a two-dimensional space called *spinor space* has such a metric, and it turns out to be of fundamental importance in physics. But its structure is outside the scope of our lectures. The point here is that we don't have SR if we just discuss vectors and tensors. We get SR when we say that we have a metric with components $\eta_{\alpha\beta}$. If we assigned other components we might get other spaces, in particular the curved spacetime of GR.

3.8 Differentiation of tensors

A function f is a $\binom{0}{0}$ tensor, and its gradient $\tilde{d}f$ is a $\binom{0}{1}$ tensor. Differentiation of a function produces a tensor of one higher (covariant) rank. We shall now see that this applies as well to differentiation of tensors of *any* rank.

Consider a $\binom{1}{1}$ tensor **T** whose components $\{T^{\alpha}{}_{\beta}\}$ are functions of position. We can write **T** as

$$\mathbf{T} = T^{\alpha}{}_{\beta}\tilde{\omega}^{\beta} \otimes \vec{e}_{\alpha}. \tag{3.61}$$

Suppose, as we did for functions, that we move along a world line with parameter τ, proper time. The rate of change of **T**,

$$\frac{d\mathbf{T}}{d\tau} = \lim_{\Delta\tau \to 0} \frac{\mathbf{T}(\tau + \Delta\tau) - \mathbf{T}(\tau)}{\Delta\tau}, \tag{3.62}$$

is not hard to calculate. Since the basis one-forms and vectors are the same everywhere (i.e. $\tilde{\omega}^{\alpha}(\tau + \Delta\tau) = \tilde{\omega}^{\alpha}(\tau)$), it follows that

$$\frac{d\mathbf{T}}{d\tau} = \left(\frac{dT^{\alpha}{}_{\beta}}{d\tau}\right)\tilde{\omega}^{\beta} \otimes \vec{e}_{\alpha}, \tag{3.63}$$

where $dT^\alpha{}_\beta/d\tau$ is the ordinary derivative of the function $T^\alpha{}_\beta$ along the world line:

$$dT^\alpha{}_\beta/d\tau = T^\alpha{}_{\beta,\gamma} U^\gamma. \tag{3.64}$$

Now, the object $d\mathbf{T}/d\tau$ is a $\binom{1}{1}$ tensor, since in Eq. (3.62) it is defined to be just the difference between two such tensors. From Eqs. (3.63) and (3.64) we have, for any vector $\vec{U}$,

$$d\mathbf{T}/d\tau = (T^\alpha{}_{\beta,\gamma}\tilde{\omega}^\beta \otimes \vec{e}_\alpha) U^\gamma, \tag{3.65}$$

from which we can deduce that

$$\nabla\mathbf{T} \equiv T^\alpha{}_{\beta,\gamma}\tilde{\omega}^\beta \otimes \tilde{\omega}^\gamma \otimes \vec{e}_\alpha \tag{3.66}$$

is a $\binom{1}{2}$ tensor. This tensor is called the gradient of $\mathbf{T}$.

We use the notation $\nabla\mathbf{T}$ rather than $\tilde{d}\mathbf{T}$ because the latter notation is usually reserved for something else. We also have a convenient notation for Eq. (3.65):

$$d\mathbf{T}/d\tau \equiv \nabla_{\vec{U}}\mathbf{T}, \tag{3.67}$$

$$\nabla_{\vec{U}}\mathbf{T} \rightarrow \{T^\alpha{}_{\beta,\gamma} U^\gamma\}. \tag{3.68}$$

This derivation made use of the fact that the basis vectors (and therefore the basis one-forms) were constant everywhere. We will find that we can't assume this in the curved spacetime of GR, and taking this into account will be our entry point into the theory!

3.9 Bibliography

Our approach to tensor analysis stresses the geometrical nature of tensors rather than the transformation properties of their components. Students who wish amplification of some of the points here can consult the early chapters of Misner *et al.* (1973) or Schutz (1980*b*). Most introductions to tensors for physicists outside relativity confine themselves to 'Cartesian' tensors, i.e. to tensor components in three-dimensional Cartesian coordinates. See, for example, the chapter in Mathews & Walker (1965).

A very complete reference for tensor analysis in the older style based upon coordinate transformations is Schouten (1954). See also Yano (1955). Books which develop that point of view for tensors in relativity include Adler *et al.* (1975), Landau & Lifshitz (1962), Robertson & Noonan (1968), and Stephani (1982).

3.10 Exercises

1(a) Given an arbitrary set of numbers $\{M_{\alpha\beta}; \alpha = 0, \ldots, 3; \beta = 0, \ldots, 3\}$ and two arbitrary sets of vector components $\{A^\mu, \mu = 0, \ldots, 3\}$ and $\{B^\nu,$

$\nu = 0, \ldots, 3\}$, show that the two expressions

$$M_{\alpha\beta}A^{\alpha}B^{\beta} \equiv \sum_{\alpha=0}^{3}\sum_{\beta=0}^{3} M_{\alpha\beta}A^{\alpha}B^{\beta}$$

and

$$\sum_{\alpha=0}^{3} M_{\alpha\alpha}A^{\alpha}B^{\alpha}$$

are not equivalent.

(b) Show that
$$A^{\alpha}B^{\beta}\eta_{\alpha\beta} = -A^0B^0 + A^1B^1 + A^2B^2 + A^3B^3.$$

2 Prove that the set of all one-forms is a vector space.

3(a) Prove, by writing out all the terms, the validity of the following
$$\tilde{p}(A^{\alpha}\vec{e}_{\alpha}) = A^{\alpha}\tilde{p}(\vec{e}_{\alpha}).$$

(b) Let the components of $\tilde{p}$ be $(-1, 1, 2, 0)$, those of $\vec{A}$ be $(2, 1, 0, -1)$ and those of $\vec{B}$ be $(0, 2, 0, 0)$. Find (i) $\tilde{p}(\vec{A})$; (ii) $\tilde{p}(\vec{B})$; (iii) $\tilde{p}(\vec{A} - 3\vec{B})$; (iv) $\tilde{p}(\vec{A}) - 3\tilde{p}(\vec{B})$.

4 Given the following vectors in $\mathcal{O}$:
$$\vec{A} \underset{\mathcal{O}}{\to} (2, 1, 1, 0),\ \vec{B} \underset{\mathcal{O}}{\to} (1, 2, 0, 0),\ \vec{C} \underset{\mathcal{O}}{\to} (0, 0, 1, 1),\ \vec{D} \underset{\mathcal{O}}{\to} (-3, 2, 0, 0),$$

(a) show that they are linearly independent;
(b) find the components of $\tilde{p}$ if
$$\tilde{p}(\vec{A}) = 1,\ \tilde{p}(\vec{B}) = -1,\ \tilde{p}(\vec{C}) = -1,\ \tilde{p}(\vec{D}) = 0;$$
(c) find the value of $\tilde{p}(\vec{E})$ for
$$\vec{E} \underset{\mathcal{O}}{\to} (1, 1, 0, 0);$$
(d) determine whether the one-forms $\tilde{p}$, $\tilde{q}$, $\tilde{r}$, and $\tilde{s}$ are linearly independent if $\tilde{q}(\vec{A}) = \tilde{q}(\vec{B}) = 0$, $\tilde{q}(\vec{C}) = 1$, $\tilde{q}(\vec{D}) = -1$, $\tilde{r}(\vec{A}) = 2$, $\tilde{r}(\vec{B}) = \tilde{r}(\vec{C}) = r(\vec{D}) = 0$, $\tilde{s}(\vec{A}) = -1$, $\tilde{s}(\vec{B}) = -1$, $\tilde{s}(\vec{C}) = \tilde{s}(\vec{D}) = 0$.

5 Justify each step leading from Eqs. (3.10a) to (3.10d).

6 Consider the basis $\{\vec{e}_{\alpha}\}$ of a frame $\mathcal{O}$ and the basis $(\bar{\lambda}^0, \bar{\lambda}^1, \bar{\lambda}^2, \bar{\lambda}^3)$ for the space of one-forms, where we have

$$\bar{\lambda}^0 \underset{\mathcal{O}}{\to} (1, 1, 0, 0),$$

$$\bar{\lambda}^1 \underset{\mathcal{O}}{\to} (1, -1, 0, 0),$$

$$\bar{\lambda}^2 \underset{\mathcal{O}}{\to} (0, 0, 1, -1),$$

$$\bar{\lambda}^3 \underset{\mathcal{O}}{\to} (0, 0, 1, 1).$$

Note that $\{\bar{\lambda}^{\beta}\}$ is *not* the basis dual to $\{\vec{e}_{\alpha}\}$.
(a) Show that $\tilde{p} \neq \tilde{p}(\vec{e}_{\alpha})\bar{\lambda}^{\alpha}$ for arbitrary $\tilde{p}$.

(b) Let $\tilde{p} \to_{\mathcal{O}} (1, 1, 1, 1)$. Find numbers l_α such that
$\tilde{p} = l_\alpha \tilde{\lambda}^\alpha$.
These are the components of $\tilde{p}$ on $\{\tilde{\lambda}^\alpha\}$.

7 Prove Eq. (3.13).

8 Draw the basis one-forms $\tilde{d}t$ and $\tilde{d}x$ of a frame $\mathcal{O}$.

9 Fig. 3.5 shows curves of equal temperature T (isotherms) of a metal plate. At the points $\mathscr{P}$ and $\mathcal{Q}$ as shown, estimate the components of the gradient $\tilde{d}T$. (Hint: the components are the contractions with the basis vectors, which can be estimated by counting the number of isotherms crossed by the vectors.)

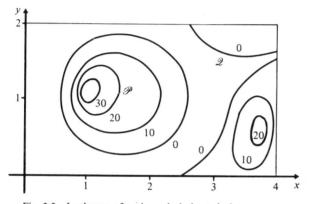

Fig. 3.5   Isotherms of an irregularly heated plate.

10(a) Given a frame $\mathcal{O}$ whose coordinates are $\{x^\alpha\}$, show that
$\partial x^\alpha / \partial x^\beta = \delta^\alpha{}_\beta$.

(b) For any two frames, we have, Eq. (3.18):
$\partial x^\beta / \partial x^{\bar{\alpha}} = \Lambda^\beta{}_{\bar{\alpha}}$.
Show that (a) and the chain rule imply
$\Lambda^\beta{}_{\bar{\alpha}} \Lambda^{\bar{\alpha}}{}_\mu = \delta^\beta{}_\mu$.
This is the inverse property again.

11 Use the notation $\partial \phi / \partial x^\alpha = \phi_{,\alpha}$ to re-write Eqs. (3.14), (3.15), and (3.18).

12 Let S be the two-dimensional plane $x = 0$ in three-dimensional Euclidean space. Let $\tilde{n} \neq 0$ be a normal one-form to S.
(a) Show that if $\vec{V}$ is a vector which is not tangent to S, then $\tilde{n}(\vec{V}) \neq 0$.
(b) Show that if $\tilde{n}(\vec{V}) > 0$, then $\tilde{n}(\vec{W}) > 0$ for any $\vec{W}$ which points toward the same side of S as $\vec{V}$ does (i.e. any $\vec{W}$ whose x component has the same sign as V^x).

(c) Show that any normal to S is a multiple of $\tilde{n}$.
(d) Generalize these statements to an arbitrary three-dimensional surface in four-dimensional spacetime.

13 Prove, by geometric or algebraic arguments, that $\tilde{d}f$ is normal to surfaces of constant f.

14 Let $\tilde{p} \rightarrow_O (1, 1, 0, 0)$ and $\tilde{q} \rightarrow_O (-1, 0, 1, 0)$ be two one-forms. Prove, by trying two vectors $\vec{A}$ and $\vec{B}$ as arguments, that $\tilde{p} \otimes \tilde{q} \neq \tilde{q} \otimes \tilde{p}$. Then find the components of $\tilde{p} \otimes \tilde{q}$.

15 Supply the reasoning leading from Eq. (3.23) to Eq. (3.24).

16(a) Prove that $\mathbf{h}_{(s)}$ defined by
$$\mathbf{h}_{(s)}(\vec{A}, \vec{B}) = \tfrac{1}{2}\mathbf{h}(\vec{A}, \vec{B}) + \tfrac{1}{2}\mathbf{h}(\vec{B}, \vec{A}) \tag{3.69}$$
is a symmetric tensor.
(b) Prove that $\mathbf{h}_{(A)}$ defined by
$$\mathbf{h}_{(A)}(\vec{A}, \vec{B}) = \tfrac{1}{2}\mathbf{h}(\vec{A}, \vec{B}) - \tfrac{1}{2}\mathbf{h}(\vec{B}, \vec{A}) \tag{3.70}$$
is an antisymmetric tensor.
(c) Find the components of the symmetric and antisymmetric parts of $\tilde{p} \otimes \tilde{q}$ defined in Exer. 14.
(d) Prove that if $\mathbf{h}$ is an antisymmetric $\binom{0}{2}$ tensor,
$$\mathbf{h}(\vec{A}, \vec{A}) = 0$$
for any vector $\vec{A}$.
(e) Find the number of independent components $\mathbf{h}_{(s)}$ and $\mathbf{h}_{(A)}$ have.

17(a) Suppose that $\mathbf{h}$ is a $\binom{0}{2}$ tensor with the property that, for *any* two vectors $\vec{A}$ and $\vec{B}$
$$\mathbf{h}(\ , \vec{A}) = \alpha \mathbf{h}(\ , \vec{B}),$$
where α is a number which may depend on $\vec{A}$ and $\vec{B}$. Show that there exist one-forms $\tilde{p}$ and $\tilde{q}$ such that
$$\mathbf{h} = \tilde{p} \otimes \tilde{q}.$$
(b) Suppose $\mathbf{T}$ is a $\binom{1}{1}$ tensor, $\tilde{\omega}$ a one-form, $\vec{v}$ a vector, and $\mathbf{T}(\tilde{\omega}; \vec{v})$ the value of $\mathbf{T}$ on $\tilde{\omega}$ and $\vec{v}$. Prove that $\mathbf{T}(\ ; \vec{v})$ is a vector and $\mathbf{T}(\tilde{\omega}; \ )$ is a one-form, i.e. that a $\binom{1}{1}$ tensor provides a map of vectors to vectors and one-forms to one-forms.

18(a) Find the one-forms mapped by the metric tensor from the vectors
$$\vec{A} \rightarrow_O (1, 0, -1, 0), \qquad \vec{B} \rightarrow_O (0, 1, 1, 0), \qquad \vec{C} \rightarrow_O (-1, 0, -1, 0),$$
$$\vec{D} \rightarrow_O (0, 0, 1, 1).$$
(b) Find the vectors mapped by the inverse of the metric tensor from the one-form $\tilde{p} \rightarrow_O (3, 0, -1, -1)$, $\tilde{q} \rightarrow_O (1, -1, 1, 1)$, $\tilde{r} \rightarrow_O (0, -5, -1, 0)$, $\tilde{s} \rightarrow_O (-2, 1, 0, 0)$.

19(a) Prove that the matrix $\{\eta^{\alpha\beta}\}$ is inverse to $\{\eta_{\alpha\beta}\}$ by performing the matrix multiplication.
(b) Derive Eq. (3.53).

20 In Euclidean three-space in Cartesian coordinates, one doesn't normally distinguish between vectors and one-forms, because their components transform identically. Prove this in two steps.

(a) Show that
$$A^{\bar{\alpha}} = \Lambda^{\bar{\alpha}}{}_{\beta} A^{\beta}$$
and
$$P_{\bar{\beta}} = \Lambda^{\alpha}{}_{\bar{\beta}} P_{\alpha}$$
are the same transformation if the matrix $\{\Lambda^{\bar{\alpha}}{}_{\beta}\}$ equals the transpose of its inverse. Such a matrix is said to be *orthogonal*.

(b) The metric of such a space has components $\{\delta_{ij}, \; i, j = 1, \ldots, 3\}$. Prove that a transformation from one Cartesian coordinate system to another must obey
$$\delta_{\bar{i}\bar{j}} = \Lambda^{k}{}_{\bar{i}} \Lambda^{l}{}_{\bar{j}} \delta_{kl}$$
and that this implies $\{\Lambda^{k}{}_{\bar{i}}\}$ is an orthogonal matrix. See Exer. 32 for the analogue of this in SR.

21(a) Let a region of the t–x plane be bounded by the lines $t = 0$, $t = 1$, $x = 0$, $x = 1$. Find the unit outward normal one-forms and their associated vectors for each of the boundary lines.

(b) Let another region be bounded by the straight lines joining the events whose coordinates are $(1, 0)$, $(1, 1)$, and $(2, 1)$. Find an outward normal for the null boundary and find its associated vector.

22 Suppose that instead of defining vectors first, we had begun by defining one-forms, aided by pictures like Fig. 3.4. Then we could have introduced vectors as linear real-valued functions of one-forms, and defined vector algebra by the analogues of Eqs. (3.6a) and (3.6b) (i.e. by exchanging arrows for tildes). Prove that, so defined, vectors form a vector space. This is another example of the duality between vectors and one-forms.

23(a) Prove that the set of all $\binom{M}{N}$ tensors for fixed M, N forms a vector space. (You must define addition of such tensors and their multiplication by numbers.)

(b) Prove that a basis for this space is the set
$$\{\underbrace{\vec{e}_{\alpha} \otimes \vec{e}_{\beta} \otimes \cdots \otimes \vec{e}_{\gamma}}_{M \text{ vectors}} \otimes \underbrace{\tilde{\omega}^{\mu} \otimes \tilde{\omega}^{\nu} \otimes \cdots \otimes \tilde{\omega}^{\lambda}}_{N \text{ one-forms}}\}$$
(You will have to define the outer product of more than two one-forms.)

24(a) Given the components of a $\binom{2}{0}$ tensor $M^{\alpha\beta}$ as the matrix
$$\begin{pmatrix} 0 & 1 & 0 & 0 \\ 1 & -1 & 0 & 2 \\ 2 & 0 & 0 & 1 \\ 1 & 0 & -2 & 0 \end{pmatrix},$$
find
(i) the components of the symmetric tensor $M^{(\alpha\beta)}$ and the antisymmetric tensor $M^{[\alpha\beta]}$;

(ii) the components of $M^\alpha{}_\beta$;

(iii) the components of $M_\alpha{}^\beta$;

(iv) the components of $M_{\alpha\beta}$.

(b) For the $\binom{1}{1}$ tensor whose components are $M^\alpha{}_\beta$, does it make sense to speak of its symmetric and antisymmetric parts? If so, define them. If not, say why.

(c) Raise an index of the metric tensor to prove

$$\eta^\alpha{}_\beta = \delta^\alpha{}_\beta.$$

25 Show that if **A** is a $\binom{2}{0}$ tensor and **B** a $\binom{0}{2}$ tensor then

$$A^{\alpha\beta}B_{\alpha\beta}$$

is frame invariant, i.e. a scalar.

26 Suppose **A** is an antisymmetric $\binom{2}{0}$ tensor, **B** a symmetric $\binom{0}{2}$ tensor, **C** an arbitrary $\binom{0}{2}$ tensor, and **D** an arbitrary $\binom{2}{0}$ tensor. Prove:

(a) $A^{\alpha\beta}B_{\alpha\beta} = 0$;

(b) $A^{\alpha\beta}C_{\alpha\beta} = A^{\alpha\beta}C_{[\alpha\beta]}$;

(c) $B_{\alpha\beta}D^{\alpha\beta} = B_{\alpha\beta}D^{(\alpha\beta)}$.

27(a) Suppose **A** is an antisymmetric $\binom{2}{0}$ tensor. Show that $\{A_{\alpha\beta}\}$, obtained by lowering indices by using the metric tensor, are components of an antisymmetric $\binom{0}{2}$ tensor.

(b) Suppose $V^\alpha = W^\alpha$. Prove that $V_\alpha = W_\alpha$.

28 Deduce Eq. (3.66) from Eq. (3.65).

29 Prove that tensor differentiation obeys the Leibniz (product) rule:

$$\nabla(\mathbf{A}\otimes\mathbf{B}) = (\nabla\mathbf{A})\otimes\mathbf{B} + \mathbf{A}\otimes\nabla\mathbf{B}.$$

30 In some frame $\mathcal{O}$, the vector fields $\vec{U}$ and $\vec{D}$ have the components:

$$\vec{U} \rightarrow (1 + t^2, t^2, \sqrt{2}t, 0),$$

$$\vec{D} \rightarrow (x, 5tx, \sqrt{2}t, 0),$$

and the scalar ρ has the value

$$\rho = x^2 + t^2 - y^2.$$

(a) Find $\vec{U}\cdot\vec{U}$, $\vec{U}\cdot\vec{D}$, $\vec{D}\cdot\vec{D}$. Is $\vec{U}$ suitable as a four-velocity field? Is $\vec{D}$?

(b) Find the spatial velocity v of a particle whose four-velocity is $\vec{U}$, for arbitrary t. What happens to it in the limits $t \rightarrow 0$, $t \rightarrow \infty$?

(c) Find U_α for all α.

(d) Find $U^\alpha{}_{,\beta}$ for all α, β.

(e) Show that $U_\alpha U^\alpha{}_{,\beta} = 0$ for all β. Show that $U^\alpha U_{\alpha,\beta} = 0$ for all β.

(f) Find $D^\beta{}_{,\beta}$.

(g) Find $(U^\alpha D^\beta)_{,\beta}$ for all α.

(h) Find $U_\alpha(U^\alpha D^\beta)_{,\beta}$ and compare with (f) above. Why are the two answers similar?

(i) Find $\rho_{,\alpha}$ for all α. Find $\rho^{,\alpha}$ for all α. (Recall that $\rho^{,\alpha} \equiv \eta^{\alpha\beta}\rho_{,\beta}$.) What are the numbers $\{\rho^{,\alpha}\}$ the components of?

(j) Find $\nabla_{\vec{U}}\rho$, $\nabla_{\vec{U}}\vec{D}$, $\nabla_{\vec{D}}\rho$, $\nabla_{\vec{D}}\vec{U}$.

31 Consider a timelike unit four-vector $\vec{U}$, and the tensor

$$\mathbf{P}_{\vec{U}} \equiv \mathbf{g} + \vec{U}\otimes\vec{U}.$$

Show that $\mathbf{P}_{\vec{U}}$ is a projection operator that projects an arbitrary vector $\vec{V}$ into one orthogonal to $\vec{U}$. That is, show that the vector $\vec{V}_\perp$ whose components are

$$V_\perp^\alpha = P^\alpha{}_\beta V^\beta = (\eta^\alpha{}_\beta + U^\alpha U_\beta) V^\beta$$

is

(a) orthogonal to $\vec{U}$,

and

(b) unaffected by $\mathbf{P}$:

$$V^\alpha{}_{\perp\perp} \equiv V^\beta_\perp P^\alpha{}_\beta = V^\alpha_\perp.$$

(c) Show that for an arbitrary nonnull vector $\vec{q}$, the tensor that projects orthogonally to it is

$$\mathbf{P}_{\vec{q}} = \mathbf{g} + \frac{\vec{q}\otimes\vec{q}}{\vec{q}\cdot\vec{q}}.$$

How does this fail for null vectors?

(d) Show that $\mathbf{P}_{\vec{U}}$ is the metric tensor for vectors perpendicular to $\vec{U}$:

$$\mathbf{P}_{\vec{U}}(\vec{V}_\perp, \vec{W}_\perp) = \mathbf{g}(\vec{V}_\perp, \vec{W}_\perp)$$
$$= \vec{V}_\perp \cdot \vec{W}_\perp.$$

32(a) From the definition $f_{\alpha\beta} = \mathbf{f}(\vec{e}_\alpha, \vec{e}_\beta)$ for the components of a $\binom{0}{2}$ tensor, prove that the transformation law is

$$f_{\bar{\alpha}\bar{\beta}} = \Lambda^\mu{}_{\bar{\alpha}}\Lambda^\nu{}_{\bar{\beta}}f_{\mu\nu}$$

and that the matrix version of this is

$$(\bar{f}) = (\Lambda)^T(f)(\Lambda),$$

where (Λ) is the matrix with components $\Lambda^\mu{}_{\bar{\alpha}}$.

(b) Since our definition of a Lorentz frame led us to deduce that the metric tensor has components $\eta_{\alpha\beta}$, this must be true in all Lorentz frames. We are thus led to a more general *definition* of a Lorentz transformation as one whose matrix $\Lambda^\mu{}_{\bar{\alpha}}$ satisfies

$$\eta_{\bar{\alpha}\bar{\beta}} = \Lambda^\mu{}_{\bar{\alpha}}\Lambda^\nu{}_{\bar{\beta}}\eta_{\mu\nu}. \qquad (3.71)$$

Prove that the matrix for a boost of velocity $v\,\vec{e}_x$ satisfies this, so that this new definition includes our older one.

(c) Suppose (Λ) and (L) are two matrices which satisfy Eq. (3.71), i.e. $(\eta) = (\Lambda)^T(\eta)(\Lambda)$ and similarly for (L). Prove that $(\Lambda)(L)$ is also the matrix of a Lorentz transformation.

33 The result of Exer. 32c establishes that Lorentz transformations form a group, represented by multiplication of their matrices. This is called the *Lorentz group*, denoted by $L(4)$ or $0(1,3)$.

(a) Find the matrices of the identity element of the Lorentz group and of the element inverse to that whose matrix is implicit in Eq. (1.12).
(b) Prove that the determinant of any matrix representing a Lorentz transformation is ± 1.
(c) Prove that those elements whose matrices have determinant $+1$ form a subgroup, while those with -1 do not.
(d) The three-dimensional orthogonal group $O(3)$ is the analogous group for the metric of three-dimensional Euclidean space. In Exer. 20b, we saw that it was represented by the orthogonal matrices. Show that the orthogonal matrices do form a group, and then show that $0(3)$ is (isomorphic to) a subgroup of $L(4)$.

34 Consider the coordinates $u = t - x$, $v = t + x$ in Minkowski space.
(a) Define $\vec{e}_u$ to be the vector connecting the events with coordinates $\{u = 1, v = 0, y = 0, z = 0\}$ and $\{u = 0, v = 0, y = 0, z = 0\}$, and analogously for $\vec{e}_v$. Show that $\vec{e}_u = (\vec{e}_t - \vec{e}_x)/2$, $\vec{e}_v = (\vec{e}_t + \vec{e}_x)/2$, and draw $\vec{e}_u$ and $\vec{e}_v$ in a spacetime diagram of the t–x plane.
(b) Show that $\{\vec{e}_u, \vec{e}_v, \vec{e}_y, \vec{e}_z\}$ are a basis for vectors in Minkowski space.
(c) Find the components of the metric tensor on this basis.
(d) Show that $\vec{e}_u$ and $\vec{e}_v$ are null and not orthogonal. (They are called a *null basis* for the t–x plane.)
(e) Compute the four one-forms $\tilde{d}u$, $\tilde{d}v$, $\mathbf{g}(\vec{e}_u, \ )$, $\mathbf{g}(\vec{e}_v, \ )$ in terms of $\tilde{d}t$ and $\tilde{d}x$.

4

Perfect fluids
in special relativity

4.1 Fluids

In many interesting situations in astrophysical GR, the source of the gravitational field can be taken to be a perfect fluid as a first approximation. In general, a 'fluid' is a special kind of *continuum*. A continuum is a collection of particles so numerous that the dynamics of individual particles cannot be followed, leaving only a description of the collection in terms of 'average' quantities: number of particles per unit volume, density of energy, density of momentum, pressure, temperature, etc. The behavior of a lake of water, and the gravitational field it generates, does not depend upon where any one particular water molecule happens to be: it depends only on the average properties of huge collections of molecules.

Nevertheless, these properties can vary from point to point in the lake: the pressure is larger at the bottom than at the top, and the temperature may vary as well. The atmosphere, another fluid, has a density that varies with position. This raises the question of how large a collection of particles to average over: it must clearly be large enough so that the individual particles don't matter, but it must be small enough so that it is relatively homogeneous: the average velocity, kinetic energy, and interparticle spacing must be the same everywhere in the collection. Such a collection is called an '*element*'. This is a somewhat imprecise but useful term for a large collection of particles that may be regarded as having a single value for such quantities as density, average velocity, and temperature. If such a collection doesn't exist (e.g. a *very* rarified gas), then the continuum approximations breaks down.

The continuum approximation assigns to each element a value of density, temperature, etc. Since the elements are regarded as 'small', this approximation is expressed mathematically by assigning to each *point* a value of density, temperature, etc. So a continuum is defined by various fields, having values at each point and at each time.

So far, this notion of a continuum embraces rocks as well as gases. A *fluid* is a continuum that 'flows': this definition is not very precise, and so the division between solids and fluids is not very well defined. Most solids will flow under high enough pressure. What makes a substance rigid? After some thought one should be able to see that rigidity comes from forces *parallel* to the interface between two elements. Two adjacent elements can push and pull on each other, but the continuum won't be rigid unless they can also prevent each other from sliding along their common boundary. A *fluid* is characterized by the weakness of such antislipping forces compared to the direct push–pull force, which is called pressure. A *perfect* fluid is defined as one in which *all* antislipping forces are zero, and the only force between neighboring fluid elements is pressure. We will soon see how to make this mathematically precise.

4.2   Dust: The number–flux vector $\vec{N}$

We will introduce the relativistic description of a fluid with the simplest one: 'dust' is defined to be a collection of particles, all of which are at rest in some one Lorentz frame. It isn't very clear how this usage of the term 'dust' evolved from the other meaning as that substance which is at rest on the windowsill, but it has become a standard usage in relativity.

The number density n. The simplest question we can ask about these particles is, how many are there per unit volume? In their rest frame, this is merely an exercise in counting the particles and dividing by the volume they occupy. By doing this in many small regions we could come up with different numbers at different points, since the particles may be distributed more densely in one area than in another. We define this *number density* to be n:

$$n \equiv \text{number density in the MCRF of the element.} \tag{4.1}$$

Clearly, n may be a function of x^i, but in the rest frame it will not depend on t. (In other frames, the Lorentz transformation may bring in a dependence on $\bar{t}$.)

What is the number density in a frame $\bar{O}$ in which the particles are not at rest? They will all have the same velocity v in $\bar{O}$. If we look at the same particles as we counted up in the rest frame, then there are clearly the same *number* of particles, but they do not occupy the same volume. Suppose they were originally in a rectangular solid of dimension $\Delta x\, \Delta y\, \Delta z$. The Lorentz contraction will reduce this to $\Delta x\, \Delta y\, \Delta z\sqrt{(1 - v^2)}$, since lengths in the direction of motion contract but lengths perpendicular do not (Fig. 4.1). Because of this, the number of particles per unit volume

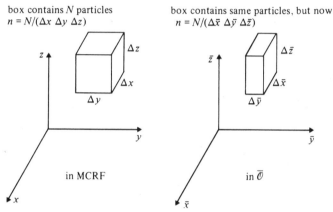

Fig. 4.1 The Lorentz contraction causes the density of particles to depend upon the frame in which it is measured.

is $[\sqrt{(1 - v^2)}]^{-1}$ times what it was in the rest frame:

$$\frac{n}{\sqrt{(1 - v^2)}} = \left\{ \begin{array}{l} \text{number density in frame in} \\ \text{which particles have velocity } v \end{array} \right\}. \tag{4.2}$$

The flux across a surface. When particles move, another question of interest is, 'how many' of them are moving in a certain direction? This is made precise by the definition of flux: *the flux of particles across a surface is the number crossing a unit area of that surface in a unit time.* This clearly depends on the inertial reference frame ('area' and 'time' are frame-dependent concepts) and on the orientation of the surface (a surface parallel to the velocity of the particles won't be crossed by any of them). In the rest frame of the dust the flux is zero, since all particles are at rest. In the frame $\bar{O}$, suppose the particles all move with velocity v in the $\bar{x}$ direction, and let us for simplicity consider a surface $\mathcal{S}$

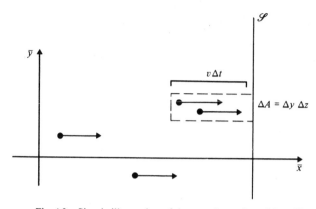

Fig. 4.2   Simple illustration of the transformation of flux: if particles move only in the x-direction, then all those within a distance $v\Delta\bar{t}$ of the surface $\mathscr{S}$ will cross $\mathscr{S}$ in the time $\Delta\bar{t}$

perpendicular to $\bar{x}$ (Fig. 4.2). The rectangular volume outlined by a dashed line clearly contains all and only those particles that will cross the area ΔA of $\mathscr{S}$ in the time $\Delta\bar{t}$. It has volume $v\,\Delta\bar{t}\,\Delta A$, and contains $(n/\sqrt{(1-v^2)})v\,\Delta\bar{t}\,\Delta A$ particles, since in this frame the number density is $n/\sqrt{(1-v^2)}$. The number crossing *per unit* time and per unit area is the flux across surfaces of constant $\bar{x}$:

$$(\text{flux})^{\bar{x}} = \frac{nv}{\sqrt{(1-v^2)}}.$$

Suppose, more generally, that the particles had a y component of velocity in $\bar{O}$ as well. Then the dashed line in Fig. 4.3 encloses all and only those particles that cross ΔA in $\mathscr{S}$ in the time $\Delta\bar{t}$. This is a 'parallelepiped', whose volume is the area of its base times its height. But its height – its

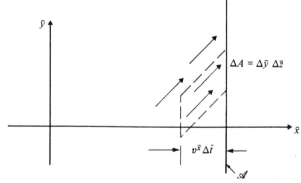

Fig. 4.3   The general situation for flux: only the x-component of the velocity carries particles across a surface of constant x.

extent in the x direction – is just $v^x \Delta \bar{t}$. Therefore we get

$$(\text{flux})^{\bar{x}} = \frac{nv^{\bar{x}}}{\sqrt{(1 - v^2)}}. \tag{4.3}$$

The number–flux four-vector $\vec{N}$. Consider the vector $\vec{N}$ defined by

◆ $\vec{N} = n\vec{U}, \tag{4.4}$

where $\vec{U}$ is the four-velocity of the particles. In a frame $\bar{O}$ in which the particles have a velocity (v^x, v^y, v^z), we have

$$\vec{U} \underset{\bar{O}}{\to} \left(\frac{1}{\sqrt{(1 - v^2)}}, \frac{v^x}{\sqrt{(1 - v^2)}}, \frac{v^y}{\sqrt{(1 - v^2)}}, \frac{v^z}{\sqrt{(1 - v^2)}} \right).$$

It follows that

$$\vec{N} \underset{\bar{O}}{\to} \left(\frac{n}{\sqrt{(1 - v^2)}}, \frac{nv^x}{\sqrt{(1 - v^2)}}, \frac{nv^y}{\sqrt{(1 - v^2)}}, \frac{nv^z}{\sqrt{(1 - v^2)}} \right). \tag{4.5}$$

Thus, in any frame, the time component of $\vec{N}$ is the number density and the spatial components are the fluxes across surfaces of the various coordinates. This is a very important conceptual result. In Galilean physics, number density was a scalar, the same in all frames (no Lorentz contraction), while flux was quite another thing: a three-vector that was frame *dependent*, since the velocities of particles are a frame-dependent notion. Our relativistic approach has unified these two notions into a single, frame-independent four-vector. This is progress in our thinking, of the most fundamental sort: the union of apparently disparate notions into a single coherent one.

It is worth reemphasizing the sense in which we use the word 'frame-independent'. The vector $\vec{N}$ is a geometrical object whose existence is independent of any frame; as a tensor, its action on a one-form to give a number is independent of any frame. Its components *do* of course depend on the frame. Since prerelativity physicists regarded the flux as a three-vector, they had to settle for it as a frame-dependent vector, in the following sense. As a three-vector it was independent of the orientation of the spatial axes in the same sense that four-vectors are independent of all frames; but the flux three-vector is different in frames that move relative to one another, since the velocity of the particles is different in different frames. To the old physicists, a flux vector had to be defined relative to some inertial frame. To a relativist, there is only *one* four-vector, and the frame dependence of the older way of looking at things came from concentrating only on a set of three of the four components of $\vec{N}$. This unification of the Galilean frame-independent number density and frame-dependent flux into a single frame-independent four-vector $\vec{N}$ is similar to the unification of 'energy' and 'momentum' into four-momentum.

One final note: it is clear that

◆ $$\vec{N} \cdot \vec{N} = -n^2, \qquad n = (-\vec{N} \cdot \vec{N})^{1/2}. \tag{4.6}$$

Thus, n is a scalar. In the same way that 'rest mass' is a scalar, even though energy and 'inertial mass' are frame dependent, here we have that n is a scalar, the 'rest density', even though number density is frame dependent. We will *always* define n to be a scalar number equal to the number density in the MCRF. We will make similar definitions for pressure, temperature, and other quantities characteristic of the fluid. These will be discussed later.

4.3   One-forms and surfaces

Number density as a timelike flux. We can complete the above discussion of the unity of number density and flux by realizing that number density can be regarded as a timelike flux. To see this, let us look at the flux across x surfaces again, this time in a *spacetime* diagram, in which we plot only $\bar{t}$ and $\bar{x}$ (Fig. 4.4). The surface $\mathscr{S}$ perpendicular to $\bar{x}$ has the world line shown. At any time $\bar{t}$ it is just one point, since we are suppressing both $\bar{y}$ and $\bar{z}$. The world lines of those particles that go through $\mathscr{S}$ in the time $\Delta \bar{t}$ are also shown. The flux is the number of world lines that cross $\mathscr{S}$ in the interval $\Delta \bar{t} = 1$. Really, since it is a two-dimensional surface, its 'world path' is three-dimensional, of which we have drawn only a section. The *flux* is the number of world lines that

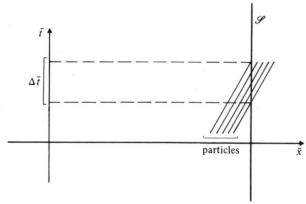

Fig. 4.4   Fig. 4.2 in a spacetime diagram, with the $\bar{y}$ direction suppressed.

cross a unit 'volume' of this three-surface: by volume we of course mean a cube of unit side – $\Delta \bar{t} = 1, \Delta \bar{y} = 1, \Delta \bar{z} = 1$. So we can define a flux as the number of world lines crossing a unit three-volume. There is no reason we cannot now define this three-volume to be an ordinary spatial volume $\Delta \bar{x} = 1, \Delta \bar{y} = 1, \Delta \bar{z} = 1$, taken at some particular time $\bar{t}$. This is shown in Fig. 4.5. Now the flux is the number crossing in the interval $\Delta \bar{x} = 1$ (since $\bar{y}$ and $\bar{z}$ are suppressed). But this is just the number 'contained' in the unit volume at the given time: the number density. So the 'timelike' flux is the number density.

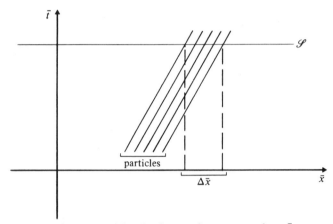

Fig. 4.5   Number density as a flux across surfaces $\bar{t} = $ const.

A one-form defines a surface. The way we described surfaces above was somewhat clumsy. To push our invariant picture further we need a somewhat more satisfactory mathematical representation of the surface that these world lines are crossing. This representation is given by one-forms. In general, a surface is defined as the solution to some equation

$$\phi(t, x, y, z) = \text{const.}$$

The gradient of the function ϕ, $\tilde{d}\phi$, is a normal one-form. In some sense, $\tilde{d}\phi$ *defines* the surface $\phi = $ const., since it uniquely determines the directions to that surface. However, any multiple of $\tilde{d}\phi$ also defines the same surface, so it is customary to use the unit-normal one-form when the surface is not null:

$$\tilde{n} \equiv \tilde{d}\phi/|\tilde{d}\phi|, \tag{4.7}$$

where

$|\tilde{d}\phi|$ is the magnitude of $\tilde{d}\phi$:

$$|\tilde{d}\phi| = |\eta^{\alpha\beta}\phi_{,\alpha}\phi_{,\beta}|^{1/2}. \tag{4.8}$$

(Do not confuse $\bar{n}$ with n, the number density in the MCRF: they are completely different, given, by historical accident, the same letter.)

As in three-dimensional vector calculus (e.g. Gauss' law), one defines the 'surface element' as the unit normal times an area element in the surface. In this case, a volume element in a three-space whose coordinates are x^α, x^β, and x^γ (for some *particular* values of α, β, and γ, all distinct) can be represented by

$$\bar{n} \, dx^\alpha \, dx^\beta \, dx^\gamma, \tag{4.9}$$

and a *unit* volume ($dx^\alpha = dx^\beta = dx^\gamma = 1$) is just $\bar{n}$. (These dxs *are* the infinitesimals that we integrate over, not the gradients.)

The flux across the surface. Recall from Gauss' law in three dimensions that the flux across a surface of, say, the electric field is just $\mathbf{E} \cdot \mathbf{n}$, the dot product of $\mathbf{E}$ with the unit normal. The situation here is exactly the same: the flux (of particles) across a surface of constant ϕ is $\langle \bar{n}, \vec{N} \rangle$. To see this, let ϕ be a coordinate, say $\bar{x}$. Then a surface of constant $\bar{x}$ has normal $\tilde{d}\bar{x}$, which is a unit normal already since $\tilde{d}\bar{x} \to_{\bar{O}} (0, 1, 0, 0)$. Then $\langle \tilde{d}\bar{x}, \vec{N} \rangle = N^\alpha (\tilde{d}\bar{x})_\alpha = N^{\bar{x}}$, which is what we have already seen is the flux across $\bar{x}$ surfaces. Clearly, had we chosen $\phi = \bar{t}$, then we would have wound up with $N^{\bar{0}}$, the number density, or flux across a surface of constant $\bar{t}$.

This is one of the first concrete physical examples of our definition of a vector as a function of one-forms into real numbers. Given the vector $\vec{N}$, we can calculate the flux across a surface by finding the unit-normal one-form for that surface, and contracting it with $\vec{N}$. We have, moreover, expressed everything frame invariantly and in a manner that separates the property of the system of particles $\vec{N}$ from the property of the surface $\bar{n}$. All of this will have many parallels in § 4.4 below.

Representation of a frame by a one-form. Before going on to discuss other properties of fluids, we should mention a useful fact. An inertial frame, which up to now has been defined by its four-velocity, can be defined also by a one-form, namely that associated with its four-velocity $\mathbf{g}(\vec{U}, \ )$. This has components

$$U_\alpha = \eta_{\alpha\beta} U^\beta$$

or, in this frame,

$$U_0 = -1, \ U_i = 0.$$

This is clearly also equal to $-\tilde{d}\bar{t}$ (since their components are equal). So we could equally well define a frame by giving $\tilde{d}t$. This has a nice picture:

đt is to be pictured as a set of surfaces of constant t, the surfaces of simultaneity. These clearly *do* define the frame, up to spatial rotations, which we usually ignore. In fact, in some sense đt is a more natural way to define the frame than $\vec{U}$. For instance, the energy of a particle whose four-momentum is $\vec{p}$ is

$$E = \langle \tilde{\mathrm{d}}t, \vec{p} \rangle = p^0. \tag{4.10}$$

There is none of the awkward minus sign that one gets in Eq. (2.35)

$$E = -\vec{p} \cdot \vec{U}.$$

4.4 Dust again: The stress–energy tensor

So far we have only discussed how many dust particles there are. But they also have energy and momentum, and it will turn out that their energy and momentum are the source of the gravitational field in GR. So we must now ask how to represent them in a frame-invariant manner. We will assume for simplicity that all the dust particles have the same rest mass m.

Energy density. In the MCRF, the energy of each particle is just m, and the number per unit volume is n. Therefore the energy per unit volume is mn. We denote this in general by ρ:

$$\rho \equiv \text{energy density in the MCRF.} \tag{4.11}$$

Thus ρ is a scalar just as n is (and m is). In our case of dust,

$$\rho = nm \text{ (dust).} \tag{4.12}$$

In more general fluids, where there is random motion of particles and hence kinetic energy of motion, even in an average rest frame, Eq. (4.12) will not be valid.

In the frame $\bar{\mathcal{O}}$ we again have that the number density is $n/\sqrt{(1-v^2)}$, but now the energy of each particle is $m/\sqrt{(1-v^2)}$, since it is moving. Therefore the energy density is $mn/(1-v^2)$:

$$\frac{\rho}{1-v^2} = \left\{ \begin{array}{l} \text{energy density in a frame in} \\ \text{which particles have velocity } v \end{array} \right\}. \tag{4.13}$$

This transformation involves *two* factors of $(1-v^2)^{1/2} = \Lambda^{\bar{0}}{}_0$, because *both* volume *and* energy transform. It is impossible, therefore, to represent energy density as some component of a vector. It is, in fact, a component of a $\binom{2}{0}$ tensor. This is most easily seen from the point of view of our definition of a tensor. To define energy requires a one-form, in order to select the 0 component of the four-vector of energy and momentum; to define a density also requires a one-form, since density is a flux across a constant-time surface. Similarly, an energy flux also requires two one-forms: one to define 'energy' and the other to define the surface. One can also speak of momentum density: again a one-form defines which component of momentum, and another one-form defines density. By analogy there is also momentum flux: the rate at which momentum crosses some surface. All these things require two one-forms. So there is a tensor **T**, called the stress–energy tensor, which has all these numbers as values when supplied with the appropriate one-forms as arguments.

Stress–energy tensor. The most convenient definition of the stress–energy tensor is in terms of its components in some (arbitrary) frame:

◆ $$\mathbf{T}(\tilde{d}x^{\alpha}, \tilde{d}x^{\beta}) = T^{\alpha\beta} \equiv \begin{Bmatrix} \text{flux of } \alpha \text{ momentum across} \\ \text{a surface of constant } x^{\beta} \end{Bmatrix}. \tag{4.14}$$

(By α momentum we mean, of course, the α component of four-momentum: $p^{\alpha} \equiv \langle \tilde{d}x^{\alpha}, \vec{p} \rangle$.) That this is truly a tensor is proved in Exer. 5, § 4.10.

Let us see how this definition fits in with our discussion above. Consider T^{00}. This is defined as the flux of 0 momentum (energy) across a surface $t = $ constant. This is just the energy density:

$$T^{00} = \text{energy density.} \tag{4.15}$$

Similarly, T^{0i} is the flux of energy across a surface $x^i = $ const:

$$T^{0i} = \text{energy flux across } x^i \text{ surface.} \tag{4.16}$$

Then T^{i0} is the flux of i momentum across a surface $t = $ const: the density of i momentum,

$$T^{i0} = i \text{ momentum density.} \tag{4.17}$$

Finally, T^{ij} is the j flux of i momentum:

$$T^{ij} = \text{flux of } i \text{ momentum across } j \text{ surface.} \tag{4.18}$$

For any particular system, giving the components of **T** in some frame defines it completely. For dust, the components of **T** in the MCRF are particularly easy. There is no motion of the particles, so all i momenta are zero and all spatial fluxes are zero. Therefore

$$(T^{00})_{\text{MCRF}} = \rho = mn,$$
$$(T^{0i})_{\text{MCRF}} = (T^{i0})_{\text{MCRF}} = (T^{ij})_{\text{MCRF}} = 0.$$

It is easy to see that the tensor $\vec{p} \otimes \vec{N}$ has exactly these components in the MCRF, where $\vec{p} = m\vec{U}$ is the four-momentum of a particle. Therefore we have

◆   Dust: $\mathbf{T} = \vec{p} \otimes \vec{N} = mn\vec{U} \otimes \vec{U} = \rho\vec{U} \otimes \vec{U}.$ $\tag{4.19}$

From this we can conclude

$$T^{\alpha\beta} = \mathbf{T}(\tilde{\omega}^\alpha, \tilde{\omega}^\beta)$$
$$= \rho\vec{U}(\tilde{\omega}^\alpha)\vec{U}(\tilde{\omega}^\beta)$$
$$= \rho U^\alpha U^\beta. \tag{4.20}$$

In the frame $\bar{O}$, where

$$\vec{U} \to \left(\frac{1}{\sqrt{(1-v^2)}}, \frac{v^x}{\sqrt{(1-v^2)}}, \cdots \right),$$

we therefore have

$$\left. \begin{aligned} T^{\bar{0}\bar{0}} &= \rho U^{\bar{0}} U^{\bar{0}} = \rho/(1-v^2), \\ T^{\bar{0}\bar{i}} &= \rho U^{\bar{0}} U^{\bar{i}} = \rho v^i/(1-v^2), \\ T^{\bar{i}\bar{0}} &= \rho U^{\bar{i}} U^{\bar{0}} = \rho v^i/(1-v^2), \\ T^{\bar{i}\bar{j}} &= \rho U^{\bar{i}} U^{\bar{j}} = \rho v^i v^j/(1-v^2). \end{aligned} \right\} \tag{4.21}$$

These are exactly what one would calculate, from first principles, for energy density, energy flux, momentum density, and momentum flux respectively. (We did the calculation for energy density above.) Notice one important point: $T^{\alpha\beta} = T^{\beta\alpha}$; that is, $\mathbf{T}$ is symmetric. This will turn out to be true in general, not just for dust.

4.5 General fluids

Until now we have dealt with the simplest possible collection of particles. To generalize this to real fluids, we have to take account of the facts that (i) besides the bulk motions of the fluid, each particle has some random velocity; and (ii) there may be various forces between particles that contribute potential energies to the total.

Definition of macroscopic quantities. The concept of a fluid element was discussed in § 4.1. For each fluid element, we go to the frame in which it is at rest (its total spatial momentum is zero). This is its MCRF. This frame is truly *momentarily* comoving: since fluid elements can be accelerated, a moment later a different inertial frame will be the MCRF. Moreover, two different fluid elements may be moving relative to one another, so that they would not have the same MCRFs. Thus, the MCRF is specific to a single fluid element, and which frame is the MCRF is a function of position and time. *All scalar quantities associated with a fluid element in relativity* (such as number density, energy density, and temperature) *are defined to be their values in the MCRF.* Thus we make the definitions displayed in Table 4.1.

Table 4.1. *Macroscopic quantities for fluids*

Symbol	Name	Definition
$\vec{U}$	Four-velocity of fluid element	Four-velocity of MCRF
n	Number density	Number of particles per unit volume in MCRF
$\vec{N}$	Flux vector	$\vec{N} \equiv n\vec{U}$
ρ	energy density	Density of *total* mass energy (rest mass, random kinetic, chemical, ...)
Π	Internal energy per particle	$\Pi \equiv (\rho/n) - m \Rightarrow \rho = n(m + \Pi)$ Thus Π is a general name for all energies other than the rest mass.
ρ_0	Rest-mass density	$\rho_0 \equiv mn$. Since m is a constant, this is the 'energy' associated with the rest mass only. Thus, $\rho = \rho_0 + n\Pi$.
T	Temperature	Usual thermodynamic definition in MCRF (see below).
p	Pressure	Usual fluid-dynamical notion in MCRF. More about this later.
S	Specific entropy	Entropy per particle (see below).

First law of thermodynamics. This law is simply a statement of conservation of energy. In the MCRF, we imagine that the fluid element is able to exchange energy with its surroundings in only two ways: by heat conduction (absorbing an amount of heat ΔQ) and by work (doing an amount of work $p\Delta V$, where V is the three-volume of the element). If we let E be the total energy of the element, then since ΔQ is energy gained and $p\Delta V$ is energy lost, we can write (assuming small changes)

$$\left. \begin{array}{l} \Delta E = \Delta Q - p\Delta V, \\ \Delta Q = \Delta E + p\Delta V. \end{array} \right\} \tag{4.22}$$

or

Now, if the element contains a total of N particles, and if this number doesn't change (i.e. no creation or destruction of particles), we can write

$$V = \frac{N}{n}, \qquad \Delta V = -\frac{N}{n^2}\Delta n. \tag{4.23}$$

Moreover, we also have (from the definition of ρ)

$$E = \rho V = \rho N/n,$$

$$\Delta E = \rho \Delta V + V \Delta \rho.$$

These two results imply

$$\Delta Q = \frac{N}{n} \Delta \rho - N(\rho + p) \frac{\Delta n}{n^2}.$$

If we write $q \equiv Q/N$, which is the heat absorbed per particle, we obtain

$$n \, \Delta q = \Delta \rho - \frac{\rho + p}{n} \Delta n. \tag{4.24}$$

Now suppose that the changes are 'infinitesimal'. It can be shown in general that a fluid's state can be given by two parameters: for instance, ρ and T or ρ and n. Everything else is a function of, say, ρ and n. That means that the right-hand side of Eq. (4.24),

$$d\rho - (\rho + p) \, dn/n,$$

depends only on ρ and n. The general theory of first-order differential equations shows that this *always* possesses an *integrating factor*: that is, there exist two functions A and B, functions only of ρ and n, such that

$$d\rho - (\rho + p) \, dn/n \equiv A \, dB$$

is an identity for all ρ and n. It is customary in thermodynamics to *define* temperature to be A/n and specific entropy to be B:

♦ $$d\rho - (\rho + p) \, dn/n = nT \, dS, \tag{4.25}$$

or, in other words,

$$\Delta q = T \, \Delta S. \tag{4.26}$$

The heat absorbed by a fluid element is proportional to its increase in entropy.

We have thus introduced T and S as convenient mathematical definitions. A full treatment would show that T is the thing normally meant by temperature, and that S is the thing used in the second law of thermodynamics, which says that the *total* entropy in any system must increase. We'll have nothing to say about the second law. Entropy appears here only because it is an integral of the first law, which is merely conservation of energy. In particular, we shall use both Eqs. (4.25) and (4.26) later.

The general stress–energy tensor. The definition of $T^{\alpha\beta}$ in Eq. (4.14) is perfectly general. Let us in particular look at it in the MCRF, where there is no bulk flow of the fluid element, and no spatial momentum in the particles. Then *in the MCRF* we have

(1) $T^{00} =$ energy density $= \rho$.
(2) $T^{0i} =$ energy flux. Although there is no motion, energy may be

transmitted by heat conduction. So T^{0i} is basically a heat-conduction term in the MCRF.

(3) T^{i0} = momentum density. Again the particles have no momentum, but if heat is being conducted, then the energy will carry momentum. We'll argue below that $T^{i0} \equiv T^{0i}$.

(4) T^{ij} = momentum flux. This is an interesting and important term. The next section gives a thorough discussion of it. It is called the *stress*.

The spatial components of **T**, T^{ij}. By definition, T^{ij} is the flux of i momentum across the j surface. Consider (Fig. 4.6) two adjacent fluid

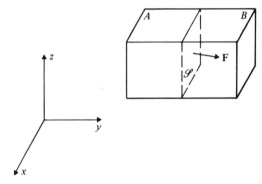

Fig. 4.6   The force **F** exerted by element A on its neighbor B may be in any direction depending on properties of the medium and any external forces.

elements, represented as cubes, having the common interface $\mathscr{S}$. In general, they exert forces on each other. Shown in the diagram is the force **F** exerted by A on B (B of course exerts an equal and opposite force on A). Since force equals the rate of change of momentum (by Newton's law, which is valid here, since we are in the MCRF where velocities are zero), A is pouring momentum into B at the rate **F** per unit time. Of course, B may or may not acquire a new velocity as a result of this new momentum it acquires; this depends upon how much momentum is put into B by its other neighbors. Obviously B's motion is the resultant of all the forces. Nevertheless, each force adds momentum to B. There is therefore a flow of momentum across $\mathscr{S}$ from A to B at the rate **F**. If $\mathscr{S}$ has area $\mathscr{A}$, then the flux of momentum across $\mathscr{S}$ is $\mathbf{F}/\mathscr{A}$. If $\mathscr{S}$ is a surface of constant x^j, then T^{ij} for fluid element A is $F^i/\mathscr{A}$.

This is a brief illustration of the meaning of T^{ij}: it represents forces between adjacent fluid elements. As mentioned before, these forces need not be perpendicular to the surfaces between the elements (i.e. viscosity or other kinds of rigidity give forces parallel to the interface). But if the forces *are* perpendicular to the interfaces, then T^{ij} will be zero unless $i = j$. (Think this through — we'll use it shortly.)

Symmetry of $T^{\alpha\beta}$ in MCRF. We now prove that **T** is a symmetric tensor. We need only prove that its components are symmetric in one frame; that implies that for any $\bar{r}, \bar{q}, \mathbf{T}(\bar{r}, \bar{q}) = \mathbf{T}(\bar{q}, \bar{r})$, which implies the symmetry of its components in any other frame. The easiest frame is the MCRF.

(a) Symmetry of T^{ij}. Consider Fig. 4.7 in which we have drawn a fluid element as a cube of side *l*. The force it exerts on a neighbor across surface (1) (a surface $x = \text{const.}$) is $F_1^i = T^{ix}l^2$, where the factor l^2 gives

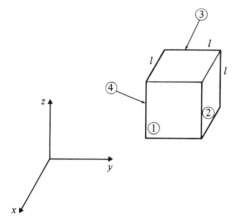

Fig. 4.7 A fluid element.

the area of the face. Here, *i* runs over 1, 2, and 3, since **F** is not necessarily perpendicular to the surface. Similarly, the force it exerts on a neighbor across (2) is $F_2^i = T^{iy}l^2$. (We shall take the limit $l \to 0$, so bear in mind that the element is small.) The element also exerts a force on its neighbor toward the $-x$ direction, which we call F_3^i. Similarly, there is F_4^i on the face looking in the negative *y* direction. The forces *on* the fluid element are, respectively, $-F_1^i, -F_2^i$, etc. The first point is that $F_3^i \approx -F_1^i$ in order that the sum of the forces on the element should vanish when $l \to 0$ (otherwise the tiny mass obtained as $l \to 0$ would have an infinite acceleration). The next point is to compute torques about the *z* axis through the center of the fluid element. (Since forces on the top and bottom of the cube don't contribute to this, we haven't considered them.) The torque due to $-\mathbf{F}_1$ is $-(\mathbf{r} \times \mathbf{F}_1)^z = -xF_1^y = -\frac{1}{2}lT^{yx}l^2$, where we have approximated force as acting at the center of the face, where $\mathbf{r} \to (l/2, 0, 0)$. The torque due to $-\mathbf{F}_3$ is the *same*, $-\frac{1}{2}l^3 T^{yx}$. The torque due to $-\mathbf{F}_2$ is $-(\mathbf{r} \times \mathbf{F}_2)^z =$

$+yF_2^x = \frac{1}{2}lT^{xy}l^2$. Similarly, the torque due to $-\mathbf{F}_4$ is the same, $\frac{1}{2}l^3 T^{xy}$. Therefore, the total torque is

$$\tau_z = l^3(T^{xy} - T^{yx}). \tag{4.27}$$

The moment of inertia of the element about the z axis is proportional to its mass times l^2, or

$$1 = \alpha\rho l^5,$$

where α is some numerical constant and ρ is the density (whether of total energy or rest mass doesn't matter in this argument). Therefore the angular acceleration is

$$\ddot{\theta} = \frac{\tau}{1} = \frac{T^{xy} - T^{yx}}{\alpha\rho l^2}. \tag{4.28}$$

Since α is a number and ρ is independent of the size of the element, as are T^{xy} and T^{yx}, this will go to infinity as $l \to 0$ unless

$$T^{xy} = T^{yx}.$$

Thus, since it is obviously not true that fluid elements are whirling around inside fluids, smaller ones whirling ever faster, we have that the stresses are always *symmetric*:

$$T^{ij} = T^{ji}. \tag{4.29}$$

Since we made no use of any property of the substance, this is true of solids as well as fluids. It is true in Newtonian theory as well as in relativity; in Newtonian theory T^{ij} are the components of a three-dimensional $\binom{2}{0}$ tensor called the stress tensor. It is familiar to any materials engineer; and it contributes its name to its relativistic generalization $\mathbf{T}$.

(b) Equality of momentum density and energy flux. Thus is much easier to demonstrate. The energy flux is the density of energy times the speed it flows at. But since energy and mass are the same, this is the density of mass times the speed it is moving at; in other words, the density of momentum. Therefore $T^{0i} = T^{i0}$.

Conservation of energy–momentum. Since $\mathbf{T}$ represents the energy and momentum content of the fluid, there must be some way of using it to express the law of conservation of energy and momentum. In fact it is reasonably easy. In Fig. 4.8 we see a cubical fluid element, seen only in cross-section (z direction suppressed). Energy can flow in across all sides. The rate of flow across face (4) is $l^2 T^{0x}(x=0)$, and across (2) is $-l^2 T^{0x}$ $(x=a)$; the second term has a minus sign, since T^{0x} represents energy flowing in the positive x direction, which is out of the volume across face (2). Similarly, energy flowing in the y direction is $l^2 T^{0y}(y=0) -$

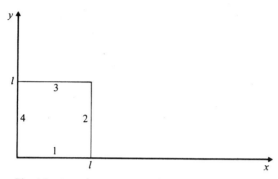

Fig. 4.8 A section $z = $ const. of a cubical fluid element.

$l^2 T^{0y}(y = l)$. The sum of these rates must be the rate of increase in the energy inside, $\partial(T^{00}l^3)/\partial t$ (statement of conservation of energy). Therefore we have

$$\frac{\partial}{\partial t} l^3 T^{00} = l^2 [T^{0x}(x = 0) - T^{0x}(x = l) + T^{0y}(y = 0)$$

$$- T^{0y}(y = l) + T^{0z}(z = 0) - T^{0z}(z = l)]. \tag{4.30}$$

Dividing by l^3 and taking the limit $l \to 0$ gives

$$\frac{\partial}{\partial t} T^{00} = -\frac{\partial}{\partial x} T^{0x} - \frac{\partial}{\partial y} T^{0y} - \frac{\partial}{\partial z} T^{0z}. \tag{4.31}$$

[In deriving this we use the definition of the derivative

$$\lim_{l \to 0} \frac{T^{0x}(x = 0) - T^{0x}(x = l)}{l} \equiv -\frac{\partial}{\partial x} T^{0x}.] \tag{4.32}$$

Eq. (4.31) can be written as

$$T^{00}{}_{,0} + T^{0x}{}_{,x} + T^{0y}{}_{,y} + T^{0z}{}_{,z} = 0$$

or

$$T^{0\alpha}{}_{,\alpha} = 0. \tag{4.33}$$

This is the statement of the law of conservation of energy.

Similarly, momentum is conserved. The same mathematics applies, with the index '0' changed to whatever spatial index corresponds to the component of momentum whose conservation is being considered. The general conservation law is, then,

◆ $$T^{\alpha\beta}{}_{,\beta} = 0. \tag{4.34}$$

This applies to any material in SR. Notice it is just a four-dimensional divergence. Its relation to Gauss' theorem, which gives an integral form of the conservation law, will be discussed later.

Conservation of particles. It may also happen that, during any flow of the fluid, the total number of particles will not change. This conservation law is derivable in the same way as Eq. (4.34) was. In particular, in Fig. 4.8 the rate of change of the number of particles in a fluid element will be due only to loss or gain across the boundaries, i.e. to net fluxes out or in. We can then write that

$$\frac{\partial}{\partial t} N^0 = - \frac{\partial}{\partial x} N^x - \frac{\partial}{\partial y} N^y - \frac{\partial}{\partial z} N^z$$

or

◆ $$N^\alpha_{,\alpha} = (nU^\alpha)_{,\alpha} = 0.$$ (4.35)

We will confine ourselves to discussing only fluids that obey this conservation law. This is hardly any restriction, since n can, if necessary, always be taken to be the density of baryons.

'Baryon', for those not familiar with high-energy physics, is a general name applied to the more massive particles in physics. The two commonest are the neutron and proton. All others are too unstable to be important in everyday physics – but when they decay they form protons and neutrons, thus conserving the total number of baryons without conserving rest mass or particle identity. Although theoretical physics suggests that baryons may not always be conserved – for instance, so-called 'grand unified theories' of the strong, weak, and electromagnetic interactions may predict a finite lifetime for the proton; and collapse to and subsequent evaporation of a black hole (see Ch. 11) will not conserve baryon number – no such phenomena have yet been observed and, in any case, are unlikely to be important in most situations.

4.6 Perfect fluids

Finally, we come to the type of fluid which is our principal subject of interest. *A perfect fluid in relativity is defined as a fluid that has no viscosity and no heat conduction in the MCRF.* It is a generalization of the 'ideal gas' of ordinary thermodynamics. It is, next to dust, the simplest kind of fluid to deal with. The two restrictions in its definition enormously simplify the stress–energy tensor, as we now see.

No heat conduction. From the definition of **T**, one sees that this immediately implies that, in the MCRF, $T^{0i} = T^{i0} = 0$. Energy can flow only if particles flow. Recall that in our discussion of the first law of thermodynamics we showed that if the number of particles were conserved, then the specific entropy was related to heat flow by Eq. (4.26). This means that in a perfect fluid, if Eq. (4.35) for conservation of particles is obeyed, then we should also have that S is a constant in time during the flow of the fluid. We shall see how this comes out of the conservation laws in a moment.

No viscosity. Viscosity is a force parallel to the interface between particles. Its absence means that the forces should always be perpendicular to the interface, i.e. that T^{ij} should be zero unless $i = j$. This means that T^{ij} should be a diagonal matrix. Moreover, it must be diagonal in *all* MCRF frames, since 'no viscosity' is a statement independent of the spatial axes. The only matrix diagonal in all frames is a multiple of the identity: all its diagonal terms are equal. Thus, an x surface will have across it only a force in the x direction, and similarly for y and z; these forces-per-unit-area are all equal, and are called the *pressure*, p. So we have $T^{ij} = p\delta^{ij}$. From six possible quantities (the number of independent elements in the 3×3 symmetric matrix T^{ij}) the zero-viscosity assumption has reduced the number of functions to one, the pressure.

Form of **T**. In the MCRF, **T** has the components we have just deduced:

$$(T^{\alpha\beta}) = \begin{pmatrix} \rho & 0 & 0 & 0 \\ 0 & p & 0 & 0 \\ 0 & 0 & p & 0 \\ 0 & 0 & 0 & p \end{pmatrix}, \tag{4.36}$$

It is not hard to show that in the MCRF

$$T^{\alpha\beta} = (\rho + p)U^\alpha U^\beta + p\eta^{\alpha\beta}. \tag{4.37}$$

For instance, if $\alpha = \beta = 0$, then $U^0 = 1$, $\eta^{00} = -1$, and $T^{\alpha\beta} = (\rho + p) - p = \rho$, as in Eq. (4.36). By trying all possible α and β you can verify that Eq. (4.37) gives Eq. (4.36). But Eq. (4.37) is a frame-invariant formula in the sense that it uniquely implies

$\blacklozenge$ $$\mathbf{T} = (\rho + p)\vec{U} \otimes \vec{U} + p\mathbf{g}^{-1}. \tag{4.38}$$

This is the stress–energy tensor of a perfect fluid.

Aside on the meaning of pressure. A comparison of Eq. (4.38) with Eq. (4.19) shows that 'dust' is the special case of a pressure-free perfect fluid. This means that a perfect fluid can be pressure free only if its particles have *no* random motion at all. Pressure arises in the random velocities of the particles. Even a gas so dilute as to be virtually collisionless has pressure. This is because pressure is the flux of momentum; whether this comes from forces or from particles crossing a boundary is immaterial.

The conservation laws. Eq. (4.34) gives us

$$T^{\alpha\beta}{}_{,\beta} = [(\rho+p)U^\alpha U^\beta + p\eta^{\alpha\beta}]_{,\beta} = 0. \qquad (4.39)$$

This gives us our first real practice with tensor calculus. There are four equations in Eq. (4.39), one for each α. First, let us also assume

$$(nU^\beta)_{,\beta} = 0 \qquad (4.40)$$

and write the first term in Eq. (4.39) as

$$[(\rho+p)U^\alpha U^\beta]_{,\beta} = \left[\frac{\rho+p}{n}U^\alpha n U^\beta\right]_{,\beta}$$

$$= nU^\beta\left(\frac{\rho+p}{n}U^\alpha\right)_{,\beta}. \qquad (4.41)$$

Moreover, $\eta^{\alpha\beta}$ is a constant matrix, so $\eta^{\alpha\beta}{}_{,\gamma} = 0$. This also implies, by the way,

$$U^\alpha{}_{,\beta}U_\alpha = 0. \qquad (4.42)$$

The proof of Eq. (4.42) is

$$U^\alpha U_\alpha = -1 \Rightarrow (U^\alpha U_\alpha)_{,\beta} = 0 \qquad (4.43)$$

or

$$(U^\alpha U^\gamma \eta_{\alpha\gamma})_{,\beta} = (U^\alpha U^\gamma)_{,\beta}\eta_{\alpha\gamma} = 2U^\alpha{}_{,\beta}U^\gamma \eta_{\alpha\gamma}. \qquad (4.44)$$

The last step follows from the symmetry of $\eta_{\alpha\beta}$, which means that $U^\alpha{}_{,\beta}U^\gamma\eta_{\alpha\gamma} = U^\alpha U^\gamma{}_{,\beta}\eta_{\alpha\gamma}$. Finally, the last expression in Eq. (4.44) converts to

$$2U^\alpha{}_{,\beta}U_\alpha,$$

which is zero by Eq. (4.43). This proves Eq. (4.42). We can make use of Eq. (4.42) in the following way. The original equation now reads, after use of Eq. (4.41),

$$nU^\beta\left(\frac{\rho+p}{n}U^\alpha\right)_{,\beta} + p_{,\beta}\eta^{\alpha\beta} = 0. \qquad (4.45)$$

From the four equations here, we can obtain one particularly useful one. Multiply by U_α and sum on α. This gives the time component of Eq. (4.45) in the MCRF:

$$nU^\beta U_\alpha\left(\frac{\rho+p}{n}U^\alpha\right)_{,\beta} + p_{,\beta}\eta^{\alpha\beta}U_\alpha = 0. \qquad (4.46)$$

The last term is just

$$p_{,\beta}U^\beta,$$

which we know to be the derivative of p along the world line of the fluid element, $dp/d\tau$. The first term gives zero when the β derivative operates on U^α (by Eq. 4.42), so we obtain (using $U^\alpha U_\alpha = -1$)

$$U^\beta \left[-n \left(\frac{\rho + p}{n} \right)_{,\beta} + p_{,\beta} \right] = 0. \tag{4.47}$$

A little algebra converts this to

$$-U^\beta \left[\rho_{,\beta} - \frac{\rho + p}{n} n_{,\beta} \right] = 0. \tag{4.48}$$

Written another way,

$$\frac{d\rho}{d\tau} - \frac{\rho + p}{n} \frac{dn}{d\tau} = 0. \tag{4.49}$$

This is to be compared with Eq. (4.25). It means

♦ $$U^\alpha S_{,\alpha} = \frac{dS}{d\tau} = 0. \tag{4.50}$$

Thus, the flow of a particle-conserving perfect fluid conserves specific entropy. This is called *adiabatic*. Because entropy is constant in a fluid element as it flows, we shall not normally need to consider it. Nevertheless, it is important to remember that the law of conservation of energy in thermodynamics is embodied in the component of the conservation equations, Eq. (4.39), parallel to U^α.

The remaining three components of Eq. (4.39) are derivable in the following way. We write, again, Eq. (4.45):

$$nU^\beta \left(\frac{\rho + p}{n} U^\alpha \right)_{,\beta} + p_{,\beta} \eta^{\alpha\beta} = 0$$

and go to the MCRF, where $U^i = 0$ but $U^i{}_{,\beta} \neq 0$. In the MCRF, the 0 component of this equation is the same as its contraction with U_α, which we have just examined. So we only need the i components:

$$nU^\beta \left(\frac{\rho + p}{n} U^i \right)_{,\beta} + p_{,\beta} \eta^{i\beta} = 0. \tag{4.51}$$

Since $U^i = 0$, the β derivative of $(\rho + p)/n$ contributes nothing, and we get

$$(\rho + p) U^i{}_{,\beta} U^\beta + p_{,\beta} \eta^{i\beta} = 0. \tag{4.52}$$

Lowering the index i makes this easier to read (and changes nothing). Since $\eta_i{}^\beta = \delta_i{}^\beta$ we get

$$(\rho + p) U_{i,\beta} U^\beta + p_{,i} = 0. \tag{4.53}$$

Finally, we recall that $U_{i,\beta} U^\beta$ is the definition of the four-acceleration a_i:

♦ $$(\rho + p) a_i + p_{,i} = 0. \tag{4.54}$$

Those familiar with nonrelativistic fluid dynamics will recognize this as the generalization of

$$\rho \mathbf{a} + \nabla p = 0, \tag{4.55}$$

where

$$\mathbf{a} = \dot{\mathbf{v}} + (\mathbf{v} \cdot \nabla)\mathbf{v}. \tag{4.56}$$

The only difference is the use of $(\rho + p)$ instead of ρ. In relativity, $(\rho + p)$ plays the role of 'inertial mass density', in that, from Eq. (4.54), the larger $(\rho + p)$, the harder it is to accelerate the object. Eq. (4.54) is essentially $\mathbf{F} = m\mathbf{a}$, with $-p_{,i}$ being the force on a fluid element. That is, p is the force a fluid element exerts on its neighbor, so $-p$ is the force on the element. But the neighbor on the opposite side of the element is pushing the other way, so only if there is a change in p across the fluid element will there be a net force causing it to accelerate. That is why $-\nabla p$ is the force.

4.7 Importance for general relativity

General relativity is a relativistic theory of gravity. We weren't able to plunge into it immediately because we lacked a good enough understanding of tensors, of fluids in SR, and of curved spaces. We have yet to study curvature (that comes next), but at this point we can look ahead and discern the vague outlines of the theory we shall study.

The first comment is on the supreme importance of **T** in GR. Newton's theory has as a source of the field the density ρ. This was understood to be the mass density, and so is closest to our ρ_0. But a theory that uses rest mass only as its source would be peculiar from a relativistic viewpoint, since rest mass and energy are interconvertible. In fact, one can show that such a theory would violate some very high-precision experiments (to be discussed later). So the source of the field should be *all* energies, the density of total mass energy T^{00}. But to have as the source of the field only one component of a tensor would give a noninvariant theory of gravity: one would need to choose a preferred frame in order to calculate T^{00}. Therefore Einstein guessed that the source of the field ought to be **T**: all stresses and pressures and momenta must also act as sources. Combining this with his insight into curved spaces led him to GR.

The second comment is about pressure, which plays a more fundamental role in GR than in Newtonian theory: first, because it is a source of the field; and second, because of its appearance in the $(\rho + p)$ term in Eq. (4.54). Consider a dense star, whose strong gravitational field requires a large pressure gradient. How large is measured by the acceleration the fluid element would have, a_i, in the absence of pressure. Given the field, and hence given a_i, the required pressure gradient is just that which would cause the opposite acceleration without gravity:

$$- a_i = \frac{p_{,i}}{\rho + p}.$$

This gives the pressure gradient $p_{,i}$. Since $(\rho + p)$ is greater than ρ, the gradient must be larger in relativity than in Newtonian theory. Moreover, since all components of **T** are sources of the gravitational field, this larger pressure adds to the gravitational field, causing even larger pressures (compared to Newtonian stars) to be required to hold the star up. For stars where $p \ll \rho$ (see below), this doesn't make much difference. But when p becomes comparable to ρ, one finds that increasing the pressure is self-defeating: *no* pressure gradient will hold the star up, and gravitational collapse must occur. This description, of course, glosses over much detailed calculation, but it shows that even by studying fluids in SR we can begin to appreciate some of the fundamental changes GR brings to gravitation.

Let us just remind ourselves of the relative sizes of p and ρ. We saw in Exer. 1, § 1.14, that $p \ll \rho$ in ordinary situations. In fact, one only gets $p \approx \rho$ for very dense material (neutron star) or material so hot that the particles move at close to the speed of light (a 'relativistic' gas).

4.8 Gauss' law

Our final topic on fluids is the integral form of the conservation laws, which are expressed in differential form in Eqs. (4.34) and (4.35). As in three-dimensional vector calculus, the conversion of a volume integral of a divergence into a surface integral is called Gauss' law. The proof of the theorem is exactly the same as in three dimensions, so we shall not derive it in detail:

◆ $$\int V^\alpha{}_{,\alpha} \, \mathrm{d}^4 x = \oint V^\alpha n_\alpha \, \mathrm{d}^3 S, \qquad (4.57)$$

where $\tilde{n}$ is the unit-normal one-form discussed in § 4.3, and $\mathrm{d}^3 S$ denotes the three-volume of the three-dimensional hypersurface bounding the four-dimensional volume of integration. The sense of the normal is that it is *outward* pointing, of course, just as in three dimensions. In Fig. 4.9 a simple volume is drawn, in order to illustrate the meaning of Eq. (4.57). The volume is bounded by four pairs of hypersurfaces, for constant t, x, y and z; only two pairs are shown, since we can only draw two dimensions easily. The normal on the t_2 surface is $\tilde{\mathrm{d}}t$. The normal on the t_1 surface

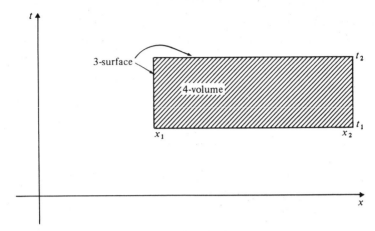

Fig. 4.9   The boundary of a region of spacetime.

is $-\tilde{d}t$, since 'outward' is clearly backwards in time. The normal on x_2 is $\tilde{d}x$, and on x_1 is $-\tilde{d}x$. So the surface integral in Eq. (4.57) is

$$\int_{t_2} V^0 \, dx \, dy \, dz + \int_{x_1} (-V^0) \, dx \, dy \, dz$$

$$+ \int_{x_2} V^x \, dt \, dy \, dz + \int_{x_1} (-V^x) \, dt \, dy \, dz$$

+ similar terms for the other surfaces in the boundary.

We can rewrite this as

$$\int [V^0(t_2) - V^0(t_1)] \, dx \, dy \, dz$$

$$+ \int [V^x(x_2) - V^x(x_1)] \, dt \, dy \, dz + \cdots. \tag{4.58}$$

If we let $\vec{V}$ be $\vec{N}$, then $N^\alpha{}_{,\alpha} = 0$ means that the above expression vanishes, which has the interpretation that change in the number of particles in the three-volume (first integral) is due to the flux across its boundaries (second and subsequent terms). If we are talking about energy conservation we replace N^α with $T^{0\alpha}$, and use $T^{0\alpha}{}_{,\alpha} = 0$. Then, obviously, a similar interpretation of Eq. (4.58) applies. Gauss' law gives an integral version of energy conservation.

4.9   Bibliography

Continuum mechanics and conservation laws are treated in most texts on GR, such as Misner *et al.* (1973). Students whose background in thermodynamics or fluid mechanics is weak are referred to Fermi (1956) and Landau & Lifshitz (1959) respectively. Apart from Exer. 25,

§ 4.10 below, we do not study much about electromagnetism, but it has a stress–energy tensor and illustrates conservation laws particularly clearly. See Landau & Lifshitz (1962) or Jackson (1975). Relativistic fluids with dissipation present their own difficulties, which reward close study. See Israel & Stewart (1980).

4.10 Exercises

1 Comment on whether the continuum approximation is likely to apply to the following physical systems: (a) planetary motions in the solar system; (b) lava flow from a volcano; (c) traffic on a major road at rush hour; (d) traffic at an intersection controlled by stop signs for each incoming road; (e) plasma dynamics.

2 Flux across a surface of constant x is often loosely called 'flux in the x direction'. Use your understanding of vectors and one-forms to argue that this is an inappropriate way of referring to a flux.

3(a) Describe how the Galilean concept of momentum is frame dependent in a manner in which the relativistic concept is not.

(b) How is this possible, since the relativistic definition is nearly the same as the Galilean one for small velocities? (Define a *Galilean* four-momentum vector.)

4 Show that the number density of dust measured by an observer whose four-velocity is $\vec{U}_{obs}$ is $-\vec{N} \cdot \vec{U}_{obs}$.

5 Complete the proof that Eq. (4.14) defines a tensor by arguing that it must be linear in both its arguments.

6 Establish Eq. (4.19) from the preceding equations.

7 Derive Eq. (4.21).

8 Argue that Eq. (4.25) can be written as an equation among one-forms, i.e.
$$\tilde{d}\rho - (\rho + p)\,\tilde{d}n/n = nT\,\tilde{d}S,$$
and similarly for Eq. (4.26). Show that the one-form $\tilde{\Delta}q$ is not a gradient, i.e. is not $\tilde{d}q$ for any function q.

9 Show that Eq. (4.34), when α is any spatial index, is just Newton's second law.

10 Take the limit of Eq. (4.35) for $|v| \ll 1$ to get
$$\partial n/\partial t + \partial(nv^i)/\partial x^i = 0.$$

11(a) Show that the matrix δ^{ij} is unchanged when transformed by a rotation of the spatial axes.

  (b) Show that any matrix which has this property is a multiple of δ^{ij}.

  12 Derive Eq. (4.37) from Eq. (4.36).

  13 Supply the reasoning in Eq. (4.44).

  14 Argue that Eq. (4.46) is the time component of Eq. (4.45) in the MCRF.

  15 Derive Eq. (4.48) from Eq. (4.47).

  16 In the MCRF, $U^i = 0$. Why can't we assume $U^i{}_{,\beta} = 0$?

  17 We have defined $a^\mu = U^\mu{}_{,\beta} U^\beta$. Go to the nonrelativistic limit (small velocity) and show that
  $$a^i = \dot{v}^i + (\mathbf{v} \cdot \boldsymbol{\nabla})v^i = Dv^i/Dt,$$
  where the operator D/Dt is the usual 'total' or 'advective' time derivative of fluid dynamics.

  18 Sharpen the discussion at the end of § 4.6 by showing that $-\boldsymbol{\nabla}p$ is actually the net force per unit volume on the fluid element in the MCRF.

  19 Show that Eq. (4.58) can be used to prove Gauss' law, Eq. (4.57).

20(a) Show that, if particles are not conserved but are generated locally at a rate ε particles per unit volume per unit time in the MCRF, then the conservation law, Eq. (4.35), becomes
  $$N^\alpha{}_{,\alpha} = \varepsilon.$$

  (b) Generalize (a) to show that if the energy and momentum of a body are not conserved (e.g. because it interacts with other systems), then there is a nonzero relativistic force four-vector F^α defined by
  $$T^{\alpha\beta}{}_{,\beta} = F^\alpha.$$
  Interpret the components of F^α in the MCRF.

  21 In an inertial frame $\mathcal{O}$ calculate the components of the stress–energy tensors of the following systems:

  (a) A group of particles all moving with the same velocity $\mathbf{v} = \beta\mathbf{e}_x$, as seen in $\mathcal{O}$. Let the rest-mass density of these particles be ρ_0, as measured in their comoving frame. Assume a sufficiently high density of particles to enable treating them as a continuum.

  (b) A ring of N similar particles of mass m rotating counter-clockwise in the x–y plane about the origin of $\mathcal{O}$, at a radius a from this point, with an angular velocity ω. The ring is a torus of circular cross-section of radius $\delta a \ll a$, within which the particles are uniformly distributed with a high enough density for the continuum approximation to apply. Do not include the stress–energy of whatever forces keep them in orbit. (Part of the calculation will relate ρ_0 of part (a) to N, a, ω, and δa.)

(c) Two such rings of particles, one rotating clockwise and the other counter-clockwise, at the same radius a. The particles do not collide or interact in any way.

22 Many physical systems may be idealized as collections of noncolliding particles (for example, black-body radiation, rarified plasmas, galaxies and globular clusters). By assuming that such a system has a random distribution of velocities at every point, with no bias in any direction in the MCRF, prove that the stress–energy tensor is that of a perfect fluid. If all particles have the same speed v and mass m, express p and ρ as functions of m, v and n. Show that a photon gas has $p = \frac{1}{3}\rho$.

23 Use the identity $T^{\mu\nu}{}_{,\nu} = 0$ to prove the following results for a bounded system (i.e. a system for which $T^{\mu\nu} = 0$ outside a bounded region of space).

(a) $\dfrac{\partial}{\partial t} \displaystyle\int T^{0\alpha}\, \mathrm{d}^3 x = 0$ (conservation of energy and momentum).

(b) $\dfrac{\partial^2}{\partial t^2} \displaystyle\int T^{00} x^i x^j\, \mathrm{d}^3 x = 2 \displaystyle\int T^{ij}\, \mathrm{d}^3 x$ (tensor virial theorem).

(c) $\dfrac{\partial^2}{\partial t^2} \displaystyle\int T^{00} (x^i x_i)^2\, \mathrm{d}^3 x = 4 \displaystyle\int T^i{}_i x^j x_j\, \mathrm{d}^3 x + 8 \displaystyle\int T^{ij} x_i x_j\, \mathrm{d}^3 x.$

24 Astronomical observations of the brightness of objects are measurements of the flux of radiation T^{0i} from the object at Earth. This problem calculates how that flux depends on the relative velocity of the object and Earth.

(a) Show that, in the rest frame $\mathcal{O}$ of a star of constant luminosity L (total energy radiated per second), the stress–energy tensor of the radiation from the star at the event $(t, x, 0, 0)$ has components $T^{00} = T^{0x} = T^{x0} = T^{xx} = L/(4\pi x^2)$. The star sits at the origin.

(b) Let $\vec{X}$ be the null vector which separates the events of emission and reception of the radiation. Show that $\vec{X} \to {}_{\mathcal{O}}(x, x, 0, 0)$ for radiation observed at the event $(x, x, 0, 0)$. Show that the stress–energy tensor of (a) has the frame-invariant form

$$\mathbf{T} = \frac{L}{4\pi} \frac{\vec{X} \otimes \vec{X}}{(\vec{U}_s \cdot \vec{X})^4},$$

where $\vec{U}_s$ is the star's four-velocity, $\vec{U}_s \to {}_{\mathcal{O}}(1, 0, 0, 0)$.

(c) Let the Earth-bound observer $\bar{\mathcal{O}}$, traveling with speed v away from the star in the x direction, measure the same radiation, again with the star on the $\bar{x}$ axis. Let $\vec{X} \to (R, R, 0, 0)$ and find R as a function of x. Express $T^{\bar{0}\bar{x}}$ in terms of R. Explain why R and $T^{\bar{0}\bar{x}}$ depend as they do on v.

25 *Electromagnetism in SR.* (This exercise is suitable only for students who have already encountered Maxwell's equations in some form.) Maxwell's equations for the electric and magnetic fields in vacuum, **E** and **B**, in three-vector notation are

$$\nabla \times \mathbf{B} - \frac{\partial}{\partial t}\mathbf{E} = 4\pi\mathbf{J},$$

$$\nabla \times \mathbf{E} + \frac{\partial}{\partial t}\mathbf{B} = 0, \tag{4.59}$$

$$\nabla \cdot \mathbf{E} = 4\pi\rho,$$

$$\nabla \cdot \mathbf{B} = 0,$$

in units where $\mu_0 = \varepsilon_0 = c = 1$. (Here ρ is the density of electric charge and **J** the current density.)

(a) An *antisymmetric* $\binom{2}{0}$ tensor **F** can be defined on spacetime by the equations $F^{0i} = E^i$ ($i = 1, 2, 3$), $F^{xy} = B^z$, $F^{yz} = B^x$, $F^{zx} = B^y$. Find from this definition all other components $F^{\mu\nu}$ in this frame and write them down in a matrix.

(b) A rotation by an angle θ about the z axis is one kind of Lorentz transformation, with the matrix

$$\Lambda^{\beta'}{}_\alpha = \begin{pmatrix} 1 & 0 & 0 & 0 \\ 0 & \cos\theta & -\sin\theta & 0 \\ 0 & \sin\theta & \cos\theta & 0 \\ 0 & 0 & 0 & 1 \end{pmatrix}.$$

Show that the new components of **F**,

$$F^{\alpha'\beta'} = \Lambda^{\alpha'}{}_\mu \Lambda^{\beta'}{}_\nu F^{\mu\nu},$$

define new electric and magnetic three-vector components (by the rule given in (a)) that are just the same as the components of the old **E** and **B** in the rotated three-space. (This shows that a spatial rotation of **F** makes a spatial rotation of **E** and **B**.)

(c) Define the current four-vector $\vec{J}$ by $J^0 = \rho$, $J^i = (J)^i$, and show that two of Maxwell's equations are just

$$F^{\mu\nu}{}_{,\nu} = 4\pi J^\mu. \tag{4.60}$$

(d) Show that the other two of Maxwell's equations are

$$F_{\mu\nu,\lambda} + F_{\nu\lambda,\mu} + F_{\lambda\mu,\nu} = 0. \tag{4.61}$$

Note that there are only *four* independent equations here. That is, choose one index value, say 0. Then the three other values (1, 2, 3) can be assigned to μ, ν, λ in *any* order, producing the same equation (up to an overall sign) each time. Try it and see: it follows from antisymmetry of $F_{\mu\nu}$.

(e) We have now expressed Maxwell's equations in tensor form. Show that conservation of charge, $J^\mu{}_{,\mu} = 0$ (recall the similar Eq. (4.35) for the number–flux vector $\vec{N}$), is implied by Eq. (4.60) above. (Hint: use antisymmetry of $F_{\mu\nu}$.)

(f) The charge density in any frame is J^0. Therefore the total charge in spacetime is $Q = \int J^0 \, dx \, dy \, dz$, where the integral extends over an entire hypersurface $t = \text{const}$. Defining $\tilde{d}t = \tilde{n}$, a unit normal for this hypersurface, show that

$$Q = \int J^\alpha n_\alpha \, dx \, dy \, dz. \tag{4.62}$$

(g) Use Gauss' law and Eq. (4.60) to show that the total charge enclosed within any closed two-surface $\mathscr{S}$ in the hypersurface $t = \text{const}$. can be determined by doing an integral over $\mathscr{S}$ itself:

$$Q = \oint_{\mathscr{S}} F^{0i} n_i \, d\mathscr{S} = \oint_{\mathscr{S}} \mathbf{E} \cdot \mathbf{n} \, d\mathscr{S},$$

where $\mathbf{n}$ is the unit normal to $\mathscr{S}$ in the hypersurface (*not* the same as $\tilde{n}$ in part (f) above).

(h) Perform a Lorentz transformation on $F^{\mu\nu}$ to a frame $\bar{O}$ moving with velocity v in the x direction relative to the frame used in (a) above. In this frame define a three-vector $\bar{\mathbf{E}}$ with components $\bar{E}^i = F^{\bar{0}\bar{i}}$, and similarly for $\bar{\mathbf{B}}$ in analogy with (a). In this way discover how $\mathbf{E}$ and $\mathbf{B}$ behave under a Lorentz transformation: they get mixed together! Thus, $\mathbf{E}$ and $\mathbf{B}$ themselves are not Lorentz invariant, but are merely components of $\mathbf{F}$, called the Faraday tensor, which is *the* invariant description of electromagnetic fields in relativity. If you think carefully, you will see that on physical grounds they *cannot* be invariant. In particular, the magnetic field is created by moving charges; but a charge moving in one frame may be at rest in another, so a magnetic field which exists in one frame may not exist in another. What is the same in *all* frames is the Faraday tensor: only its components get transformed.

5

Preface to curvature

5.1 On the relation of gravitation to curvature

Until now we have discussed only SR. In SR, forces have played a background role, and we have never introduced gravitation explicitly as a possible force. One ingredient of SR is the existence of inertial frames that fill all of spacetime: all of spacetime can be described by a single frame, all of whose coordinate points are always at rest relative to the origin, and all of whose clocks run at the same rate relative to the origin's clock. From the fundamental postulates we were led to the idea of the interval Δs^2, which gives an invariant geometrical meaning to certain physical statements. For example, a timelike interval between two events is the time elapsed on a clock which passes through the two events; a spacelike interval is the length of a rod that joins two events in a frame in which they are simultaneous. The mathematical function that calculates the interval is the metric, and so the metric of SR is defined physically by lengths of rods and readings of clocks. This is the power of SR and one reason for the elegance and compactness of tensor notation in it (for instance the replacement of 'number density' and 'flux' by $\vec{N}$). On a piece of paper on which one had plotted all the events and world lines of interest in some coordinate system, it would always be possible to define *any* metric by just giving its components $g_{\alpha\beta}$ as some arbitrarily chosen set of functions of the coordinates. But this arbitrary metric would be useless in doing physical calculations. The usefulness of $\eta_{\alpha\beta}$ is its close relation to experiment, and our derivation of it drew heavily on the experiments.

This closeness to experiment is, of course, a test. Since $\eta_{\alpha\beta}$ makes certain predictions about rods and clocks, one can ask for their verification. In particular, is it *possible* to construct a frame in which the clocks all run at the same rate? This is a crucial question, and we shall show that in a nonuniform gravitational field the answer, experimentally, is no. In this sense, gravitational fields are incompatible with *global* SR: the ability to construct a global inertial frame. We shall see that in small regions of spacetime – regions small enough that nonuniformities of the gravitational forces are too small to measure – one can always construct a 'local' SR frame. In this sense, we shall have to build local SR into a more general theory. The first step is the proof that clocks don't all run at the same rate in a gravitational field.

The gravitational redshift experiment. Let us first imagine performing an idealized experiment, first suggested by Einstein. (i) Let a tower of height h be constructed on the surface of Earth, as in Fig. 5.1. Begin with a

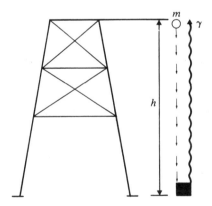

Fig. 5.1 A mass m is dropped from a tower of height h. The total mass at the bottom is converted into energy and returned to the top as a photon. Perpetual motion will be performed unless the photon loses as much energy in climbing as the mass gained in falling. Light is therefore redshifted as it climbs in a gravitational field.

particle of rest mass m at the top of the tower. (ii) The particle is dropped and falls freely with acceleration g. It reaches the ground with velocity $v = (2gh)^{1/2}$, so its total energy E, as measured by an experimenter on the ground, is $m + \frac{1}{2}mv^2 + 0(v^4) = m + mgh + 0(v^4)$. (iii) The experimenter on the ground has some magical method of changing all this energy into a single photon of the same energy, which he directs upwards. (Such a process does not violate conservation laws, since Earth absorbs the photon's momentum but not its energy, just as it does for a bouncing rubber ball. The student sceptical of 'magic' should show how the argument proceeds if only a fraction ε of the energy is converted into a photon.) (iv) Upon its arrival at the top of the tower with energy E', the photon is again magically changed into a particle of rest mass $m' = E'$. It must be that $m' = m$; otherwise, perpetual motion could result by the gain in energy obtained by operating such an experiment. So we are led by our abhorrence of the injustice of perpetual motion to *predict* that $E' = m$ or, for the photon,

$$\blacklozenge \qquad \frac{E'}{E} = \frac{h\nu'}{h\nu} = \frac{m}{m + mgh + 0(v^4)} = 1 - gh + 0(v^4). \qquad (5.1)$$

We predict that a photon climbing in Earth's gravitational field will lose energy (not surprisingly) and will consequently be redshifted.

Although our thought experiment is too idealized to be practical, it is possible to measure the redshift predicted by Eq. (5.1) directly. This was first done by Pound & Rebka (1960) and improved by Pound & Snider (1965). The experiment used the Mossbauer effect to obtain great precision in the measurement of the difference $\nu' - \nu$ produced in a photon climbing a distance $h = 22.5$ m. Eq. (5.1) was verified to approximately 1% precision. A detailed description of the experiment may be found in Misner *et al.* (1973).

This experimental verification is comforting from the point of view of energy conservation. But it is the death-blow to our chances of finding a simple, special-relativistic theory of gravity, as we shall now show.

Nonexistence of a Lorentz frame at rest on Earth. If SR is to be valid in a gravitational field, it is a natural first guess to assume that the 'laboratory' frame at rest on Earth is a Lorentz frame. The following argument, due originally to Schild (1967), easily shows this assumption to be false. In Fig. 5.2 we draw a spacetime diagram in this hypothetical frame, in which the one spatial dimension plotted is the vertical one. Consider light as a wave, and look at two successive 'crests' of the wave as they move upward in the Pound–Rebka–Snider experiment. The top and bottom of the tower have vertical world lines in this diagram, since they are at rest. Light is shown moving on a wiggly line, and it is purposely drawn curved in some arbitrary way. This is to allow for the possibility that gravity may act on light in an unknown way, deflecting it from a null path. But

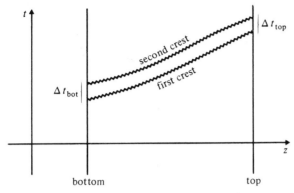

Fig. 5.2 In a time-independent gravitational field, two successive 'crests' of an electromagnetic wave must travel identical paths. Because of the redshift (Eq. (5.1)) the time between them at the top is larger than at the bottom. An observer at the top therefore 'sees' a clock at the bottom running slowly.

no matter how light is affected by gravity the effect must be the same on both wave crests, since the gravitational field does not change from one time to another. Therefore the two crests' paths are *congruent*, and we conclude from the hypothetical Minkowski geometry that $\Delta t_{top} = \Delta t_{bottom}$. On the other hand, the time between two crests is simply the reciprocal of the measured frequency $\Delta t = 1/\nu$. Since the Pound–Rebka–Snider experiment establishes that $\nu_{bottom} > \nu_{top}$, we know that $\Delta t_{top} > \Delta t_{bottom}$. The conclusion from Minkowski geometry is wrong, and the reference frame at rest on Earth is not a Lorentz frame.

Is this the end, then, of SR? Not quite. We have shown that a particular frame is not inertial. We have not shown that there are *no* inertial frames. In fact there are certain frames which are inertial in a restricted sense, and in the next paragraph we shall use another physical argument to find them.

The principle of equivalence. One important property of an inertial frame is that a particle at rest in it stays at rest if no forces act on it. In order to use this, we have to have an idea of what a force is. Ordinarily, gravity is regarded as a force, but gravity is distinguished from all other forces in a remarkable way: all bodies given the same initial velocity follow the same trajectory in a gravitational field, regardless of their internal composition. With all other forces, some bodies are affected and others are not: electromagnetism affects charged particles but not neutral ones, and the trajectory of a charged particle depends on the ratio of its charge to its mass, which is not the same for all particles. Similarly, the other two basic forces in physics – the so-called 'strong' and 'weak' interactions – affect different particles differently. With all these forces, it would always be possible to define experimentally the trajectory of a particle unaffected by the force, i.e. a particle that remained at rest in an inertial frame. But, with gravity, this does not work. Attempting to define an inertial frame at rest on Earth, then, is vacuous, since *no* free particle (not even a photon) could possibly be a physical 'marker' for it.

But there is a frame in which particles do keep a uniform velocity. This is a frame which falls freely in the gravitational field. Since this frame accelerates at the same rate as free particles do (at least the low-velocity particles to which Newtonian gravitational physics applies), it follows that all such particles will maintain a uniform velocity relative to this frame. This frame is at least a candidate for an inertial frame. In the next section we will show that photons are not redshifted in this frame, which makes it an even better candidate. Einstein built GR by taking the hypothesis that these frames are inertial.

The argument we have just made, that freely falling frames are inertial, will perhaps be more familiar to the student if it is turned around. Consider, in empty space free of gravity, a uniformly accelerating rocket ship. From the point of view of an observer inside, it appears that there is a gravitational field in the rocket: objects dropped accelerate toward the rear of the ship, all with the same acceleration, independent of their internal composition.[1] Moreover, an object held stationary relative to the ship has 'weight' equal to the force required to keep it accelerating with the ship. Just as in 'real' gravity, this force is proportional to the mass of the object. A true inertial frame is one which falls freely toward the rear of the ship, at the same acceleration as particles. From this it can be seen that uniform gravitational fields are equivalent to frames that accelerate uniformly relative to inertial frames. This is the *principle of equivalence* between gravity and acceleration, and is a cornerstone of Einstein's theory. In more modern terminology, what we have described is called the *weak equivalence principle*, 'weak' because it refers only to gravity. We shall later use the *strong equivalence principle*, which says that one can discover how all the other forces of nature behave in a gravitational field by postulating that their laws in a freely falling inertial frame are identical to their laws in SR, i.e. when there are no gravitational fields.

---

1   This has been tested experimentally to extremely high precision in the so-called Eötvös experiment. See Dicke (1964).

Before we return to the proof that freely falling frames are inertial, even for photons, we must make two important observations. The first is that our arguments are valid only locally – since the gravitational field of Earth is not uniform, particles some distance away do not remain at uniform velocity in a particular freely falling frame. We shall discuss this in some detail below. The second point is that there are of course an infinity of freely falling frames at any point. They differ in their velocities and in the orientation of their spatial axes, but they all accelerate relative to Earth at the same rate.

The redshift experiment again. Let us now take a different point of view on the Pound–Rebka–Snider experiment. Let us view it in a freely falling frame, which we have seen has at least some of the characteristics of an inertial frame. Let us take the particular frame that is at rest when the photon begins its upward journey and falls freely after that. Since the photon rises a distance h, it takes time $\Delta t = h$ to arrive at the top. In this time, the frame has acquired velocity gh downward relative to the experimental apparatus. So the photon's frequency relative to the freely falling frame can be obtained by the redshift formula

$$\frac{\nu(\text{freely falling})}{\nu'(\text{apparatus at top})} = \frac{1+gh}{\sqrt{(1-g^2h^2)}} = 1 + gh + 0(v^4). \qquad (5.2)$$

From Eq. (5.1) we see that if we neglect terms of higher order (as we did to derive Eq. (5.1)), then we get ν(photon emitted at bottom) = ν(in freely falling frame when photon arrives at top). So there is *no* redshift in a freely falling frame. This gives us a sound basis for postulating that the freely falling frame is an inertial frame.

Local inertial frames. The above discussion makes one suggest that the gravitational redshift experiment really does not render SR and gravity incompatible. Perhaps one simply has to realize that the frame at rest on Earth is not inertial and the freely falling one – in which there is no redshift and so Fig. 5.2 leads to no contradiction – is the true inertial frame. Unfortunately this doesn't completely save SR, for the simple reason that the freely falling frames on different sides of Earth fall in different directions: there is *no* single global frame which is everywhere freely falling in Earth's gravitational field and which is still rigid, in that the distances between its coordinate points are constant in time. It is still impossible to construct a *global* inertial frame, and so the most we can salvage is a *local* inertial frame, which we now describe.

Consider a freely falling frame in Earth's gravitational field. An inertial frame in SR fills all of spacetime, but this freely falling frame would not be inertial if it were extended too far horizontally, because then it would not be falling vertically. In Fig. 5.3 the frame is freely falling at *B*, but

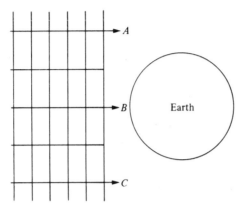

Fig. 5.3   A rigid frame cannot fall freely in the Earth's field and still remain rigid.

at *A* and *C* is not moving on the trajectory of a test particle. Moreover, since the acceleration of gravity changes with height, the frame cannot remain inertial if extended over too large a vertical distance; if it were falling with particles at one height, it would not be at another. Finally, the frame can have only a limited extent in time as well, since, as it falls, both the above limitations become more severe due to the frame's approaching closer to Earth. All of these limitations are due to nonuniformities in the gravitational field. Insofar as nonuniformities can be neglected, the freely falling frame can be regarded as inertial. *Any* gravitational field can be regarded as uniform over a small enough region of space and time, and so one can always set up *local* inertial frames. They are analogous to the MCRFs of fluids: in this case the frame is inertial in only a small region for a small time. How small depends on (a) the strength of the nonuniformities of the gravitational field, and (b) the sensitivity of whatever experiment is being used to detect noninertial properties of the frame. Since *any* nonuniformity is, in principle, detectable, a frame can only be regarded mathematically as inertial in a vanishingly small region. But for current technology, the freely falling frames near the surface of Earth can be regarded as inertial to a high accuracy. We will be more quantitative in a later chapter. For now, we just emphasize the mathematical notion that any theory of gravity must admit *local inertial frames*: frames that, at a point, are inertial frames of SR.

Tidal forces. Nonuniformities in gravitational fields are called tidal forces, since they are the ones that raise tides. (If Earth were in a uniform gravitational field, it would fall freely and have no tides. Tides bulge due to the *difference* of the Moon's and Sun's gravitational fields across the diameter of Earth.) We have seen that these tidal forces prevent the construction of global inertial frames. It is therefore these forces that are regarded as the fundamental manifestation of gravity in GR.

The role of curvature. The world lines of free particles have been our probe of the possibility of constructing inertial frames. In SR, two such world lines which begin parallel to each other remain parallel, no matter how far they are extended. This is exactly the property that straight lines have in Euclidean geometry. It is natural, therefore, to discuss the *geometry* of spacetime as defined by the world lines of free particles. In these terms, Minkowski space is a *flat* space, because it obeys Euclid's parallelism axiom. It is not a Euclidean space, however, since its metric is different: photons travel on straight world lines of zero proper length. So SR has a flat, non-Euclidean geometry.

Now, in a nonuniform gravitational field, the world lines of two nearby particles which begin parallel do not generally remain parallel. Gravitational spacetime is therefore not flat. In Euclidean geometry, when one drops the parallelism axiom, one gets a curved space. For example, the surface of a sphere is curved. Locally straight lines on a sphere extend to great circles, and two great circles always intersect. Nevertheless, sufficiently near to any point, one can pretend that the geometry is flat: the map of a town can be represented on a flat sheet of paper without significant distortion, while a similar attempt for the whole globe fails completely. The sphere is thus locally flat. This is true for all so-called Riemannian[2] spaces: they all are locally flat, but the locally straight lines (called *geodesics*) do not usually remain parallel.

Einstein's important advance was to see the similarity between Reimannian spaces and gravitational physics. He identified the trajectories of freely falling particles with the geodesics of a curved geometry: they are locally straight since spacetime admits local inertial frames in which those trajectories are straight lines, but globally they do not remain parallel.

We shall follow Einstein and look for a theory of gravity which uses a curved spacetime to represent the effects of gravity on particles' trajectories. To do this we shall clearly have to study the mathematics of

---

2   B. Riemann (1826–66) was the first to publish a detailed study of the consequences of dropping Euclid's parallelism axiom.

curvature. The simplest introduction is actually to study curvilinear coordinate systems in a flat space, where our intuition is soundest. We shall see that this will develop nearly all the mathematical concepts we need, and the step to a curved space will be simple. So for the rest of this chapter we will study the Euclidean plane: no more SR (for the time being!) and no more indefinite inner products. What we are after in this chapter is parallelism, not metrics. This approach has the added bonus of giving a more sensible derivation to such often-mysterious formulae as the expression for ∇^2 in polar coordinates!

5.2   Tensor algebra in polar coordinates

Consider the Euclidean plane. The usual coordinates are x and y. Sometimes polar coordinates $\{r, \theta\}$ are convenient:

$$\left. \begin{aligned} r &= (x^2 + y^2)^{1/2}, & x &= r \cos \theta, \\ \theta &= \arctan \frac{y}{x}, & y &= r \sin \theta. \end{aligned} \right\} \tag{5.3}$$

Small increments Δr and $\Delta \theta$ are produced by Δx and Δy according to

$$\left. \begin{aligned} \Delta r &= \frac{x}{r} \Delta x + \frac{y}{r} \Delta y = \cos \theta \, \Delta x + \sin \theta \, \Delta y, \\ \Delta \theta &= -\frac{y}{r^2} \Delta x + \frac{x}{r^2} \Delta y = -\frac{1}{r} \sin \theta \, \Delta x + \frac{1}{r} \cos \theta \, \Delta y, \end{aligned} \right\} \tag{5.4}$$

which are valid to first order.

It is also possible to use other coordinate systems. Let us denote a general coordinate system by ξ and η:

$$\left. \begin{aligned} \xi &= \xi(x, y), & \Delta \xi &= \frac{\partial \xi}{\partial x} \Delta x + \frac{\partial \xi}{\partial y} \Delta y, \\ \eta &= \eta(x, y), & \Delta \eta &= \frac{\partial \eta}{\partial x} \Delta x + \frac{\partial \eta}{\partial y} \Delta y. \end{aligned} \right\} \tag{5.5}$$

In order for (ξ, η) to be good coordinates, it is necessary that any two distinct points (x_1, y_1) and (x_2, y_2) be assigned different pairs (ξ_1, η_1) and (ξ_2, η_2), by Eq. (5.5). For instance, the definitions $\xi = x$, $\eta = 1$ would not give good coordinates, since the distinct points $(x = 1, y = 2)$ and $(x = 1, y = 3)$ both have $(\xi = 1, \eta = 1)$. Mathematically, this implies that if $\Delta \xi = \Delta \eta = 0$ in Eq. (5.5), then the points must be the same, or $\Delta x = \Delta y = 0$. This will be true if the determinant of Eq. (5.5) is nonzero,

$$\det \begin{pmatrix} \partial \xi / \partial x & \partial \xi / \partial y \\ \partial \eta / \partial x & \partial \eta / \partial y \end{pmatrix} \neq 0. \tag{5.6}$$

This determinant is called the *Jacobian* of the coordinate transformation,

Eq. (5.5). If the Jacobian vanishes at a point, the transformation is said to be *singular* there.

Vectors and one-forms. The old way of defining a vector is to say that it transforms under an *arbitrary* coordinate transformation in the way that the displacement transforms. That is, a vector $\overrightarrow{\Delta r}$ can be represented[3] as a displacement $(\Delta x, \Delta y)$, or in polar coordinates $(\Delta r, \Delta \theta)$, or in general $(\Delta \xi, \Delta \eta)$. Then it is clear that for *small* $(\Delta x, \Delta y)$ we have (from Eq. (5.5))

$$\begin{pmatrix} \Delta \xi \\ \\ \Delta \eta \end{pmatrix} = \begin{pmatrix} \dfrac{\partial \xi}{\partial x} & \dfrac{\partial \xi}{\partial y} \\ \dfrac{\partial \eta}{\partial x} & \dfrac{\partial \eta}{\partial y} \end{pmatrix} \begin{pmatrix} \Delta x \\ \\ \Delta y \end{pmatrix}. \tag{5.7}$$

By defining the matrix of transformation

$$(\Lambda^{\alpha'}{}_{\beta}) = \begin{pmatrix} \partial \xi / \partial x & \partial \xi / \partial y \\ \partial \eta / \partial x & \partial \eta / \partial y \end{pmatrix}, \tag{5.8}$$

we can write the transformation for an arbitrary $\vec{V}$ in the same manner as in SR

$$V^{\alpha'} = \Lambda^{\alpha'}{}_{\beta} V^{\beta}, \tag{5.9}$$

where unprimed indices refer to (x, y) and primed to (ξ, η), and where indices can only take the values 1 and 2. A vector can be defined as something whose components transform according to Eq. (5.9). There is a more sophisticated and natural way, however. This is the modern way, which we now introduce.

Consider a scalar field ϕ. Given coordinates (ξ, η) it is always possible to form the derivatives $\partial \phi / \partial \xi$ and $\partial \phi / \partial \eta$. We *define* the one-form $\tilde{d}\phi$ to be the geometrical object whose components are

$$\tilde{d}\phi \rightarrow (\partial \phi / \partial \xi, \partial \phi / \partial \eta) \tag{5.10}$$

in the (ξ, η) coordinate system. This is a general definition of an infinity of one-forms, each formed from a different scalar field. The transformation of components is automatic from the chain rule for partial derivatives:

$$\frac{\partial \phi}{\partial \xi} = \frac{\partial x}{\partial \xi} \frac{\partial \phi}{\partial x} + \frac{\partial y}{\partial \xi} \frac{\partial \phi}{\partial y}, \tag{5.11}$$

---

3   We shall denote Euclidean vectors by arrows, and we shall use Greek letters for indices (numbered 1 and 2) to denote the fact that the sum is over all possible (i.e. both) values.

and similarly for $\partial\phi/\partial\eta$. This can be written as

$$\begin{pmatrix} \partial\phi/\partial\xi \\ \partial\phi/\partial\eta \end{pmatrix} = \begin{pmatrix} \partial x/\partial\xi & \partial y/\partial\xi \\ \partial x/\partial\eta & \partial y/\partial\eta \end{pmatrix} \begin{pmatrix} \partial\phi/\partial x \\ \partial\phi/\partial y \end{pmatrix} \tag{5.12}$$

Then the transformation matrix is

$$(\Lambda^{\alpha}{}_{\beta'}) = \begin{pmatrix} \partial x/\partial\xi & \partial y/\partial\xi \\ \partial x/\partial\eta & \partial y/\partial\eta \end{pmatrix}. \tag{5.13}$$

Thus, in the modern view we first define one-forms. This is more natural than the old way, in which a *single* vector $(\Delta x, \Delta y)$ was defined and others were obtained by analogy. Here a whole *class* of one-forms is defined in terms of derivatives, and the transformation properties of one-forms follow automatically.

Now a vector is defined as a linear function of one-forms into real numbers. The implications of this will be explored in the next paragraph. First we just note that all this is the same as we had in SR, so that vectors do in fact obey the transformation law, Eq. (5.9). It is of interest to see explicitly that $(\Lambda^{\alpha}{}_{\beta})$ and $(\Lambda^{\alpha}{}_{\beta'})^{T}$ are inverses to each other. (We need the transpose $(\Lambda^{\alpha}{}_{\beta'})^{T}$ here. This came out in SR but wasn't used because the transpose of the Lorentz transformation equaled the original matrix, which was symmetric. In general, $(\Lambda^{\alpha}{}_{\beta'})$ is not symmetric.) The product of the matrices is

$$\begin{pmatrix} \partial\xi/\partial x & \partial\xi/\partial y \\ \partial\eta/\partial x & \partial\eta/\partial y \end{pmatrix} \begin{pmatrix} \partial x/\partial\xi & \partial x/\partial\eta \\ \partial y/\partial\xi & \partial y/\partial\eta \end{pmatrix}$$

$$= \begin{pmatrix} \dfrac{\partial\xi}{\partial x}\dfrac{\partial x}{\partial\xi} + \dfrac{\partial\xi}{\partial y}\dfrac{\partial y}{\partial\xi} & \dfrac{\partial\xi}{\partial x}\dfrac{\partial x}{\partial\eta} + \dfrac{\partial\xi}{\partial y}\dfrac{\partial y}{\partial\eta} \\ \dfrac{\partial\eta}{\partial x}\dfrac{\partial x}{\partial\xi} + \dfrac{\partial\eta}{\partial y}\dfrac{\partial y}{\partial\xi} & \dfrac{\partial\eta}{\partial x}\dfrac{\partial x}{\partial\eta} + \dfrac{\partial\eta}{\partial y}\dfrac{\partial y}{\partial\eta} \end{pmatrix}. \tag{5.14}$$

By the chain rule this matrix is

$$\begin{pmatrix} \partial\xi/\partial\xi & \partial\xi/\partial\eta \\ \partial\eta/\partial\xi & \partial\eta/\partial\eta \end{pmatrix} = \begin{pmatrix} 1 & 0 \\ 0 & 1 \end{pmatrix}, \tag{5.15}$$

where the equality follows from the definition of a partial derivative.

Curves and vectors. The usual notion of a curve is of a connected series of points in the plane. This we shall call a *path*, and reserve the word curve for a parametrized path. That is, we shall follow modern mathematical terminology and define a *curve* as a mapping of an interval of the real line into a path in the plane. What this means is that a curve is a path with a real number associated with each point on the path. This number is called the parameter s. Each point has coordinates which may then be expressed as a function of s:

◆   Curve: $\{\xi = f(s), \eta = g(s), a \leqslant s \leqslant b\}$ (5.16)

defines a curve in the plane. If we were to change the parameter (but not the points) to $s' = s'(s)$, which is a function of the old s, then we would have

$$\{\xi = f'(s'), \eta = g'(s'), a' \leqslant s' \leqslant b'\}, \tag{5.17}$$

where f' and g' are *new* functions, and where $a' = s'(a)$, $b' = s'(b)$. This is, mathematically, a *new* curve, even though its *image* (the points of the plane that it passes through) is the same. So there are an infinite number of curves having the same path.

The derivative of a scalar field ϕ along the curve is $\mathrm{d}\phi/\mathrm{d}s$. This depends on s, so by changing the parameter, one changes the derivative. One can write this as

$$\mathrm{d}\phi/\mathrm{d}s = \langle \tilde{\mathrm{d}}\phi, \vec{V} \rangle, \tag{5.18}$$

where $\vec{V}$ is the vector whose components are $(\mathrm{d}\xi/\mathrm{d}s, \mathrm{d}\eta/\mathrm{d}s)$. This vector depends only on the curve, while $\tilde{\mathrm{d}}\phi$ depends only on ϕ. Therefore $\vec{V}$ is a vector characteristic of the curve, called the *tangent* vector. (It clearly lies tangent to curve: see Fig. 5.4.) So a vector may be regarded as a

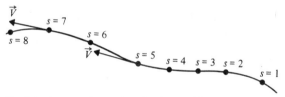

Fig. 5.4   A curve, its parametrization, and its tangent vector.

thing which produces $\mathrm{d}\phi/\mathrm{d}s$, given ϕ. This leads to the most modern view, that the tangent vector to the curve should be *called* $\mathrm{d}/\mathrm{d}s$. Some relativity texts occasionally use this notation. For our purposes, however, we shall just let $\vec{V}$ be the tangent vector whose components are $(\mathrm{d}\xi/\mathrm{d}s, \mathrm{d}\eta/\mathrm{d}s)$. Notice that a path in the plane has, at any point, an infinity of tangents, all of them parallel but differing in length. These are to be regarded as vectors tangent to *different* curves, curves that have different parameters at that point. A curve has a *unique* tangent, since the path *and* parameter are given. Moreover, even curves that have identical tangents at a point may not be identical elsewhere. From the Taylor expansion $\xi(s+1) \approx \xi(s) + \mathrm{d}\xi/\mathrm{d}s$, we see that $\vec{V}(s)$ stretches approximately from s to $s+1$ along the curve.

Now, it is clear that under a coordinate transformation s does not change (its definition had nothing to do with coordinates) but the components of $\vec{V}$ will, since by the chain rule

$$\begin{pmatrix} d\xi/ds \\ d\eta/ds \end{pmatrix} = \begin{pmatrix} \partial\xi/\partial x & \partial\xi/\partial y \\ \partial\eta/\partial x & \partial\eta/\partial y \end{pmatrix} \begin{pmatrix} dx/ds \\ dy/ds \end{pmatrix}. \tag{5.19}$$

This is the same transformation law as we had for vectors earlier, Eq. (5.7).

To sum up, the modern view is that a vector is a *tangent* to some curve, and is the function that gives $d\phi/ds$ when it takes $\tilde{d}\phi$ as an argument. Having said this, we are now in a position to do polar coordinates more thoroughly.

Polar coordinate basis one-forms and vectors. The bases of the coordinates are clearly

$$\vec{e}_{\alpha'} = \Lambda^\beta{}_{\alpha'}\vec{e}_\beta,$$

or

$$\vec{e}_r = \Lambda^x{}_r\vec{e}_x + \Lambda^y{}_r\vec{e}_y \tag{5.20}$$

$$= \frac{\partial x}{\partial r}\vec{e}_x + \frac{\partial y}{\partial r}\vec{e}_y$$

$$= \cos\theta\,\vec{e}_x + \sin\theta\,\vec{e}_y, \tag{5.21}$$

and, similarly,

$$\vec{e}_\theta = \frac{\partial x}{\partial \theta}\vec{e}_x + \frac{\partial y}{\partial \theta}\vec{e}_y$$

$$= -r\sin\theta\,\vec{e}_x + r\cos\theta\,\vec{e}_y. \tag{5.22}$$

Notice in this that we have used, among others,

$$\Lambda^x{}_r = \frac{\partial x}{\partial r}. \tag{5.23}$$

Similarly, to transform the other way we would need

$$\Lambda^r{}_x = \frac{\partial r}{\partial x}. \tag{5.24}$$

The transformation matrices are exceedingly simple: just keeping track of which index is up and which is down gives the right derivative to use.

The basis one-forms are, analogously,

$$\tilde{d}\theta = \frac{\partial \theta}{\partial x}\tilde{d}x + \frac{\partial \theta}{\partial y}\tilde{d}y,$$

$$= -\frac{1}{r}\sin\theta\,\tilde{d}x + \frac{1}{r}\cos\theta\,\tilde{d}y. \tag{5.25}$$

(Notice the similarity to ordinary calculus, Eq. (5.4).) Similarly, we find

$$\tilde{d}r = \cos \theta \; \tilde{d}x + \sin \theta \; \tilde{d}y. \tag{5.26}$$

We can draw pictures of the bases at various points (Fig. 5.5). Drawing the basis vectors is no problem. Drawing the basis one-forms is most easily done by drawing surfaces of constant r and θ for $\tilde{d}r$ and $\tilde{d}\theta$. These surfaces have different orientations in different places.

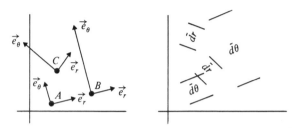

Fig. 5.5   Basis vectors and one-forms for polar coordinates.

There is a point of great importance to note here: the bases change from point to point. For the vectors, the basis vectors at A in Fig. 5.5 are not parallel to those at C. This is because they point in the direction of increasing coordinate, which changes from point to point. Moreover, the lengths of the bases are not constant. For example, from Eq. (5.22) we find

$$|\vec{e}_\theta|^2 = = \vec{e}_\theta \cdot \vec{e}_\theta = r^2 \sin^2 \theta + r^2 \cos^2 \theta = r^2, \tag{5.27a}$$

so that $\vec{e}_\theta$ increases in magnitude as one gets further from the origin. The reason is that the basis vector $\vec{e}_\theta$, having components $(0, 1)$ with respect to r and θ, has essentially a θ displacement of one unit, i.e. one radian. It must be longer to do this at large radii than at small. So we do not have a *unit* basis. It is easy to verify that

$$|\vec{e}_r| = 1, \qquad |\tilde{d}r| = 1, \qquad |\tilde{d}\theta| = r^{-1}. \tag{5.27b}$$

Again, $|\tilde{d}\theta|$ gets larger near $r = 0$ because a given vector can span a larger range of θ near the origin than farther away.

Metric tensor. The dot products above were all calculated by knowing the metric in Cartesian coordinates x, y:

$$\vec{e}_x \cdot \vec{e}_x = \vec{e}_y \cdot \vec{e}_y = 1, \qquad \vec{e}_x \cdot \vec{e}_y = 0;$$

or, put in tensor notation,

$$\mathbf{g}(\vec{e}_\alpha, \vec{e}_\beta) = \delta_{\alpha\beta} \quad \text{in Cartesian coordinates.} \tag{5.28}$$

What are the components of $\mathbf{g}$ in polar coordinates? Simply

$$g_{\alpha'\beta'} = \mathbf{g}(\vec{e}_{\alpha'}, \vec{e}_{\beta'}) = \vec{e}_{\alpha'} \cdot \vec{e}_{\beta'} \tag{5.29}$$

or, by Eq. (5.27),

$$g_{rr} = 1, \qquad g_{\theta\theta} = r^2, \qquad\qquad (5.30a)$$

and, from Eqs. (5.21) and (5.22),

$$g_{r\theta} = 0. \qquad\qquad (5.30b)$$

So we can write the components of **g** as

$$(g_{\alpha\beta})_{polar} = \begin{pmatrix} 1 & 0 \\ 0 & r^2 \end{pmatrix}, \qquad\qquad (5.31)$$

A convenient way of displaying the components of **g** and at the same time showing the coordinates is the line element, which is the magnitude of an arbitrary 'infinitesimal' displacement $d\vec{l}$:

◆ $$\begin{aligned} d\vec{l} \cdot d\vec{l} = ds^2 &= |dr\, \vec{e}_r + d\theta\, \vec{e}_\theta|^2 \\ &= dr^2 + r^2\, d\theta^2. \end{aligned} \qquad\qquad (5.32)$$

Do *not* confuse dr and $d\theta$ here with the basis one-forms $\tilde{d}r$ and $\tilde{d}\theta$. The things in this equation are components of $d\vec{l}$ in polar coordinates, and 'd' simply means 'infinitesimal Δ'.

There is another way of deriving Eq. (5.32) which is instructive. Recall Eq. (3.26) in which a general $\binom{0}{2}$ tensor is written as a sum over basis $\binom{0}{2}$ tensors $\tilde{d}x^\alpha \otimes \tilde{d}x^\beta$. For the metric this is

$$\mathbf{g} = g_{\alpha\beta}\, \tilde{d}x^\alpha \otimes \tilde{d}x^\beta = \tilde{d}r \otimes \tilde{d}r + r^2\, \tilde{d}\theta \otimes \tilde{d}\theta.$$

Although this has a superficial resemblance to Eq. (5.32), it is different: it is an operator which, when supplied with the vector $d\vec{l}$, whose components are dr and $d\theta$, gives Eq. (5.32). Unfortunately, the two expressions resemble each other rather too closely because of the confusing way notation has evolved in this subject. Most texts and research papers still use the 'old-fashioned' form in Eq. (5.32) for displaying the components of the metric, and we follow the same practice.

The metric has an inverse:

$$\begin{pmatrix} 1 & 0 \\ 0 & r^2 \end{pmatrix}^{-1} \equiv \begin{pmatrix} 1 & 0 \\ 0 & r^{-2} \end{pmatrix}. \qquad\qquad (5.33)$$

So we have $g^{rr} = 1$, $g^{r\theta} = 0$, $g^{\theta\theta} = 1/r^2$. This enables us to make the mapping between one-forms and vectors. For instance, if ϕ is a scalar field and $\tilde{d}\phi$ is its gradient, then the vector $\vec{d}\phi$ has components

$$(\vec{d}\phi)^\alpha = g^{\alpha\beta}\phi_{,\beta}, \qquad\qquad (5.34)$$

or

$$\begin{aligned} (\vec{d}\phi)^r = g^{r\beta}\phi_{,\beta} &= g^{rr}\phi_{,r} + g^{r\theta}\phi_{,\theta} \\ &= \partial\phi/\partial r; \end{aligned} \qquad\qquad (5.35a)$$

$$(\vec{d}\phi)^\theta = g^{\theta r}\phi_{,r} + g^{\theta\theta}\phi_{,\theta}$$

$$= \frac{1}{r^2}\frac{\partial\phi}{\partial\theta}. \tag{5.35b}$$

So, while $(\phi_{,r}, \phi_{,\theta})$ are components of a one-form, the vector gradient has components $(\phi_{,r}, \phi_{,\theta}/r^2)$. Even though we are in Euclidean space, vectors generally have different components from their associated one-forms. Cartesian coordinates are the only coordinates in which the components are the same.

5.3 Tensor calculus in polar coordinates

The fact that the basis vectors of polar coordinates are not constant everywhere, leads to some problems when one tries to differentiate vectors. For instance, consider the simple vector $\vec{e}_x$, which is a constant vector field, the same at any point. In polar coordinates it has components $\vec{e}_x \rightarrow (\Lambda^r{}_x, \Lambda^\theta{}_x) = (\cos\theta, -r^{-1}\sin\theta)$. These are clearly not constant, even though $\vec{e}_x$ is. The reason is that they are components on a nonconstant basis. If we were to differentiate them with respect to, say, θ, we would most certainly *not* get $\partial\vec{e}_x/\partial\theta$, which must be identically zero. So, from this example, one sees that differentiating the components of a vector does not necessarily give the derivative of the vector. One must also differentiate the nonconstant basis vectors. This is the key to the understanding of curved coordinates and, indeed, of curved spaces. We shall now make these ideas systematic.

Derivatives of basis vectors. Since $\vec{e}_x$ and $\vec{e}_y$ are constant vector fields, we easily find that

$$\frac{\partial}{\partial r}\vec{e}_r = \frac{\partial}{\partial r}(\cos\theta\,\vec{e}_x + \sin\theta\,\vec{e}_y) = 0, \tag{5.36a}$$

$$\frac{\partial}{\partial\theta}\vec{e}_r = \frac{\partial}{\partial\theta}(\cos\theta\,\vec{e}_x + \sin\theta\,\vec{e}_y)$$

$$= -\sin\theta\,\vec{e}_x + \cos\theta\,\vec{e}_y = \frac{1}{r}\vec{e}_\theta. \tag{5.36b}$$

These have a simple geometrical picture, shown in Fig. 5.6. At two nearby points, A and B, $\vec{e}_r$ must point directly away from the origin, and so in slightly different directions. The derivative of $\vec{e}_r$ with respect to θ is just the difference between $\vec{e}_r$ at A and B divided by $\Delta\theta$. The difference in this case is clearly a vector parallel to $\vec{e}_\theta$, which then makes Eq. (5.36b) reasonable.

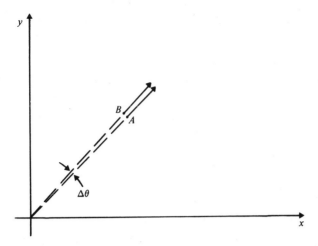

Fig. 5.6   Change in $\vec{e}_r$ when θ changes by $\Delta\theta$.

Similarly,

$$\frac{\partial}{\partial r}\vec{e}_\theta = \frac{\partial}{\partial r}(-r\sin\theta\,\vec{e}_x + r\cos\theta\,\vec{e}_y)$$

$$= -\sin\theta\,\vec{e}_x + \cos\theta\,\vec{e}_y = \frac{1}{r}\vec{e}_\theta, \tag{5.37a}$$

$$\frac{\partial}{\partial\theta}\vec{e}_\theta = r\cos\theta\,\vec{e}_x = r\sin\theta\,\vec{e}_y = -r\vec{e}_r. \tag{5.37b}$$

The student is encouraged to draw a picture similar to Fig. 5.6 to explain these formulas.

Derivatives of general vectors. Let us go back to the derivative of $\vec{e}_x$. Since

$$\vec{e}_x = \cos\theta\,\vec{e}_r - \frac{1}{r}\sin\theta\,\vec{e}_\theta, \tag{5.38}$$

we have

$$\frac{\partial}{\partial\theta}\vec{e}_x = \frac{\partial}{\partial\theta}(\cos\theta)\,\vec{e}_r + \cos\theta\frac{\partial}{\partial\theta}(\vec{e}_r)$$

$$- \frac{\partial}{\partial\theta}\left(\frac{1}{r}\sin\theta\right)\vec{e}_\theta - \frac{1}{r}\sin\theta\frac{\partial}{\partial\theta}(\vec{e}_\theta) \tag{5.39}$$

$$= -\sin\theta\,\vec{e}_r + \cos\theta\left(\frac{1}{r}\vec{e}_\theta\right)$$

$$- \frac{1}{r}\cos\theta\,\vec{e}_\theta - \frac{1}{r}\sin\theta(-r\vec{e}_r). \tag{5.40}$$

To get this we used Eqs. (5.36) and (5.37). Simplifying gives

$$\frac{\partial}{\partial \theta} \vec{e}_x = 0,$$ (5.41)

just as one should have. Now, in Eq. (5.39) the first and third terms come from differentiating the *components* of $\vec{e}_x$ on the polar coordinate basis; the other two terms are the derivatives of the polar basis vectors themselves, and are necessary for cancelling out the derivatives of the components.

A general vector $\vec{V}$ has components (V^r, V^θ) on the polar basis. Its derivative, by analogy with Eq. (5.39), is

$$\frac{\partial \vec{V}}{\partial r} = \frac{\partial}{\partial r}(V^r \vec{e}_r + V^\theta \vec{e}_\theta)$$

$$= \frac{\partial V^r}{\partial r} \vec{e}_r + V^r \frac{\partial \vec{e}_r}{\partial r} + \frac{\partial V^\theta}{\partial r} \vec{e}_\theta + V^\theta \frac{\partial \vec{e}_\theta}{\partial r},$$

and similarly for $\partial \vec{V}/\partial \theta$. Written in index notation, this becomes

$$\frac{\partial \vec{V}}{\partial r} = \frac{\partial}{\partial r}(V^\alpha \vec{e}_\alpha) = \frac{\partial V^\alpha}{\partial r} \vec{e}_\alpha + V^\alpha \frac{\partial \vec{e}_\alpha}{\partial r}.$$

(Here α runs of course over r and θ.)

This shows explicitly that the derivative of $\vec{V}$ is more than just the derivative of its components V^α. Now, since r is just one coordinate, we can generalize the above equation to

$$\frac{\partial \vec{V}}{\partial x^\beta} = \frac{\partial V^\alpha}{\partial x^\beta} \vec{e}_\alpha + V^\alpha \frac{\partial \vec{e}_\alpha}{\partial x^\beta},$$ (5.42)

where, now, x^β can be either r or θ for $\beta = 1$ or 2.

The Christoffel symbols. The final term in Eq. (5.42) is obviously of great importance. Since $\partial \vec{e}_\alpha/\partial x^\beta$ is itself a vector, it can be written as a linear combination of the basis vectors; we introduce the symbol $\Gamma^\mu{}_{\alpha\beta}$ to denote the coefficients in this combination:

$$\frac{\partial \vec{e}_\alpha}{\partial x^\beta} = \Gamma^\mu{}_{\alpha\beta} \vec{e}_\mu.$$ (5.43)

The interpretation of $\Gamma^\mu{}_{\alpha\beta}$ is that it is the μth component of $\partial \vec{e}_\alpha/\partial x^\beta$. It needs three indices: one (α) gives the basis vector being differentiated; the second (β) gives the coordinate with respect to which it is being differentiated; and the third (μ) denotes the component of the resulting derivative vector. These things, $\Gamma^\mu{}_{\alpha\beta}$, are so useful that they have been given a name: the Christoffel symbols. The question of whether or not they are components of tensors we postpone until much later.

We have of course already calculated them for polar coordinates. From Eqs. (5.36) and (5.37) we find

$$
\left.
\begin{aligned}
&(1)\quad \frac{\partial \vec{e}_r}{\partial r}=0 \Rightarrow \Gamma^{\mu}{}_{rr}=0 \quad \text{for all } \mu, \\[2mm]
&(2)\quad \frac{\partial \vec{e}_r}{\partial \theta}=\frac{1}{r}\vec{e}_\theta \Rightarrow \Gamma^{r}{}_{r\theta}=0, \qquad \Gamma^{\theta}{}_{r\theta}=\frac{1}{r}, \\[2mm]
&(3)\quad \frac{\partial \vec{e}_\theta}{\partial r}=\frac{1}{r}\vec{e}_\theta \Rightarrow \Gamma^{r}{}_{\theta r}=0, \qquad \Gamma^{\theta}{}_{\theta r}=\frac{1}{r}, \\[2mm]
&(4)\quad \frac{\partial \vec{e}_\theta}{\partial \theta}=-r\vec{e}_r \Rightarrow \Gamma^{r}{}_{\theta\theta}=-r, \qquad \Gamma^{\theta}{}_{\theta\theta}=0.
\end{aligned}
\right\}
\tag{5.44}
$$

In the definition, Eq. (5.43), all indices must refer to the same coordinate system. Thus, although we computed the derivatives of $\vec{e}_r$ and $\vec{e}_\theta$ by using the constancy of $\vec{e}_x$ and $\vec{e}_y$,, the Cartesian bases do not in the end make any appearance in Eq. (5.44). The Christoffel symbols' importance is that they enable one to express these derivatives without using any other coordinates than polar.

The covariant derivative. Using the definition of the Christoffel symbols, Eq. (5.43), the derivative in Eq. (5.42) becomes

$$
\frac{\partial \vec{V}}{\partial x^\beta}=\frac{\partial V^\alpha}{\partial x^\beta}\vec{e}_\alpha + V^\alpha \Gamma^{\mu}{}_{\alpha\beta}\vec{e}_\mu.
\tag{5.45}
$$

In the last term there are two sums, on α and μ. Relabeling the dummy indices will help here: we change μ to α and α to μ and get

$$
\frac{\partial \vec{V}}{\partial x^\beta}=\frac{\partial V^\alpha}{\partial x^\beta}\vec{e}_\alpha + V^\mu \Gamma^{\alpha}{}_{\mu\beta}\vec{e}_\alpha.
\tag{5.46}
$$

The reason for the relabeling was that, now, $\vec{e}_\alpha$ can be factored out of both terms:

$$
\frac{\partial \vec{V}}{\partial x^\beta}=\left(\frac{\partial V^\alpha}{\partial x^\beta}+V^\mu \Gamma^{\alpha}{}_{\mu\beta}\right)\vec{e}_\alpha.
\tag{5.47}
$$

So the vector field $\partial \vec{V}/\partial x^\beta$ has components

$$
\partial V^\alpha/\partial x^\beta + V^\mu \Gamma^{\alpha}{}_{\mu\beta}.
\tag{5.48}
$$

Recall our original notation for the partial derivative, $\partial V^\alpha/\partial x^\beta = V^\alpha{}_{,\beta}$. We keep this notation and define a *new* symbol:

◆ $$V^\alpha{}_{;\beta} \equiv V^\alpha{}_{,\beta}+V^\mu \Gamma^{\alpha}{}_{\mu\beta}. \tag{5.49}$$

Then, with this shorthand semicolon notation, we have

◆ $$\partial \vec{V}/\partial x^\beta = V^\alpha{}_{;\beta}\,\vec{e}_\alpha, \tag{5.50}$$

a very compact way of writing Eq. (5.47).

Now $\partial \vec{V}/\partial x^\beta$ is a vector field if we regard β as a given fixed number. But there are two values that β *can* have, and so we can also regard $\partial \vec{V}/\partial x^\beta$ as being associated with a $\binom{1}{1}$ tensor field which maps the vector $\vec{e}_\beta$ into the vector $\partial \vec{V}/\partial x^\beta$, as in Exer. 17, § 3.10. This tensor field is called the *covariant derivative* of $\vec{V}$, denoted, naturally enough, as $\nabla \vec{V}$. Then its components are

$$(\nabla \vec{V})^\alpha{}_\beta = (\nabla_\beta \vec{V})^\alpha = V^\alpha{}_{;\beta}. \tag{5.51}$$

On a Cartesian basis the components are just $V^\alpha{}_{,\beta}$. On the curvilinear basis, however, the derivatives of the basis vectors must be taken into account, and we get that $V^\alpha{}_{;\beta}$ are the components of $\nabla \vec{V}$ in whatever coordinate system the Christoffel symbols in Eq. (5.49) refer to. The significance of this statement should not be underrated, as it is the foundation of all our later work. There is a single $\binom{1}{1}$ tensor called $\nabla \vec{V}$. In Cartesian coordinates its components are $\partial V^\alpha/\partial x^\beta$. In general coordinates $\{x^{\mu'}\}$ its components are called $V^{\alpha'}{}_{;\beta'}$ and can be obtained in either of two equivalent ways: (i) compute them directly in $\{x^{\mu'}\}$ using Eq. (5.49) and a knowledge of what the $\Gamma^{\alpha'}{}_{\mu'\beta'}$s are in these coordinates; or (ii) obtain them by the usual tensor transformation laws from Cartesian to $\{x^{\mu'}\}$.

What is the covariant derivative of a scalar? The covariant derivative differs from the partial derivative with respect to the coordinates only because the basis vectors change. But a scalar does not depend on the basis vectors, so its covariant derivative is the same as its partial derivative, which is its gradient:

$$\nabla_\alpha f = \partial f/\partial x^\alpha ; \qquad \nabla f = \tilde{d} f. \tag{5.52}$$

Divergence and Laplacian. Before doing any more theory, let us link this up with things we have seen before. In Cartesian coordinates the divergence of a vector V^α is $V^\alpha{}_{,\alpha}$. This is the scalar obtained by contracting $V^\alpha{}_{,\beta}$ on its two indices. Since contraction is a frame-invariant operation, the divergence of $\vec{V}$ can be calculated in other coordinates $\{x^{\mu'}\}$ also by contracting the components of $\nabla \vec{V}$ on their two indices. This results in a scalar with the value $V^{\alpha'}{}_{;\alpha'}$. It is important to realize that this is the *same* number as $V^\alpha{}_{,\alpha}$ in Cartesian coordinates:

$$V^\alpha{}_{,\alpha} \equiv V^{\beta'}{}_{;\beta'}, \tag{5.53}$$

where unprimed indices refer to Cartesian coordinates and primed to the arbitrary system.

For polar coordinates (dropping primes for convenience here)

$$V^\alpha{}_{;\alpha} = \frac{\partial V^\alpha}{\partial x^\alpha} + \Gamma^\alpha{}_{\mu\alpha} V^\mu.$$

Now, from Eq. (5.44) we can calculate

$$\left.\begin{aligned} \Gamma^\alpha{}_{r\alpha} &= \Gamma^r{}_{rr} + \Gamma^\theta{}_{r\theta} = \frac{1}{r}, \\ \Gamma^\alpha{}_{\theta\alpha} &= \Gamma^r{}_{\theta r} + \Gamma^\theta{}_{\theta\theta} = 0. \end{aligned}\right\} \tag{5.54}$$

Therefore we have

$$V^\alpha{}_{;\alpha} = \frac{\partial V^r}{\partial r} + \frac{\partial V^\theta}{\partial \theta} + \frac{1}{r} V^r,$$

$$= \frac{1}{r} \frac{\partial}{\partial r} (r V^r) + \frac{\partial}{\partial \theta} V^\theta. \tag{5.55}$$

This may be a familiar formula to the student. What is probably more familiar is the Laplacian, which is the divergence of the gradient. But we only have the divergence of vectors, and the gradient is a one-form. Therefore we must first convert the one-form to a vector. Thus, given a scalar ϕ, we have the vector gradient (see Eq. (5.52) and the last part of § 5.2 above) with components $(\phi_{,r}, \phi_{,\theta}/r^2)$. Using these as the components of the vector in the divergence formula, Eq. (5.55) gives

$$\nabla \cdot \nabla \phi \equiv \nabla^2 \phi = \frac{1}{r} \frac{\partial}{\partial r} \left(r \frac{\partial \phi}{\partial r} \right) + \frac{1}{r^2} \frac{\partial^2 \phi}{\partial \theta^2}. \tag{5.56}$$

This is the Laplacian in plane polar coordinates. It is, of course, identically equal to

$$\nabla^2 \phi = \frac{\partial^2 \phi}{\partial x^2} + \frac{\partial^2 \phi}{\partial y^2}. \tag{5.57}$$

Derivatives of one-forms and tensors of higher types. Since a scalar ϕ depends on no basis vectors, its derivative $\tilde{d}\phi$ is the same as its covariant derivative $\nabla\phi$. We shall almost always use the symbol $\nabla\phi$. To compute the derivative of a one-form (which as for a vector won't be simply the derivatives of its components), we use the property that a one-form and a vector give a scalar. Thus, if $\tilde{p}$ is a one-form and $\vec{V}$ is an arbitrary vector, then for fixed β, $\nabla_\beta \tilde{p}$ is also a one-form, $\nabla_\beta \vec{V}$ is a vector, and $\langle \tilde{p}, \vec{V} \rangle \equiv \phi$ is a scalar. In any (arbitrary) coordinate system this scalar is just

$$\phi = p_\alpha V^\alpha. \tag{5.58}$$

Therefore $\nabla_\beta \phi$ is, by the product rule for derivatives,

$$\nabla_\beta \phi = \phi_{,\beta} = \frac{\partial p_\alpha}{\partial x^\beta} V^\alpha + p_\alpha \frac{\partial V^\alpha}{\partial x^\beta}. \tag{5.59}$$

But we can use Eq. (5.49) to replace $\partial V^\alpha / \partial x^\beta$ in favor of $V^\alpha{}_{;\beta}$, which are the components of $\nabla_\beta \vec{V}$:

$$\nabla_\beta \phi = \frac{\partial p_\alpha}{\partial x^\beta} V^\alpha + p_\alpha V^\alpha{}_{;\beta} - p_\alpha V^\mu \Gamma^\alpha{}_{\mu\beta}. \tag{5.60}$$

Rearranging terms, and relabeling dummy indices in the term that contains the Christoffel symbol, gives

$$\nabla_\beta \phi = \left(\frac{\partial p_\alpha}{\partial x^\beta} - p_\mu \Gamma^\mu{}_{\alpha\beta} \right) V^\alpha + p_\alpha V^\alpha{}_{;\beta}. \tag{5.61}$$

Now, every term in this equation except the one in parentheses is *known* to be the component of a tensor, for an arbitrary vector field $\vec{V}$. Therefore, since multiplication and addition of components always gives new tensors, it must be true that the term in parentheses is also the component of a tensor. This is, of course, the covariant derivative of $\tilde{p}$:

$$\blacklozenge \qquad (\nabla_\beta \tilde{p})_\alpha \equiv (\nabla \tilde{p})_{\alpha\beta} \equiv p_{\alpha;\beta} = p_{\alpha,\beta} - p_\mu \Gamma^\mu{}_{\alpha\beta}. \tag{5.62}$$

Then Eq. (5.61) reads

$$\blacklozenge \qquad \nabla_\beta (p_\alpha V^\alpha) = p_{\alpha;\beta} V^\alpha + p_\alpha V^\alpha{}_{;\beta}. \tag{5.63}$$

Thus covariant differentiation obeys the same sort of product rule as Eq. (5.59). It *must* do this, since in Cartesian coordinates ∇ is just partial differentiation of components, so Eq. (5.63) reduces to Eq. (5.59).

Let us compare the two formulae we have:

$$V^\alpha{}_{;\beta} = V^\alpha{}_{,\beta} + V^\mu \Gamma^\alpha{}_{\mu\beta}, \tag{5.49}$$

$$p_{\alpha;\beta} = p_{\alpha,\beta} - p_\mu \Gamma^\mu{}_{\alpha\beta}. \tag{5.62}$$

There are certain similarities and certain differences. If one remembers that the derivative index β is the *last* one on Γ, then the other indices are the only ones they can be without raising and lowering with the metric. The only thing to watch is the sign difference. It may help to remember that $\Gamma^\alpha{}_{\mu\beta}$ was related to derivatives of the basis vectors, for then it is reasonable that $-\Gamma^\mu{}_{\alpha\beta}$ be related to derivatives of the basis one-forms. The change in sign means that the basis one-forms change 'oppositely' to basis vectors, which makes sense when one realizes that the contraction $\langle \tilde{\omega}^\alpha, \vec{e}_\beta \rangle = \delta^\alpha{}_\beta$ is a *constant* whose derivative must be zero.

The same procedure that led to Eq. (5.62) would lead to the following:

$$\blacklozenge \qquad \nabla_\beta T_{\mu\nu} = T_{\mu\nu,\beta} - T_{\alpha\nu} \Gamma^\alpha{}_{\mu\beta} - T_{\mu\alpha} \Gamma^\alpha{}_{\nu\beta}; \tag{5.64}$$

$$\blacklozenge \qquad \nabla_\beta A^{\mu\nu} = A^{\mu\nu}{}_{,\beta} + A^{\alpha\nu} \Gamma^\mu{}_{\alpha\beta} + A^{\mu\alpha} \Gamma^\nu{}_{\alpha\beta}; \tag{5.65}$$

$$\blacklozenge \qquad \nabla_\beta B^\mu{}_\nu = B^\mu{}_{\nu,\beta} + B^\alpha{}_\nu \Gamma^\mu{}_{\alpha\beta} - B^\mu{}_\alpha \Gamma^\alpha{}_{\nu\beta}. \tag{5.66}$$

Inspect these closely: they are *very* systematic. Simply throw in one Γ term for each index; a raised index is treated like a vector and a lowered one like a one-form. The geometrical meaning of Eq. (5.64) is that $\nabla_\beta T_{\mu\nu}$ is a component of the $\binom{0}{3}$ tensor $\nabla \mathbf{T}$, where $\mathbf{T}$ is a $\binom{0}{2}$ tensor. Similarly, in Eq. (5.65), $\mathbf{A}$ is a $\binom{2}{0}$ tensor and $\nabla \mathbf{A}$ is a $\binom{2}{1}$ tensor with components $\nabla_\beta A^{\mu\nu}$.

5.4   Christoffel symbols and the metric

The formalism developed above has not used any properties of the metric tensor to derive covariant derivatives. But the metric must be involved somehow, because it can convert a vector into a one-form, and so it must have something to say about the relationship between their derivatives. In particular, in Cartesian coordinates the components of the one-form and its related vector are *equal*, and since ∇ is just differentiation of components, the components of the covariant derivatives of the one-form and vector must be equal. This means that if $\tilde{V}$ is an arbitrary vector and $\tilde{V} = \mathbf{g}\,(\vec{V},\ )$ is its related one-form, then in Cartesian coordinates

$$\nabla_\beta \tilde{V} = \mathbf{g}\,(\nabla_\beta \vec{V},\ ). \tag{5.67}$$

But Eq. (5.67) is a tensor equation, so it must be valid in *all* coordinates. We conclude that

$$V_{\alpha;\beta} = g_{\alpha\mu} V^\mu{}_{;\beta}, \tag{5.68}$$

which is the component representation of Eq. (5.67).

If the above argument in words wasn't satisfactory, let us go through it again in equations. Let unprimed indices α, β, γ, ... denote Cartesian coordinates and primed indices α', β', γ', ... denote *arbitrary* coordinates.

We begin with the statement

$$V_{\alpha'} = g_{\alpha'\mu'} V^{\mu'}, \tag{5.69}$$

valid in any coordinate system. But in Cartesian coordinates

$$g_{\alpha\mu} = \delta_{\alpha\mu}, \qquad V_\alpha = V^\alpha.$$

Now, also in Cartesian coordinates, the Christoffel symbols vanish, so

$$V_{\alpha;\beta} = V_{\alpha,\beta} \quad \text{and} \quad V^\alpha{}_{;\beta} = V^\alpha{}_{,\beta}.$$

Therefore we conclude

$$V_{\alpha;\beta} = V^\alpha{}_{;\beta}$$

in Cartesian coordinates only. To convert this into an equation valid in all coordinate systems, we note that in Cartesian coordinates

$$V^\alpha{}_{;\beta} = g_{\alpha\mu} V^\mu{}_{;\beta},$$

so that again in Cartesian coordinates we have

$$V_{\alpha;\beta} = g_{\alpha\mu} V^\mu{}_{;\beta}.$$

But now this equation *is* a tensor equation, so its validity in one coordinate system implies its validity in all. This is just Eq. (5.68) again:

$$V_{\alpha';\beta'} = g_{\alpha'\mu'} V^{\mu'}{}_{;\beta'}. \tag{5.70}$$

This result has far-reaching implications. If we take the β' covariant derivative of Eq. (5.69) we find

$$V_{\alpha';\beta'} = g_{\alpha'\mu';\beta'} V^{\mu'} + g_{\alpha'\mu'} V^{\mu'}{}_{;\beta'}.$$

Comparison of this with Eq. (5.70) shows (since $\vec{V}$ is an arbitrary vector) that we must have

◆ $$g_{\alpha'\mu';\beta'} \equiv 0 \tag{5.71}$$

in all coordinate systems. This is a consequence of Eq. (5.67). In Cartesian coordinates

$$g_{\alpha\mu;\beta} \equiv g_{\alpha\mu,\beta} = \delta_{\alpha\mu,\beta} \equiv 0$$

is a trivial identity. However, in other coordinates it is not obvious, so we shall work it out as a check on the consistency of our formalism.

Using Eq. (5.64) gives (now unprimed indices are general)

$$g_{\alpha\beta;\mu} = g_{\alpha\beta,\mu} - \Gamma^{\nu}{}_{\alpha\mu} g_{\nu\beta} - \Gamma^{\nu}{}_{\beta\mu} g_{\alpha\nu}. \tag{5.72}$$

In polar coordinates let us work out a few examples. Let $\alpha = r, \beta = r, \mu = r$:

$$g_{rr;r} = g_{rr,r} - \Gamma^{\nu}{}_{rr} g_{\nu r} - \Gamma^{\nu}{}_{rr} g_{r\nu}.$$

Since $g_{rr,r} = 0$ and $\Gamma^{\nu}{}_{rr} = 0$ for all ν, this is trivially zero. Not so trivial is $\alpha = \theta, \beta = \theta, \mu = r$:

$$g_{\theta\theta;r} = g_{\theta\theta,r} - \Gamma^{\nu}{}_{\theta r} g_{\nu\theta} - \Gamma^{\nu}{}_{\theta r} g_{\theta\nu}.$$

With $g_{\theta\theta} = r^2$, $\Gamma^{\theta}{}_{\theta r} = 1/r$ and $\Gamma^{r}{}_{\theta r} = 0$, this becomes

$$g_{\theta\theta;r} = (r^2)_{,r} - \frac{1}{r}(r^2) - \frac{1}{r}(r^2) = 0.$$

So it works, almost magically. But it is important to realize that it is not magic: it follows directly from the facts that $g_{\alpha\beta,\mu} = 0$ in Cartesian coordinates and that $g_{\alpha\beta;\mu}$ are the components of the *same* tensor ∇g in arbitrary coordinates.

Perhaps it is useful to pause here to get some perspective on what we have just done. We introduced covariant differentiation in arbitrary coordinates by using our understanding of parallelism in Euclidean space. We then showed that the metric of Euclidean space is covariantly constant: Eq. (5.71). When we go on to curved (Riemannian) spaces we will have to discuss parallelism much more carefully, but Eq. (5.71) will *still* be true, and therefore so will all its consequences, such as those we now go on to describe.

Calculating the Christoffel symbols from the metric. The vanishing of Eq. (5.72) leads to an extremely important result. One sees that Eq. (5.72) can be used to determine $g_{\alpha\beta,\mu}$ in terms of $\Gamma^\mu{}_{\alpha\beta}$. It turns out that the reverse is also true, that $\Gamma^\mu{}_{\alpha\beta}$ can be expressed in terms of $g_{\alpha\beta,\mu}$. This gives an easy way to derive the Christoffel symbols. To show this we first prove a result of some importance in its own right: *in any coordinate system* $\Gamma^\mu{}_{\alpha\beta} \equiv \Gamma^\mu{}_{\beta\alpha}$. To prove this symmetry consider an arbitrary scalar field ϕ. Its first derivative $\nabla\phi$ is a one-form with components $\phi_{,\beta}$. Its second covariant derivative $\nabla\nabla\phi$ has components $\phi_{,\beta;\alpha}$ and is a $\binom{0}{2}$ tensor. In Cartesian coordinates these components are

$$\phi_{,\beta,\alpha} \equiv \frac{\partial}{\partial x^\alpha} \frac{\partial}{\partial x^\beta} \phi$$

and we see that they are symmetric in α and β, since partial derivatives commute. But if a tensor is symmetric in one basis it is symmetric in all bases. Therefore

$$\phi_{,\beta;\alpha} = \phi_{,\alpha;\beta} \tag{5.73}$$

in *any* basis. Using the definition, Eq. (5.62) gives

$$\phi_{,\beta,\alpha} - \phi_{,\mu}\Gamma^\mu{}_{\beta\alpha} = \phi_{,\alpha,\beta} - \phi_{,\mu}\Gamma^\mu{}_{\alpha\beta}$$

in any coordinate system. But again we have

$$\phi_{,\alpha,\beta} = \phi_{,\beta,\alpha}$$

in *any* coordinates, which leaves

$$\Gamma^\mu{}_{\alpha\beta}\phi_{,\mu} = \Gamma^\mu{}_{\beta\alpha}\phi_{,\mu}$$

for arbitrary ϕ. This proves the assertion

◆ $\Gamma^\mu{}_{\alpha\beta} = \Gamma^\mu{}_{\beta\alpha}$ in any coordinate system.   (5.74)

We use this to invert Eq. (5.72) by some advanced index gymnastics. We write three versions of Eq. (5.72) with different permutations of indices:

$$g_{\alpha\beta,\mu} = \Gamma^\nu{}_{\alpha\mu}g_{\nu\beta} + \Gamma^\nu{}_{\beta\mu}g_{\alpha\nu},$$
$$g_{\alpha\mu,\beta} = \Gamma^\nu{}_{\alpha\beta}g_{\nu\mu} + \Gamma^\nu{}_{\mu\beta}g_{\alpha\nu},$$
$$-g_{\beta\mu,\alpha} = -\Gamma^\nu{}_{\beta\alpha}g_{\nu\mu} - \Gamma^\nu{}_{\mu\alpha}g_{\beta\nu}.$$

We add these up and group terms, using the symmetry of **g**, $g_{\beta\nu} = g_{\nu\beta}$:

$$g_{\alpha\beta,\mu} + g_{\alpha\mu,\beta} - g_{\beta\mu,\alpha}$$
$$= (\Gamma^\nu{}_{\alpha\mu} - \Gamma^\nu{}_{\alpha\mu})g_{\nu\beta} + (\Gamma^\nu{}_{\alpha\beta} - \Gamma^\nu{}_{\beta\alpha})g_{\nu\mu} + (\Gamma^\nu{}_{\beta\mu} + \Gamma^\nu{}_{\mu\beta})g_{\alpha\nu}.$$

In this equation the first two terms on the right vanish by the symmetry, Eq. (5.74), of Γ, and we get

$$g_{\alpha\beta,\mu} + g_{\alpha\mu,\beta} - g_{\beta\mu,\alpha} = 2g_{\alpha\nu}\Gamma^\nu{}_{\beta\mu}.$$

We are almost there. Dividing by 2, multiplying by $g^{\alpha\gamma}$ (with summation implied on α) and using

$$g^{\alpha\gamma}g_{\alpha\nu} \equiv \delta^\gamma{}_\nu$$

gives

◆ $$\tfrac{1}{2}g^{\alpha\gamma}(g_{\alpha\beta,\mu} + g_{\alpha\mu,\beta} - g_{\beta\mu,\alpha}) = \Gamma^\gamma{}_{\beta\mu}.$$ (5.75)

This is the expression of the Christoffel symbols in terms of the partial derivatives of the components of $\mathbf{g}$. In polar coordinates, for example,

$$\Gamma^\theta{}_{r\theta} = \tfrac{1}{2}g^{\alpha\theta}(g_{\alpha r,\theta} + g_{\alpha\theta,r} - g_{r\theta,\alpha}).$$

Since $g^{r\theta} = 0$ and $g^{\theta\theta} = r^{-2}$ we have

$$\Gamma^\theta{}_{r\theta} = \frac{1}{2r^2}(g_{\theta r,\theta} + g_{\theta\theta,r} - g_{r\theta,\theta})$$

$$= \frac{1}{2r^2}g_{\theta\theta,r} = \frac{1}{2r^2}(r^2)_{,r} = \frac{1}{r}.$$

This is the same value for $\Gamma^\theta{}_{r\theta}$, as we derived earlier. This method of computing $\Gamma^\alpha{}_{\beta\mu}$ is so useful that it is well worth committing Eq. (5.75) to memory. It will be exactly the same in curved spaces.

5.5 The tensorial nature of $\Gamma^\alpha{}_{\beta\mu}$

Since $\vec{e}_\alpha$ is a vector, $\nabla\vec{e}_\alpha$ is a $\binom{1}{1}$ tensor whose components are $\Gamma^\mu{}_{\alpha\beta}$. Here α is fixed and μ and β are the component indices: changing α changes the tensor $\nabla\vec{e}_\alpha$, while changing μ or β changes only the component under discussion. So it is possible to regard μ and β as component indices and α as a label giving the particular tensor referred to. There is one such tensor for each basis vector $\vec{e}_\alpha$. However, this is not terribly useful, since under a change of coordinates the basis changes and the important quantities in the new system are the *new* tensors $\nabla\vec{e}_{\beta'}$ which are obtained from the old ones $\nabla\vec{e}_\alpha$ in a complicated way: they are *different* tensors, not just different components of the same tensor. So the set $\Gamma^\mu{}_{\alpha\beta}$ in one frame is not obtained by a simple tensor transformation from the set $\Gamma^{\mu'}{}_{\alpha'\beta'}$ of another frame. The easiest example of this is Cartesian coordinates, where $\Gamma^\alpha{}_{\beta\mu} \equiv 0$, while they are not zero in other frames. So in many books it is said that $\Gamma^\mu{}_{\alpha\beta}$ are not components of tensors. As we have seen, this is not strictly true: $\Gamma^\mu{}_{\alpha\beta}$ are the (μ,β) components of a *set* of tensors $\nabla\vec{e}_\alpha$. But there is no single tensor whose components are $\Gamma^\mu{}_{\alpha\beta}$, so expressions like $\Gamma^\mu{}_{\alpha\beta}V^\alpha$ are not components of a single tensor, either. The combination

$$V^\beta{}_{,\alpha} + V^\mu\Gamma^\beta{}_{\mu\alpha}$$

is a component of a single tensor $\nabla\vec{V}$.

5.6 Noncoordinate bases

In this whole discussion we have generally assumed that the non-Cartesian basis vectors were generated by a coordinate transformation from (x, y) to some (ξ, η). However, as we shall show below, not every field of basis vectors can be obtained in this way, and we shall have to look carefully at our results to see which need modification (few actually do). We will almost never use non-coordinate bases in our work in this course, but one frequently encounters them in the standard references on curved coordinates in flat space, so we should pause to take a brief look at them now.

Polar coordinate basis. The basis vectors for our polar coordinate system were defined by

$$\vec{e}_{\alpha'} = \Lambda^{\beta}{}_{\alpha'} \vec{e}_{\beta},$$

where primed indices refer to polar coordinates and unprimed to Cartesian. Moreover, we had

$$\Lambda^{\beta}{}_{\alpha'} = \partial x^{\beta} / \partial x^{\alpha'},$$

where we regard the Cartesian coordinates $\{x^{\beta}\}$ as functions of the polar coordinates $\{x^{\alpha'}\}$. We found that

$$\vec{e}_{\alpha'} \cdot \vec{e}_{\beta'} \equiv g_{\alpha'\beta'} \neq \delta_{\alpha'\beta'},$$

i.e. that these basis vectors are *not* unit vectors.

Polar unit basis. Often it is convenient to work with *unit* vectors. A simple set of unit vectors derived from the polar coordinate basis is:

$$\vec{e}_{\hat{r}} = \vec{e}_r, \quad \vec{e}_{\hat{\theta}} = \frac{1}{r} \vec{e}_{\theta}, \tag{5.76}$$

with a corresponding unit one-form basis

$$\tilde{\omega}^{\hat{r}} = \tilde{d}r, \quad \tilde{\omega}^{\hat{\theta}} = r\tilde{d}\theta. \tag{5.77}$$

The student should verify that

$$\left.\begin{aligned} \vec{e}_{\hat{\alpha}} \cdot \vec{e}_{\hat{\beta}} &\equiv g_{\hat{\alpha}\hat{\beta}} = \delta_{\hat{\alpha}\hat{\beta}}, \\ \tilde{\omega}^{\hat{\alpha}} \cdot \tilde{\omega}^{\hat{\beta}} &\equiv g^{\hat{\alpha}\hat{\beta}} = \delta^{\hat{\alpha}\hat{\beta}} \end{aligned}\right\} \tag{5.78}$$

so these constitute orthonormal bases for the vectors and one-forms. Our notation, which is fairly standard, is to use a 'caret' or 'hat', ^, above an index to denote an orthornormal basis. Now, the question arises, do there exist coordinates (ξ, η) such that

$$\vec{e}_{\hat{r}} = \vec{e}_{\xi} = \frac{\partial x}{\partial \xi} \vec{e}_x + \frac{\partial y}{\partial \xi} \vec{e}_y \tag{5.79a}$$

and

$$\vec{e}_{\hat{\theta}} = \vec{e}_{\eta} = \frac{\partial x}{\partial \eta} \vec{e}_x + \frac{\partial y}{\partial \eta} \vec{e}_y ? \tag{5.79b}$$

If so, then $\{\vec{e}_{\hat{r}}, \vec{e}_{\hat{\theta}}\}$ are the basis for the coordinates (ξ, η) and so can be called a coordinate basis; if such (ξ, η) can be shown not to exist then these vectors are a noncoordinate basis. The question is actually more easily answered if we look at the basis one-form. Thus, we seek (ξ, η) such that

$$\left.\begin{aligned} \tilde{\omega}^{\hat{r}} &= \tilde{d}\xi = \frac{\partial \xi}{\partial x} \tilde{d}x + \frac{\partial \xi}{\partial y} \tilde{d}y, \\ \tilde{\omega}^{\hat{\theta}} &= \tilde{d}\eta = \frac{\partial \eta}{\partial x} \tilde{d}x + \frac{\partial \eta}{\partial y} \tilde{d}y. \end{aligned}\right\} \tag{5.80}$$

Since we know $\tilde{\omega}^{\hat{r}}$ and $\tilde{\omega}^{\hat{\theta}}$ in terms of $\tilde{d}r$ and $\tilde{d}\theta$ we have, from Eqs. (5.25) and (5.26),

$$\left.\begin{aligned} \tilde{\omega}^{\hat{r}} &= \tilde{d}r = \cos\theta \, \tilde{d}x + \sin\theta \, \tilde{d}y, \\ \tilde{\omega}^{\hat{\theta}} &= r \, \tilde{d}\theta = -\sin\theta \, \tilde{d}x + \cos\theta \, \tilde{d}y. \end{aligned}\right\} \tag{5.81}$$

(The orthonormality of $\tilde{\omega}^{\hat{r}}$ and $\tilde{\omega}^{\hat{\theta}}$ are obvious here.) Thus if (ξ, η) exist we have

$$\frac{\partial \eta}{\partial x} = -\sin\theta, \qquad \frac{\partial \eta}{\partial y} = \cos\theta. \tag{5.82}$$

By looking at the mixed partial derivatives we deduce

$$\frac{\partial}{\partial y}\frac{\partial \eta}{\partial x} = -\frac{\partial \sin\theta}{\partial y} = \frac{\partial}{\partial x}\frac{\partial \eta}{\partial y} = \frac{\partial \cos\theta}{\partial x}, \tag{5.83}$$

which implies

$$\frac{\partial}{\partial y}(-\sin\theta) = \frac{\partial}{\partial x}(\cos\theta) \tag{5.84}$$

or

$$\frac{\partial}{\partial y}\left(\frac{y}{\sqrt{(x^2+y^2)}}\right) + \frac{\partial}{\partial x}\left(\frac{x}{\sqrt{(x^2+y^2)}}\right) = 0.$$

This is certainly *not* true. Therefore ξ and η do *not* exist: we have a noncoordinate basis. (If this manner of proof is surprising, try it on $\tilde{d}r$ and $\tilde{d}\theta$ themselves.)

In textbooks that deal with vector calculus in curvilinear coordinates, almost all use the unit orthonormal basis rather than the coordinate basis. Thus, for polar coordinates, if a vector has components in the *coordinate* basis *PC*,

$$\vec{V} \xrightarrow[PC]{} (a, b) = \{V^\alpha\}, \tag{5.85}$$

then it has components in the *orthonormal* basis *PO*

$$\vec{V} \xrightarrow[PO]{} (a, rb) = \{V^{\hat{a}}\}. \tag{5.86}$$

So if, for example, the books calculate the divergence of the vector, they obtain, instead of our Eq. (5.85),

$$\nabla \cdot V = \frac{1}{r}\frac{\partial}{\partial r}(r\,V^{\hat{r}}) + \frac{1}{r}\frac{\partial}{\partial \theta}\,V^{\hat{\theta}}. \tag{5.87}$$

The difference between Eqs. (5.55) and (5.87) is purely a matter of the basis for $\vec{V}$.

General remarks on noncoordinate bases. The principal differences between coordinate and noncoordinate bases arise from the following. Consider an arbitrary scalar field ϕ and the number $\tilde{d}\phi(\vec{e}_\mu)$, where $\vec{e}_\mu$ is a basis vector of some arbitrary basis. We have used the notation

$$\tilde{d}\phi(\vec{e}_\mu) = \phi_{,\mu}. \tag{5.88}$$

Now, if $\vec{e}_\mu$ is a member of a coordinate basis, then $\tilde{d}\phi(\vec{e}_\mu) = \partial\phi/\partial x^\mu$ and we have, as defined in an earlier chapter,

$$\phi_{,\mu} = \frac{\partial\phi}{\partial x^\mu} : \text{coordinate basis.} \tag{5.89}$$

But if no coordinates exist for $\{\vec{e}_\mu\}$, then Eq. (5.89) must fail. For example, if we let Eq. (5.88) *define* $\phi_{,\hat{\mu}}$, then we have

$$\phi_{,\hat{\theta}} = \frac{1}{r}\frac{\partial\phi}{\partial\theta}. \tag{5.90}$$

In general, we get

$$\nabla_{\hat{\alpha}}\phi \equiv \phi_{,\hat{\alpha}} = \Lambda^\beta{}_{\hat{\alpha}}\nabla_\beta\phi = \Lambda^\beta{}_{\hat{\alpha}}\frac{\partial\phi}{\partial x^\beta} \tag{5.91}$$

for any coordinate system $\{x^\beta\}$ and noncoordinate basis $\{\vec{e}_{\hat{\alpha}}\}$. It is thus convenient to continue with the notation, Eq. (5.88), and to make the rule that $\phi_{,\mu} = \partial\theta/\partial x^\mu$ *only* in a coordinate basis.

The Christoffel symbols may be defined just as before

$$\nabla_{\hat{\beta}}\vec{e}_{\hat{\alpha}} = \Gamma^{\hat{\mu}}{}_{\hat{\alpha}\hat{\beta}}\,\vec{e}_{\hat{\mu}}, \tag{5.92}$$

but now

$$\nabla_{\hat{\beta}} = \Lambda^\alpha{}_{\hat{\beta}}\frac{\partial}{\partial x^\alpha}, \tag{5.93}$$

where $\{x^\alpha\}$ is any coordinate system and $\{\vec{e}_{\hat{\beta}}\}$ any basis (coordinate or not). Now, however, one *cannot* prove that $\Gamma^{\hat{\mu}}{}_{\hat{\alpha}\hat{\beta}} = \Gamma^{\hat{\mu}}{}_{\hat{\beta}\hat{\alpha}}$, since that proof used $\phi_{,\hat{\alpha},\hat{\beta}} = \phi_{,\hat{\beta},\hat{\alpha}}$, which was true in a coordinate basis (partial derivatives commute) but is not true otherwise. Hence, also, Eq. (5.75) for $\Gamma^{\mu}{}_{\alpha\beta}$ in terms of $g_{\alpha\beta,\gamma}$ applies only coordinate systems. More general expressions are worked out in Exer. 20, § 5.9.

What is the general reason for the nonexistence of coordinates for a basis? If $\{\tilde{\omega}^{\bar{\alpha}}\}$ is a coordinate one-form basis, then its relation to another one $\{\tilde{d}x^{\alpha}\}$ is

$$\tilde{\omega}^{\bar{\alpha}} = \Lambda^{\bar{\alpha}}{}_{\beta}\,\tilde{d}x^{\beta} = \frac{\partial x^{\bar{\alpha}}}{\partial x^{\beta}}\,\tilde{d}x^{\beta}. \tag{5.94}$$

The key point is that $\Lambda^{\bar{\alpha}}{}_{\beta}$, which is generally a function of position, must actually be the partial derivative $\partial x^{\bar{\alpha}}/\partial x^{\beta}$ everywhere. Thus we have

$$\frac{\partial}{\partial x^{\gamma}}\Lambda^{\bar{\alpha}}{}_{\beta} = \frac{\partial^{2}x^{\bar{\alpha}}}{\partial x^{\gamma}\partial x^{\beta}} = \frac{\partial^{2}x^{\bar{\alpha}}}{\partial x^{\beta}\partial x^{\gamma}} = \frac{\partial}{\partial x^{\beta}}\Lambda^{\bar{\alpha}}{}_{\gamma}. \tag{5.95}$$

These 'integrability conditions' must be satisfied by all the elements $\Lambda^{\bar{\alpha}}{}_{\beta}$ in order for $\tilde{\omega}^{\bar{\alpha}}$ to be a coordinate basis. Clearly, one can always choose a transformation matrix for which this fails, thereby generating a noncoordinate basis.

Noncoordinate bases in this book. We shall not have occasion to use such bases very often. Mainly, it is important to understand that they exist, that not every basis is derivable from a coordinate system. The algebra of coordinate bases is simpler in almost every respect. One may ask why the standard treatments of curvilinear coordinates in vector calculus, then, stick to orthonormal bases. The reason is that in such a basis in Euclidean space the metric has components $\delta_{\alpha\beta}$, so the form of the dot product and the equality of vector and one-form components carry over directly from Cartesian coordinates (which have the *only* orthonormal coordinate basis!). In order to gain the simplicity of coordinate bases for vector and tensor calculus, one has to spend time learning the difference between vectors and one-forms!

5.7 Looking ahead

The work we have done in this chapter has developed almost all the notation and concepts we will need in our study of curved spaces and spacetimes. It is particularly important that the student understands §§ 5.2–5.4 because the mathematics of curvature will be developed by analogy with the development here. What we have to add to all this is a discussion of parallelism, of how to measure the extent to which the Euclidean parallelism axiom fails. This measure is the famous Riemann tensor.

5.8 Bibliography

The Eötvös and Pound–Rebka–Snider experiments, and other experimental fundamentals underpinning GR, are discussed by Dicke (1964), Misner *et al.* (1973), Shapiro (1980), and Will (1972, 1979, 1981), as well as by some authors in the collections by Bertotti (1974, 1977). See Hoffmann (1983) for a less mathematical discussion of the motivation for introducing curvature.

The mathematics of curvilinear coordinates is developed from a variety of points of view in: Abraham & Marsden (1978), Dodson & Poston (1977), Hicks (1965), Lovelock & Rund (1975), Schutz (1980b), and Spivak (1979).

5.9 Exercises

1   Repeat the argument that led to Eq. (5.1) under more realistic assumptions: suppose a fraction ε of the kinetic energy of the mass at the bottom can be converted into a photon and sent back up, the remaining energy staying at ground level in a useful form. Devise a perpetual motion engine if Eq. (1.1) is violated.

2   Explain why a *uniform* external gravitational field would raise no tides on Earth.

3(a) Show that the coordinate transformation $(x, y) \to (\xi, \eta)$ with $\xi = x$ and $\eta = 1$ violates Eq. (5.6).

(b) Are the following coordinate transformations good ones? Compute the Jacobian and list any points at which the transformations fail.
   (i) $\xi = (x^2 + y^2)^{1/2}$, $\eta = \arctan(y/x)$;
   (ii) $\xi = \ln x$, $\eta = y$;
   (iii) $\xi = \arctan(y/x)$, $\eta = (x^2 + y^2)^{-1/2}$.

4   A curve is defined by $\{x = f(\lambda),\ y = g(\lambda),\ 0 \le \lambda \le 1\}$. Show that the tangent vector $(dx/d\lambda, dy/d\lambda)$ does actually lie tangent to the curve.

5   Sketch the following curves. Which have the same paths? Find also their tangent vectors where the parameter equals zero.
   (a) $x = \sin \lambda, y = \cos \lambda$; (b) $x = \cos(2\pi t^2)$, $y = \sin(2\pi t^2 + \pi)$; (c) $x = s, y = s + 4$; (d) $x = s^2, y = -(s-2)(s+2)$; (e) $x = \mu, y = 1$.

6   Justify the pictures in Fig. 5.5.

7   Calculate all elements of the transformation matrices $\Lambda^{\alpha'}{}_\beta$ and $\Lambda^\mu{}_{\nu'}$ for the transformation from Cartesian (x, y) – the unprimed indices – to polar (r, θ) – the primed indices.

8(a) (Uses the result of Exer. 7.) Let $f = x^2 + y^2 + 2xy$, and in Cartesian coordinates $\vec{V} \to (x^2 + 3y, y^2 + 3x)$, $\vec{W} \to (1, 1)$. Compute f as a function of r and θ, and find the components of $\vec{V}$ and $\vec{W}$ on the polar basis, expressing them as functions of r and θ.

(b) Find the components of $\mathrm{d}f$ in Cartesian coordinates and obtain them in polars (i) by direct calculation in polars, and (ii) by transforming components from Cartesian.

(c) (i) Use the metric tensor in polar coordinates to find the polar components of the one-forms $\tilde{V}$ and $\tilde{W}$ associated with $\vec{V}$ and $\vec{W}$. (ii) Obtain the polar components of $\vec{V}$ and $\vec{W}$ by transformation of their Cartesian components.

9 Draw a diagram similar to Fig. 5.6 to explain Eq. (5.37).

10 Prove that $\nabla \vec{V}$, defined in Eq. (5.51), is a $\binom{1}{1}$ tensor.

11 (Uses the result of Exer. 7 and 8.) For the vector field $\vec{V}$ whose Cartesian components are $(x^2 + 3y, y^2 + 3x)$, compute: (a) $V^\alpha{}_{,\beta}$ in Cartesian; (b) the transformation $\Lambda^{\mu'}{}_\alpha \Lambda^\beta{}_{\nu'} V^\alpha{}_{,\beta}$ to polars; (c) the components $V^{\mu'}{}_{;\nu'}$ directly in polars using the Christoffel symbols, Eq. (5.44), in Eq. (5.49); (d) the divergence $V^\alpha{}_{,\alpha}$ using your results in (a); (e) the divergence $V^{\mu'}{}_{;\mu'}$ using your results in either (b) or (c); (f) the divergence $V^{\mu'}{}_{;\mu'}$ using Eq. (5.55) directly.

12 For the one-form field $\tilde{p}$ whose Cartesian components are $(x^2 + 3y, y^2 + 3x)$, compute: (a) $p_{\alpha,\beta}$ in Cartesian; (b) the transformation $\Lambda^\alpha{}_{\mu'} \Lambda^\beta{}_{\nu'} p_{\alpha,\beta}$ to polars; (c) the components $p_{\mu';\nu'}$ directly in polars using the Christoffel symbols, Eq. (5.44), in Eq. (5.62).

13 For those who have done both Exer. 11 and Exer. 12, show in polars that $g_{\mu'\alpha'} V^{\alpha'}{}_{;\nu'} = p_{\mu';\nu'}$.

14 For the tensor whose polar components are $(A^{rr} = r^2, A^{r\theta} = r \sin\theta, A^{\theta r} = r \cos\theta, A^{\theta\theta} = \tan\theta)$, compute Eq. (5.65) in polars for all possible indices.

15 For the vector whose polar components are $(V^r = 1, V^\theta = 0)$, compute in polars all components of the second covariant derivative $V^\alpha{}_{;\mu;\nu}$. (Hint: to find the second derivative, treat the first derivative $V^\alpha{}_{;\mu}$ as any $\binom{1}{1}$ tensor: Eq. (5.66).)

16 Fill in all the missing steps leading from Eq. (5.74) to Eq. (5.75).

17 Discover how each expression $V^\beta{}_{,\alpha}$ and $V^\mu \Gamma^\beta{}_{\mu\alpha}$ separately transforms under a change of coordinates (for $\Gamma^\beta{}_{\mu\alpha}$, begin with Eq. (5.43)). Show that neither is the standard tensor law, but that their *sum* does obey the standard law.

18 Verify Eq. (5.78).

19   Verify that the calculation from Eq. (5.81) to (5.84), when repeated for $\vec{d}r$ and $\vec{d}\theta$, shows them to be a coordinate basis.

20   For a noncoordinate basis $\{\vec{e}_{\mu}\}$, define $\nabla_{\vec{e}_{\mu}} \vec{e}_{\nu} - \nabla_{\vec{e}_{\nu}} \vec{e}_{\mu} \equiv c^{\alpha}{}_{\mu\nu} \vec{e}_{\alpha}$ and use this in place of Eq. 5.74 to generalize Eq. (5.75).

21   Consider the $x-t$ plane of an inertial observer in SR. A certain uniformly accelerated observer wishes to set up an orthonormal coordinate system. By Exer. 21, § 2.9, his world line is

$$t(\lambda) = a \sinh \lambda, \qquad x(\lambda) = a \cosh \lambda, \tag{5.96}$$

where a is a constant and $a\lambda$ is his proper time (clock time on his wrist watch).

(a)   Show that the spacelike line described by Eq. (5.96) with a as the variable parameter and λ fixed is orthogonal to his world line where they intersect. Changing λ in Eq. (5.96) then generates a *family* of such lines.

(b)   Show that Eq. (5.96) defines a transformation from coordinates (t, x) to coordinates (λ, a) which form an *orthogonal* coordinate system. Draw these coordinates and show that they cover only one half of the original $t-x$ plane. Show that the coordinates are bad on the lines $|x| = |t|$, so they really cover two disjoint quadrants.

(c)   Find the metric tensor and all the Christoffel symbols in this coordinate system. This observer will do a perfectly good job provided that he always uses Christoffel symbols appropriately and sticks to events in his quadrant. In this sense, SR admits accelerated observers. The right-hand quadrant in these coordinates is sometimes called *Rindler space*, and the boundary lines $x = \pm t$ bear some resemblance to the black-hole horizons we will study later.

22   Show that if $U^{\alpha}\nabla_{\alpha}V^{\beta} = W^{\beta}$, then $U^{\alpha}\nabla_{\alpha}V_{\beta} = W_{\beta}$.

6

Curved manifolds

6.1 Differentiable manifolds and tensors

The mathematical concept of a curved space begins (but does not end) with the idea of a *manifold*. A manifold is essentially a continuous space which looks locally like Euclidean space. To the concept of a manifold is added the idea of curvature itself. The introduction of curvature into a manifold will be the subject of subsequent sections. First we study the idea of a manifold, which one can regard as just a fancy word for 'space'.

Manifolds. The surface of a sphere is a manifold. So is any m-dimensional 'hyperplane' in an n-dimensional Euclidean space ($m \leq n$). More abstractly, the set of all rigid rotations of Cartesian coordinates in three-dimensional Euclidean space will be shown below to be a manifold. Basically, a manifold is any set that can be continuously parametrized. The number of independent parameters is the *dimension* of the manifold, and the parameters themselves are the *coordinates* of the manifold. Consider the examples just mentioned. The surface of a sphere is 'parametrized' by two coordinates θ and ϕ. The m-dimensional 'hyperplane' has m Cartesian coordinates, and the set of all rotations can be parametrized by the three 'Euler angles', which in effect give the direction of the axis of rotation (two parameters for this) and the amount of rotation (one parameter). So the set of rotations is a manifold: each point is a particular rotation, and the coordinates are the three parameters. It is a three-dimensional manifold. Mathematically, the association of points with the values of their parameters can be thought of as a mapping of points of a manifold into points of the Euclidean space of the correct dimension. This is the meaning of the fact that a manifold looks locally like Euclidean space: it is 'smooth' and has a certain number of dimensions. It must be stressed that the large-scale topology of a manifold may be very different from Euclidean space: the surface of a torus is not Euclidean, even topologically. But locally the correspondence is good: a small patch of the surface of a torus can be mapped 1–1 into the plane tangent to it. This is the way to think of a manifold: it is a space with coordinates, that locally looks Euclidean but that globally can warp, bend, and do almost anything (as long as it stays continuous).

Differential structure. We shall really only consider 'differentiable manifolds'. These are spaces that are continuous and differentiable. Roughly, this means it is possible to define a scalar field ϕ at each point of the manifold and be sure that it can be differentiated everywhere. The surface of a sphere is differentiable everywhere. That of a cone is differentiable except at its apex. Nearly all manifolds of use in physics are differentiable almost everywhere. The curved spacetimes of GR certainly are. The assumption of differentiability immediately means that we can define one-forms and vectors. That is, in a certain coordinate system on the manifold, the members of the set $\{\phi_{,\alpha}\}$ are the components of the one-form $\tilde{d}\phi$; and any set of the form $\{a\phi_{,\alpha} + b\psi_{,\alpha}\}$, where a and b are functions, is also a one-form field. Similarly, every curve (with parameter, say, λ) has a tangent vector $\vec{V}$ defined as the linear function that takes the one-form $\tilde{d}\phi$ into the derivative of ϕ along the curve, $d\phi/d\lambda$:

$$\langle \tilde{d}\phi, \vec{V} \rangle = \vec{V}(\tilde{d}\phi) = \nabla_{\vec{V}}\phi = d\phi/d\lambda. \tag{6.1}$$

Any linear combination of vectors is also a vector. Using the vectors and one-forms so defined, we can build up the whole set of tensors of type $\binom{M}{N}$, just as we did in SR. Since we have not yet picked out any $\binom{0}{2}$ tensor to serve as the metric, there is not yet any correspondence between forms and vectors. Everything else, however, is exactly as we had in SR and in polar coordinates. All of this comes only from differentiability, so the set of all tensors is said to be part of the 'differential structure' of the manifold. We will not have much occasion to use that term.

Review. It is useful here to review the fundamentals of tensor algebra. We can summarize the following rules.

(1) A tensor *field* defines a tensor at every point.

(2) Vectors and one-forms are linear operators on each other, producing real numbers. The linearity means:

$$\langle \tilde{p}, a\vec{V} + b\vec{W} \rangle = a\langle \tilde{p}, \vec{V} \rangle + b\langle \tilde{p}, \vec{W} \rangle,$$
$$\langle a\tilde{p} + b\tilde{q}, \vec{V} \rangle = a\langle \tilde{p}, \vec{V} \rangle + b\langle \tilde{q}, \vec{V} \rangle,$$

where a and b are any scalar fields.

(3) Tensors are similarly linear operators on one-forms and vectors, producing real numbers.

(4) If two tensors of the same type have equal components in a given basis, they have equal components in all bases and are said to be identical (or equal, or the same). In particular, if a tensor's components are all zero in one basis they are zero in all, and the tensor is said to be zero.

(5) A number of manipulations of components of tensor fields are called 'permissible tensor operations' because they produce components of new tensors:

  (i) Multiplication by a scalar field produces components of a new tensor of the same type.

  (ii) Addition of components of two tensors of the same type gives components of a new tensor of the same type. (In particular, only tensors of the same type can be equal.)

  (iii) Multiplication of components of two tensors of arbitrary type gives components of a new tensor of the sum of the types, the outer product of the two tensors.

  (iv) Covariant differentiation (to be discussed later) of the components of a tensor of type $\binom{N}{M}$ gives components of a tensor of type $\binom{N}{M+1}$.

  (v) Contraction on a pair of indices of the components of a tensor of type $\binom{N}{M}$ produces components of a tensor of type $\binom{N-1}{M-1}$. (Contraction is only defined between an upper and lower index.)

(6) If an equation is formed using components of tensors combined only by the permissible tensor operations, and if the equation is true in one basis, then it is true in any other. This is a very useful result. It comes from the fact that the equation (from (5) above) is simply an equality between components of two tensors of the same type, which (from (4)) is then true in any system.

6.2   Riemannian manifolds

So far we have not introduced a metric onto the manifold. Indeed, on certain manifolds a metric would be unnecessary or inconvenient for whichever problem is being considered. But in our case the metric is absolutely fundamental, since it will carry the information about the rates at which clocks run and the distances between points, just as it does in SR. A differentiable manifold on which a symmetric $\binom{0}{2}$ tensor field **g** has been singled out to act as the metric at each point is called a Riemannian manifold. (Strictly speaking, only if the metric is positive-definite – that is, $\mathbf{g}(\vec{V}, \vec{V}) > 0$ for all $\vec{V} \neq 0$ – is it called Riemannian; indefinite metrics, like SR and GR, are called pseudo-Riemannian. This is a distinction that we won't bother to make.) It is important to understand that in picking out a metric we 'add' structure to the manifold; we shall see that the metric completely defines the curvature of the manifold. Thus, by our choosing one metric **g** the manifold gets a certain curvature (perhaps that of a sphere), while a different **g**′ would give it a different curvature (perhaps an ellipsoid of revolution). The differentiable manifold itself is 'primitive': an amorphous collection of points, arranged locally like the points of Euclidean space, but not having any distance relation or shape specified. Giving the metric **g** gives it a specific shape, as we shall see. From now on we shall study Riemannian manifolds, on which a metric **g** is assumed to be defined at every point.

(For completeness we should remark that it is in fact possible to define the notion of curvature on a manifold without introducing a metric (so-called 'affine' manifolds). Some texts actually approach the subject this way. But since the metric is essential in GR, we shall simply study those manifolds whose curvature is defined by a metric.)

The metric and local flatness. The metric, of course, provides a mapping between vectors and one-forms at every point. Thus, given a vector field $\vec{V}(\mathcal{P})$ (which notation means that $\vec{V}$ depends on the position $\mathcal{P}$, where $\mathcal{P}$ is any point), there is a unique one-form field $\tilde{V}(\mathcal{P}) = \mathbf{g}(\vec{V}(\mathcal{P}),\ )$. The components of **g** are called $g_{\alpha\beta}$; the components of the inverse matrix are called $g^{\alpha\beta}$. The metric permits raising and lowering of indices in the same way as in SR, which means

$$V_\alpha = g_{\alpha\beta} V^\beta.$$

In general, $\{g_{\alpha\beta}\}$ will be complicated functions of position, so it will not be true that there would be a simple relation between, say, V_0 and V^0 in an arbitrary coordinate system.

Since we wish to study general curved manifolds, we have to allow any coordinate system. In SR we only studied Lorentz (inertial) frames because they were simple. But because gravity prevents such frames from being global, we shall have to allow all coordinates, and hence all coordinate transformations, that are nonsingular. (Nonsingular means, as in § 5.2, that the matrix of the transformation, $\Lambda^{\alpha'}{}_{\beta} \equiv \partial x^{\alpha'}/\partial x^{\beta}$, has an inverse.) Now, the matrix $(g_{\alpha\beta})$ is a symmetric matrix by definition. It is a well-known theorem of matrix algebra (see Exer. 3, § 6.9) that a transformation matrix can always be found that will make any symmetric matrix into a diagonal matrix with each entry on the main diagonal either $+1$, -1, or zero. The number of $+1$ entries equals the number of positive eigenvalues of $(g_{\alpha\beta})$, while the number of -1 entries is the number of negative eigenvalues. So if we choose **g** originally to have three positive eigenvalues and one negative, then we can always find a $\Lambda^{\alpha'}{}_{\beta}$ to make the metric components become

$$(g_{\alpha'\beta'}) = \begin{pmatrix} -1 & 0 & 0 & 0 \\ 0 & 1 & 0 & 0 \\ 0 & 0 & 1 & 0 \\ 0 & 0 & 0 & 1 \end{pmatrix} \equiv (\eta_{\alpha\beta}). \tag{6.2}$$

From now on we will use $\eta_{\alpha\beta}$ to denote *only* the matrix in Eq. (6.2), which is of course the metric of SR.

There are two remarks that must be made here. The first is that Eq. (6.2) relied on choosing **g** to have the appropriately signed eigenvalues. The sum of the diagonal elements in Eq. (6.2) is called the *signature* of the metric. For SR and GR it is $+2$. Thus, the fact that we have previously deduced from physical arguments that one can always construct a *local* inertial frame at any event, finds its mathematical representation in Eq. (6.2), that the metric can be transformed into $\eta_{\alpha\beta}$ at that point. This in turn implies that the metric has to have signature $+2$ if it is to describe a spacetime with gravity.

The second remark is that the matrix $\Lambda^{\alpha'}{}_{\beta}$ that produces Eq. (6.2) at every point may *not* be a coordinate transformation. That is, the set $\{\tilde{\omega}^{\alpha'} = \Lambda^{\alpha'}{}_{\beta}\, \tilde{d}x^{\beta}\}$ may not be a coordinate basis. By our earlier discussion of noncoordinate bases, it would be a coordinate transformation only if Eq. (5.95) holds:

$$\frac{\partial \Lambda^{\alpha'}{}_{\beta}}{\partial x^{\gamma}} = \frac{\partial \Lambda^{\alpha'}{}_{\gamma}}{\partial x^{\beta}}.$$

In a general gravitational field this will be impossible, because otherwise it would imply the existence of coordinates for which Eq. (6.2) is true everywhere: a global Lorentz frame. However, having found a basis at a particular point $\mathcal{P}$ for which Eq. (6.2) is true, it is possible to find coordinates such that, in the neighborhood of $\mathcal{P}$, Eq. (6.2) is 'nearly' true. This is embodied in the following theorem, whose (rather long) proof is at the end of this section. Choose any point $\mathcal{P}$ of the manifold. A coordinate system $\{x^\alpha\}$ can be found whose origin is at $\mathcal{P}$ and in which:

$$g_{\alpha\beta}(x^\mu) = \eta_{\alpha\beta} + 0[(x^\mu)^2]. \tag{6.3}$$

That is, the metric near $\mathcal{P}$ is approximately that of SR, differences being of second order in the coordinates. From now on we shall refer to such coordinate systems as 'local Lorentz frames' or 'local inertial frames'. Eq. (6.3) can be rephrased in a somewhat more precise way as:

◆ $g_{\alpha\beta}(\mathcal{P}) = \eta_{\alpha\beta}$ for all α, β; $\qquad\qquad$ (6.4)

◆ $\dfrac{\partial}{\partial x^\gamma} g_{\alpha\beta}(\mathcal{P}) = 0$ for all α, β, γ; $\qquad\qquad$ (6.5)

but

◆ $\dfrac{\partial^2}{\partial x^\gamma \, \partial x^\mu} g_{\alpha\beta}(\mathcal{P}) \neq 0$

for at least some values of α, β, γ, and μ if the manifold is not exactly flat.

The existence of local Lorentz frames is merely the statement that any curved space has a flat space 'tangent' to it at any point. Recall that straight lines in flat spacetime are the world lines of free particles; the absence of first-derivative terms (Eq. (6.5)) in the metric of a curved spacetime will mean that free particles are moving on lines that are locally straight in this coordinate system. This makes such coordinates very useful for us, since the equations of physics will be nearly as simple in them as in flat spacetime, and if constructed by the rules of § 6.1 will be valid in any coordinate system. The proof of this theorem is at the end of this section, and is worth studying.

Lengths and volumes. The metric of course gives a way to define lengths of curves. Let $d\vec{x}$ be a small vector displacement on some curve. Then $d\vec{x}$ has squared length $ds^2 = g_{\alpha\beta} \, dx^\alpha \, dx^\beta$. (Recall that we call this the line element of the metric.) If we take the absolute value of this and take its square root, we get a measure of length: $dl \equiv |g_{\alpha\beta} \, dx^\alpha \, dx^\beta|^{1/2}$. Then integrating it gives

$$l = \int_{\substack{\text{along} \\ \text{curve}}} |g_{\alpha\beta} \, dx^\alpha \, dx^\beta|^{1/2} \tag{6.6}$$

$$= \int_{\lambda_0}^{\lambda_1} \left| g_{\alpha\beta} \frac{dx^\alpha}{d\lambda} \frac{dx^\beta}{d\lambda} \right|^{1/2} d\lambda, \tag{6.7}$$

where λ is the parameter of the curve (whose endpoints are λ_0 and λ_1). But since the tangent vector $\vec{V}$ has components $V^\alpha = dx^\alpha/d\lambda$, we finally have:

♦
$$l = \int_{\lambda_0}^{\lambda_1} |\vec{V} \cdot \vec{V}|^{1/2} d\lambda \tag{6.8}$$

as the length of the arbitrary curve.

The computation of volumes is very important for integration in spacetime. Here, we mean by 'volume' the four-dimensional volume element we used for integrations in Gauss' law in § 4.4. Let us go to a local Lorentz frame, where we know that a small four-dimensional region has four-volume $dx^0 dx^1 dx^2 dx^3$, where $\{x^\alpha\}$ are the coordinates which at this point give the nearly Lorentz metric, Eq. (6.3). In *any* other coordinate system $\{x^{\alpha'}\}$ it is a well-known result of the calculus of several variables that:

$$dx^0 dx^1 dx^2 dx^3 = \frac{\partial(x^0, x^1, x^2, x^3)}{\partial(x^{0'}, x^{1'}, x^{2'}, x^{3'})} dx^{0'} dx^{1'} dx^{2'} dx^{3'}, \tag{6.9}$$

where the factor $\partial(\ )/\partial(\ )$ is the Jacobian of the transformation from $\{x^{\alpha'}\}$ to $\{x^\alpha\}$, as defined in § 5.2:

$$\frac{\partial(x^0, x^1, x^2, x^3)}{\partial(x^{0'}, x^{1'}, x^{2'}, x^{3'})} = \det \begin{pmatrix} \partial x^0/\partial x^{0'} & \partial x^0/\partial x^{1'} & \cdots \\ \partial x^1/\partial x^{0'} & & \\ \vdots & & \end{pmatrix}$$

$$= \det(\Lambda^\alpha{}_{\beta'}). \tag{6.10}$$

This would be a rather tedious way to calculate the Jacobian, but there is an easier way using the metric. In matrix terminology, the transformation of the metric components is

$$(g) = (\Lambda)(\eta)(\Lambda)^T, \tag{6.11}$$

where (g) is the matrix of $g_{\alpha\beta}$, (η) of $\eta_{\alpha\beta}$, etc., and where 'T' denotes transpose. It follows that the determinants satisfy

$$\det(g) = \det(\Lambda) \det(\eta) \det(\Lambda^T). \tag{6.12}$$

But for any matrix

$$\det(\Lambda) = \det(\Lambda^T), \tag{6.13}$$

and we can easily see from Eq. (6.2) that

$$\det(\eta) = -1. \tag{6.14}$$

Therefore we get

$$\det(g) = -[\det(\Lambda)]^2. \tag{6.15}$$

Now we introduce the notation

$$g \equiv \det(g_{\alpha'\beta'}), \tag{6.16}$$

which enables us to conclude from Eq. (6.15) that

$$\det(\Lambda^\alpha{}_{\beta'}) = (-g)^{1/2}. \tag{6.17}$$

Thus, from Eq. (6.9) we get

◆ $$dx^0 \, dx^1 \, dx^2 \, dx^3 = [-\det(g_{\alpha'\beta'})]^{1/2} \, dx^{0'} \, dx^{1'} \, dx^{2'} \, dx^{3'}$$
$$= (-g)^{1/2} \, dx^{0'} \, dx^{1'} \, dx^{2'} \, dx^{3'}. \tag{6.18}$$

This is a very useful result. It is also conceptually an important result because it is the first example of a kind of argument we will frequently employ, an argument that uses locally flat coordinates to generalize our flat-space concepts to analogous ones in curved space. In this case we began with $dx^0 \, dx^1 \, dx^2 \, dx^3 = d^4x$ in a locally flat coordinate system. We argue that this volume element at $\mathcal{P}$ must be the volume physically measured by rods and clocks, since the space is the same as Minkowski space in this small region. We then find that the value of this expression in arbitrary coordinates $\{x^{\mu'}\}$ is Eq. (6.18), $(-g)^{1/2} \, d^4x'$, which is thus the expression for the true volume in a curved space at any point in any coordinates.

It should not be surprising that the metric comes into it, of course, since the metric measures lengths. One only need remember that in any coordinates the square root of the negative of the determinant of $(g_{\alpha\beta})$ is the thing to multiply by d^4x to get the true, or *proper*, volume element.

Perhaps it would be helpful to quote an example from three dimensions. Here proper volume is $(g)^{1/2}$, since the metric is positive-definite (Eq. (6.14) would have a + sign). In spherical coordinates the line element is $dl^2 = dr^2 + r^2 \, d\theta^2 + r^2 \sin^2 \theta \, d\phi^2$, so the metric is

$$(g_{ij}) = \begin{pmatrix} 1 & 0 & 0 \\ 0 & r^2 & 0 \\ 0 & 0 & r^2 \sin^2 \theta \end{pmatrix}. \tag{6.19}$$

Its determinant is $r^4 \sin^2 \theta$, so $(g)^{1/2} \, d^3x'$ is

$$r^2 \sin \theta \, dr \, d\theta \, d\phi, \tag{6.20}$$

which we know is the correct volume element in these coordinates.

Proof of the local-flatness theorem. Let $\{x^\alpha\}$ be an arbitrary given coordinate system and $\{x^{\alpha'}\}$ the one which is desired: it reduces to the inertial system at a certain fixed point $\mathcal{P}$. (A point in this four-dimensional manifold is, of course, an event.) Then there is some relation

$$x^\alpha = x^\alpha(x^{\mu'}), \tag{6.21}$$

$$\Lambda^\alpha{}_{\mu'} = \partial x^\alpha/\partial x^{\mu'}. \tag{6.22}$$

Expanding $\Lambda^\alpha_{\mu'}$ in a Taylor series about $\mathcal{P}$ (whose coordinates are $x_0^{\mu'}$) gives

$$\Lambda^\alpha{}_{\mu'}(\vec{x}') = \Lambda^\alpha{}_{\mu'}(\mathcal{P}) + (x^{\gamma'} - x_0^{\gamma'})\frac{\partial \Lambda^\alpha{}_{\mu'}}{\partial x^{\gamma'}}$$

$$+ \tfrac{1}{2}(x^{\gamma'} - x_0^{\gamma'})(x^{\lambda'} - x_0^{\lambda'})\frac{\partial^2 \Lambda^\alpha{}_{\mu'}}{\partial x^{\lambda'}\partial x^{\gamma'}} + \cdots,$$

$$= \Lambda^\alpha{}_{\mu'}|_{\mathcal{P}} + (x^{\gamma'} - x_0^{\gamma'})\frac{\partial^2 x^\alpha}{\partial x^{\gamma'}\partial x^{\mu'}}\bigg|_{\mathcal{P}}$$

$$+ \tfrac{1}{2}(x^{\gamma'} - x_0^{\gamma'})(x^{\lambda'} - x_0^{\lambda'})\frac{\partial^3 x^\alpha}{\partial x^{\lambda'}\partial x^{\gamma'}\partial x^{\mu'}}\bigg|_{\mathcal{P}} + \cdots. \tag{6.23}$$

Expanding the metric in the same way gives

$$g_{\alpha\beta}(\vec{x}') = g_{\alpha\beta}|_{\mathcal{P}} + (x^{\gamma'} - x_0^{\gamma'})\frac{\partial g_{\alpha\beta}}{\partial x^{\gamma'}}\bigg|_{\mathcal{P}}$$

$$+ \tfrac{1}{2}(x^{\gamma'} - x_0^{\gamma'})(x^{\lambda'} - x_0^{\lambda'})\frac{\partial^2 g_{\alpha\beta}}{\partial x^{\lambda'}\partial x^{\gamma'}}\bigg|_{\mathcal{P}} + \cdots. \tag{6.24}$$

We put these into the transformation,

$$g_{\mu'\nu'} = \Lambda^\alpha{}_{\mu'}\Lambda^\beta{}_{\nu'}g_{\alpha\beta}, \tag{6.25}$$

to obtain

$$g_{\mu'\nu'}(\vec{x}') = \Lambda^\alpha{}_{\mu'}|_{\mathcal{P}}\Lambda^\beta{}_{\nu'}|_{\mathcal{P}}g_{\alpha\beta}|_{\mathcal{P}}$$

$$+ (x^{\gamma'} - x_0^{\gamma'})[\Lambda^\alpha{}_{\mu'}|_{\mathcal{P}}\Lambda^\beta{}_{\nu'}|_{\mathcal{P}}g_{\alpha\beta,\gamma}|_{\mathcal{P}}$$

$$+ \Lambda^\alpha{}_{\mu'}|_{\mathcal{P}}g_{\alpha\beta}|_{\mathcal{P}}\, \partial^2 x^\beta/\partial x^{\gamma'}\partial x^{\nu'}|_{\mathcal{P}}$$

$$+ \Lambda^\beta{}_{\nu'}|_{\mathcal{P}}g_{\alpha\beta}|_{\mathcal{P}}\, \partial^2 x^\alpha/\partial x^{\gamma'}\partial x^{\mu'}|_{\mathcal{P}}]$$

$$+ \tfrac{1}{2}(x^{\gamma'} - x_0^{\gamma'})(x^{\lambda'} - x_0^{\lambda'})[\cdots]. \tag{6.26}$$

Now, we do not know the transformation, Eq. (6.21), but we can define it by its Taylor expansion. Let us count the number of free variables we have for this purpose. The matrix $\Lambda^\alpha{}_{\mu'}|_{\mathcal{P}}$ has 16 numbers, all of which are freely specifiable. The array $\{\partial^2 x^\alpha/\partial x^{\gamma'}\partial x^{\mu'}|_{\mathcal{P}}\}$ has $4 \times 10 = 40$ free numbers (not $4 \times 4 \times 4$, since it is *symmetric* in γ' and μ'). The array $\{\partial^3 x^\alpha/\partial x^{\lambda'}\partial x^{\gamma'}\partial x^{\mu'}|_{\mathcal{P}}\}$ has $4 \times 20 = 80$ free variables, since symmetry on *all* rearrangements of λ', γ' and μ' gives only 20 independent arrangements (the general expression for three indices is $n(n+1)(n+2)/3!$, where n is the number of values each index can take, four in our case). On the other hand, $g_{\alpha\beta}|_{\mathcal{P}}$, $g_{\alpha\beta,\gamma}|_{\mathcal{P}}$ and $g_{\alpha\beta,\gamma'\mu'}|_{\mathcal{P}}$ are all given initially. They have, respectively, 10, $10 \times 4 = 40$, and $10 \times 10 = 100$ independent numbers for a fully general metric. The first question is, can we satisfy Eq. (6.4),

$$g_{\mu'\nu'}|_{\mathscr{P}} = \eta_{\mu'\nu'}? \qquad (6.27)$$

This can be written as

$$\eta_{\mu'\nu'} = \Lambda^{\alpha}{}_{\mu'}|_{\mathscr{P}} \Lambda^{\beta}{}_{\nu'}|_{\mathscr{P}} g_{\alpha\beta}|_{\mathscr{P}}. \qquad (6.28)$$

These are ten equations, and to satisfy them we have 16 free values in $\Lambda^{\alpha}{}_{\mu'}|_{\mathscr{P}}$. They can indeed, therefore, be satisfied, leaving six elements of $\Lambda^{\alpha}{}_{\mu'}|_{\mathscr{P}}$ unspecified. These six correspond to the six degrees of freedom in the Lorentz transformations that preserve the form of the metric $\eta_{\mu'\nu'}$. That is, one can boost by a velocity v (three free parameters) or rotate by an angle θ around a direction defined by two other angles. These total six degrees of freedom in $\Lambda^{\alpha}{}_{\mu'}|_{\mathscr{P}}$ that leave the local inertial frame inertial.

The next question is, can we choose the 40 free numbers $\partial \Lambda^{\alpha}{}_{\mu'}/\partial x^{\gamma'}|_{\mathscr{P}}$ in Eq. (6.26) in such a way as to satisfy the 40 equations, Eq. (6.5),

$$g_{\alpha'\beta',\mu'}|_{\mathscr{P}} = 0? \qquad (6.29)$$

Since 40 equals 40, the answer is yes, just barely. Given the matrix $\Lambda^{\alpha}{}_{\mu'}|_{\mathscr{P}}$, there is one and only one way to arrange the coordinates near $\mathscr{P}$ such that $\Lambda^{\alpha}{}_{\mu',\gamma'}|_{\mathscr{P}}$ has the right values to make $g_{\alpha'\beta',\mu'}|_{\mathscr{P}} = 0$. So there is no extra freedom other than that with which to make local Lorentz transformations.

The final question is, can we make this work at higher order? Can we find 80 numbers $\Lambda^{\alpha}{}_{\mu',\gamma'\lambda'}|_{\mathscr{P}}$ which can make the 100 numbers $g_{\alpha'\beta',\mu'\lambda'}|_{\mathscr{P}} = 0$? The answer, since $80 < 100$, is no. There are, in the general metric, 20 'degrees of freedom' among the second derivatives $g_{\alpha'\beta',\mu'\lambda'}|_{\mathscr{P}}$ ($100 - 80 = 20$): in general, at least 20 of these components will not vanish.

Therefore we see that a general metric is characterized at any point $\mathscr{P}$ not so much by its value at $\mathscr{P}$ (which can always be made to be $\eta_{\alpha\beta}$), nor by its first derivatives there (which can be made zero), but by the 20 second derivatives there which cannot be made to vanish. These 20 numbers will be seen to be components of a tensor which represents the curvature; this we shall show later. In a *flat* space, of course, all 20 vanish. In a general space they do not.

6.3 Covariant differentiation

We now look at the subject of differentiation. By definition, the derivative of a vector field involves the difference between vectors at two different points (in the limit as the points come together). In a curved space the notion of the difference between vectors at different points must be handled with care, since in between the points the space is curved and the idea of vectors at the two points, pointing in the 'same' direction, is fuzzy. However, the local flatness of the Riemannian manifold helps us out. We only need to compare vectors in the limit as they get infinitesimally close together, and we know that we can construct a coordinate system at any point which is as close to being flat as we would like in this same limit. So in a small region the manifold looks flat, and it is then natural to say that the derivative of a vector whose components are constant in this coordinate system is zero at that point. That is, we say that the derivatives of the basis vectors of the locally inertial coordinate system are zero at $\mathcal{P}$.

Let us emphasize that this is a *definition* of the covariant derivative. For us, its justification is in the physics: the local inertial frame is a frame in which everything is locally like SR, and in SR the derivatives of these basis vectors are zero. This definition immediately leads to the fact that in these coordinates at this point, the covariant derivative of a vector has components given by the partial derivatives of the components (that is, the Christoffel symbols vanish):

$$V^{\alpha}{}_{;\beta} = V^{\alpha}{}_{,\beta} \quad \text{at } \mathcal{P} \text{ in this frame.} \tag{6.30}$$

This is of course also true for any other tensor, including the metric:

$$g_{\alpha\beta;\gamma} = g_{\alpha\beta,\gamma} = 0 \quad \text{at } \mathcal{P}.$$

(The second equality is just Eq. (6.5).) Now, the equation $g_{\alpha\beta;\gamma} = 0$ is true in one frame (the locally inertial one), and is a valid tensor equation; therefore it is true in *any* basis:

$$\blacklozenge \qquad g_{\alpha\beta;\gamma} = 0 \quad \text{in any basis.} \tag{6.31}$$

This is a very important result, and comes directly from our definition of the covariant derivative. Recalling § 5.4, we see that *if* we have $\Gamma^{\mu}{}_{\alpha\beta} = \Gamma^{\mu}{}_{\beta\alpha}$, then Eq. (6.31) leads to Eq. (5.75) for *any* metric:

$$\blacklozenge \qquad \Gamma^{\alpha}{}_{\mu\nu} = \tfrac{1}{2} g^{\alpha\beta} (g_{\beta\mu,\nu} + g_{\beta\nu,\mu} - g_{\mu\nu,\beta}). \tag{6.32}$$

It is left to Exer. 5, § 6.9, to demonstrate, by repeating the flat-space argument now in the locally inertial frame, that $\Gamma^{\mu}{}_{\beta\alpha}$ is indeed symmetric in any coordinate system, so that Eq. (6.32) is correct in any coordinates. We assumed at the start that at $\mathcal{P}$ in a locally inertial frame, $\Gamma^{\alpha}{}_{\mu\nu} = 0$. But, importantly, the derivatives of $\Gamma^{\alpha}{}_{\mu\nu}$ at $\mathcal{P}$ in this frame are not all zero generally, since they involve $g_{\alpha\beta,\gamma\mu}$. This means that even though coordinates can be found in which $\Gamma^{\alpha}{}_{\mu\nu} = 0$ at a point, these symbols do not generally vanish elsewhere. This differs from flat space, where a coordinate system exists in which $\Gamma^{\alpha}{}_{\mu\nu} = 0$ everywhere. So we can see that at any given point, the difference between a general manifold and a flat one manifests itself in the derivatives of the Christoffel symbols.

Eq. (6.32) means that, given $g_{\alpha\beta}$, one can calculate $\Gamma^{\alpha}{}_{\mu\nu}$ everywhere. One can therefore calculate all covariant derivatives, given **g**. To review the formulas:

◆ $\qquad V^{\alpha}{}_{;\beta} = V^{\alpha}{}_{,\beta} + \Gamma^{\alpha}{}_{\mu\beta} V^{\mu},$ \hfill (6.33)

◆ $\qquad \mathcal{P}_{\alpha;\beta} = \mathcal{P}_{\alpha,\beta} - \Gamma^{\mu}{}_{\alpha\beta} \mathcal{P}_{\mu},$ \hfill (6.34)

◆ $\qquad T^{\alpha\beta}{}_{;\gamma} = T^{\alpha\beta}{}_{,\gamma} + \Gamma^{\alpha}{}_{\mu\gamma} T^{\mu\beta} + \Gamma^{\beta}{}_{\mu\gamma} T^{\alpha\mu}.$ \hfill (6.35)

Divergence formula. Quite often one deals with the divergence of vectors. Given an arbitrary vector field V^{α}, its divergence is defined by Eq. (5.52),

$$V^{\alpha}{}_{;\alpha} = V^{\alpha}{}_{,\alpha} + \Gamma^{\alpha}{}_{\mu\alpha} V^{\mu}. \tag{6.36}$$

This formula involves a sum in the Christoffel symbol, which, from Eq. (6.32), is

$$\begin{aligned}
\Gamma^{\alpha}{}_{\mu\alpha} &= \tfrac{1}{2} g^{\alpha\beta} (g_{\beta\mu,\alpha} + g_{\beta\alpha,\mu} - g_{\mu\alpha,\beta}) \\
&= \tfrac{1}{2} g^{\alpha\beta} (g_{\beta\mu,\alpha} - g_{\mu\alpha,\beta}) + \tfrac{1}{2} g^{\alpha\beta} g_{\beta\alpha,\mu}.
\end{aligned} \tag{6.37}$$

This has had its terms rearranged to simplify it: notice that the term in parentheses is antisymmetric in α and β, while it is contracted on α and β with $g^{\alpha\beta}$, which is symmetric. The first term therefore vanishes and we find

$$\Gamma^{\alpha}{}_{\mu\alpha} = \tfrac{1}{2} g^{\alpha\beta} g_{\alpha\beta,\mu}. \tag{6.38}$$

Since $(g^{\alpha\beta})$ is the inverse matrix of $(g_{\alpha\beta})$, it can be shown (see Exer. 7, § 6.9) that the derivative of the determinant g of the matrix $(g_{\alpha\beta})$ is

◆ $\qquad g_{,\mu} = g g^{\alpha\beta} g_{\beta\alpha,\mu}.$ \hfill (6.39)

Using this in Eq. (6.38), one finds

◆ $\qquad \Gamma^{\alpha}{}_{\mu\alpha} = (\sqrt{-g})_{,\mu} / \sqrt{-g}.$ \hfill (6.40)

Then we can write the divergence, Eq. (6.36), as

$$V^{\alpha}{}_{;\alpha} = V^{\alpha}{}_{,\alpha} + \frac{1}{\sqrt{-g}} V^{\alpha} (\sqrt{-g})_{,\alpha} \tag{6.41}$$

or

◆ $\qquad V^{\alpha}{}_{;\alpha} = \dfrac{1}{\sqrt{-g}} (\sqrt{-g}\, V^{\alpha})_{,\alpha}.$ \hfill (6.42)

This is a very much easier formula to use than Eq. (6.36). It is also important for Gauss' law, where we integrate the divergence over a volume (using, of course, the proper volume element):

$$\int V^{\alpha}{}_{;\alpha} \sqrt{-g}\ \mathrm{d}^4 x = \int (\sqrt{-g}\, V^{\alpha})_{,\alpha}\ \mathrm{d}^4 x. \tag{6.43}$$

Since the final term involves simple partial derivatives, the mathematics of Gauss' law applies to it, just as in SR (§ 4.8):

$$\int (\sqrt{-g}\,V^\alpha)_{,\alpha}\, \mathrm{d}^4x = \oint V^\alpha n_\alpha \sqrt{-g}\, \mathrm{d}^3S. \tag{6.44}$$

This means

$$\blacklozenge \qquad \int V^\alpha{}_{;\alpha}\sqrt{-g}\, \mathrm{d}^4x = \oint V^\alpha n_\alpha \sqrt{-g}\, \mathrm{d}^3S. \tag{6.45}$$

So Gauss' law does apply on a curved manifold, in the form given by Eq. (6.45). One needs to integrate the divergence over proper volume and to use the *proper surface element*, $n_\alpha \sqrt{-g}\, \mathrm{d}^3S$, in the surface integral.

6.4 Parallel-transport, geodesics and curvature

Until now, we have used the local-flatness theorem to develop as much mathematics on curved manifolds as possible without considering the curvature explicitly. Indeed, we have yet to give a precise mathematical definition of curvature. It is important to distinguish two different kinds of curvature: intrinsic and extrinsic. Consider, for example, a cylinder. Since a cylinder is round in one direction, one thinks of it as curved. This is its *extrinsic* curvature: the curvature it has in relation to the flat three-dimensional space it is part of. On the other hand, a cylinder can be made by rolling a flat piece of paper without tearing or crumpling it, so the *intrinsic* geometry is that of the original paper: it is flat. This means that the distance in the surface of the cylinder between any two points is the same as it was in the original paper; parallel lines remain parallel when continued; in fact, *all* of Euclid's axioms hold for the surface of a cylinder. A two-dimensional 'ant' confined to that surface would decide it was flat; only its global topology is funny, in that going in a certain direction in a straight line brings him back to where he started. The *intrinsic* geometry of an n-dimensional manifold considers only the relationships between its points on paths that remain in the manifold (for the cylinder, in the two-dimensional surface). The *extrinsic* curvature of the cylinder comes from considering it as a surface in a space of higher dimension, and asking about the curvature of lines that stay in the surface compared with 'straight' lines that go off it. So *extrinsic* curvature relies on the notion of a higher-dimensional space. In this book, when we talk about the curvature of spacetime, we talk about its *intrinsic* curvature, since it is clear that all world lines are confined to remain in spacetime. Whether or not there is a higher-dimensional flat space in which our four-dimensional space is a mere surface is of no interest here, since we apparently can't enter it. The only thing of interest to us is the intrinsic geometry of spacetime.

The cylinder, as we have just seen, is intrinsically flat; a sphere, on the other hand, has an intrinsically curved surface. To see this, consider Fig. 6.1, in which two neighboring lines begin at *A* and *B* perpendicular to the equator, and hence are parallel. When continued as locally straight lines they follow the arc of great circles, and the two lines meet at the pole *P*. Parallel lines, when continued, do not remain parallel, so the space is not flat.

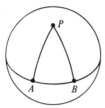

Fig. 6.1   A spherical triangle *APB*.

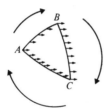

Fig. 6.2   A 'triangle' made of curved lines in flat space.

There is an even more striking illustration of the curvature of the sphere. Consider, first, flat space. In Fig. 6.2 a closed path in flat space is drawn, and, starting at *A*, at each point a vector is drawn parallel to the one at the previous point. This construction is carried around the loop from *A* to *B* to *C* and back to *A*. The vector finally drawn at *A* is, of course, parallel to the original one. A completely different thing happens on a sphere! Consider the path shown in Fig. 6.3. Remember, we are drawing the vector as it is seen to a two-dimensional ant on the sphere, so it must always be tangent to the sphere. Aside from that, each vector is drawn as parallel as possible to the previous one. In this loop, *A* and *C* are on the equator 90° apart, and *B* is at the pole. Each arc is the arc of a great circle, and each is 90° long. At *A* we choose the vector parallel to the equator. Each new vector is therefore drawn perpendicular to the arc *AB*. When we get to *B*, the vectors are tangent to *BC*. So, going from *B* to *C*, we keep drawing tangents to *BC*. These are perpendicular

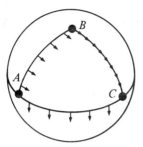

Fig. 6.3   Parallel transport around a spherical triangle.

to the equator of *C*, and so, from *C* to *A* the new vectors remain perpendicular to the equator. Thus the vector field has rotated 90° in this construction! Despite the fact that each vector is drawn parallel to its neighbor, the closed loop has caused a discrepancy. Since this doesn't happen in flat space, it must be an effect of the sphere's curvature.

This result has radical implications: on a curved manifold it simply isn't possible to define globally parallel vector fields. One can still define local parallelism, for instance how to move a vector from one point to another, keeping it parallel and of the same length. But the result of such 'parallel transport' from point *A* to point *B* depends on the path taken. One therefore cannot assert that a vector at *A* is or is not parallel to (or the same as) a certain vector at *B*.

Parallel-transport. The construction we have just made on the sphere is called parallel-transport. If a vector field $\vec{V}$ is defined at every point of a curve, as in Fig. 6.4, and if the vectors $\vec{V}$ at infinitesimally close points of the curve are parallel and of equal length, then $\vec{V}$ is said to be parallel-transported along the curve. It is easy to write down an equation for this. If $\vec{U} = d\vec{x}/d\lambda$ is the tangent to the curve (λ being the parameter along it; $\vec{U}$ is not necessarily normalized), then in a locally inertial

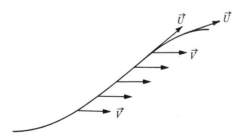

Fig. 6.4   Parallel transport of $\vec{V}$ along $\vec{U}$.

coordinate system at a point $\mathcal{P}$ the components of $\vec{V}$ must be constant along the curve at $\mathcal{P}$:

$$\frac{\mathrm{d}V^{\alpha}}{\mathrm{d}\lambda} = 0 \quad \text{at } \mathcal{P}. \tag{6.46}$$

This can be written as:

$$\frac{\mathrm{d}V^{\alpha}}{\mathrm{d}\lambda} = U^{\beta}V^{\alpha}{}_{,\beta} = U^{\beta}V^{\alpha}{}_{;\beta} = 0 \quad \text{at } \mathcal{P}. \tag{6.47}$$

The first equality is the definition of the derivative of a function (in this case V^{α}) along the curve; the second equality comes from the fact that $\Gamma^{\alpha}{}_{\mu\nu} = 0$ at $\mathcal{P}$ in these coordinates. But the third equality is a frame-invariant expression and holds in any basis, so it can be taken as a frame-invariant *definition of the parallel-transport of $\vec{V}$ along $\vec{U}$*:

◆ $$\quad U^{\beta}V^{\alpha}{}_{;\beta} = 0 \Leftrightarrow \frac{\mathrm{d}}{\mathrm{d}\lambda}\vec{V} = \nabla_{\vec{U}}\vec{V} = 0. \tag{6.48}$$

The last step uses the notation for the derivative along $\vec{U}$ introduced in Eq. (3.67).

Geodesics. The most important curves in flat space are straight lines. One of Euclid's axioms is that two straight lines that are initially parallel remain parallel when extended. What does he mean by 'extended'? He *doesn't* mean 'continued in such a way that the distance between them remains constant', because even then they could both bend. What he means is that each line keeps going in the direction it has been going in. More precisely, the tangent to the curve at one point is parallel to the tangent at the previous point. In fact, a straight line in Euclidean space is the *only* curve that parallel-transports its own tangent vector! In a curved space, we can also draw lines that are 'as nearly straight as possible' by demanding parallel-transport of the tangent vector. These are called *geodesics*:

◆ $$\quad \{\vec{U} \text{ is tangent to a geodesic}\} \Leftrightarrow \nabla_{\vec{U}}\vec{U} = 0. \tag{6.49}$$

(Note that in a locally inertial system these lines *are* straight.) In component notation:

$$U^{\beta}U^{\alpha}{}_{;\beta} = U^{\beta}U^{\alpha}{}_{,\beta} + \Gamma^{\alpha}{}_{\mu\beta}U^{\mu}U^{\beta} = 0. \tag{6.50}$$

Now, if we let λ be the parameter of the curve, then $U^{\alpha} = \mathrm{d}x^{\alpha}/\mathrm{d}\lambda$ and $U^{\beta}\partial/\partial x^{\beta} = \mathrm{d}/\mathrm{d}\lambda$:

◆ $$\quad \frac{\mathrm{d}}{\mathrm{d}\lambda}\left(\frac{\mathrm{d}x^{\alpha}}{\mathrm{d}\lambda}\right) + \Gamma^{\alpha}{}_{\mu\beta}\frac{\mathrm{d}x^{\mu}}{\mathrm{d}\lambda}\frac{\mathrm{d}x^{\beta}}{\mathrm{d}\lambda} = 0. \tag{6.51}$$

Since the Christoffel symbols $\Gamma^{\alpha}{}_{\mu\beta}$ are known functions of the coordinates

$\{x^{\alpha}\}$, this is a nonlinear (quasi-linear), second-order differential equation for $x^{\alpha}(\lambda)$. It has a unique solution when initial conditions at $\lambda = \lambda_0$ are given: $x_0^{\alpha} = x^{\alpha}(\lambda_0)$ and $U_0^{\alpha} = (dx^{\alpha}/d\lambda)_{\lambda_0}$. So, by giving an initial position (x_0^{α}) and an initial direction (U_0^{α}), one gets a unique geodesic.

Recall that if we change parameter we change, mathematically speaking, the curve (though not the points it passes through). Now, if λ is a parameter of a geodesic (so that Eq. (6.51) is satisfied), and if we define a new parameter

$$\phi = a\lambda + b, \tag{6.52}$$

where a and b are *constants* (not depending on position on the curve), then ϕ is also a parameter in which Eq. (6.51) is satisfied:

$$\frac{d^2 x^{\alpha}}{d\phi^2} + \Gamma^{\alpha}{}_{\mu\beta} \frac{dx^{\mu}}{d\phi} \frac{dx^{\beta}}{d\phi} = 0.$$

Generally speaking, *only* linear transformations of λ like Eq. (6.52) will give new parameters in which the geodesic equation is satisfied. A parameter like λ and ϕ above is called an *affine* parameter. A curve having the same path as a geodesic but parametrized by a nonaffine parameter is, strictly speaking, not a geodesic curve.

A geodesic is also a curve of *extremal length*: between any two points, its length is unchanged to first order in small changes in the curve. The student is urged to prove this by using Eq. (6.7), finding the Euler–Lagrange equations for it to be an extremal for fixed λ_0 and λ_1, and showing that these reduce to Eq. (6.51) when Eq. (6.32) is used. This is a very instructive exercise. One can also show that proper distance along the geodesic is itself an affine parameter. (See Exer. 13–15, § 6.9.)

6.5 The curvature tensor

At last we are in a position to give a mathematical description of the intrinsic curvature of a manifold. We go back to the curious example of the parallel-transport of a vector around a closed loop, and take it as our *definition* of curvature. Let us imagine in our manifold a very small closed loop (Fig. 6.5) whose four sides are the coordinate

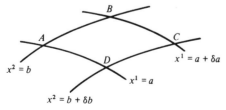

Fig. 6.5   Small section of a coordinate grid.

lines $x^1 = a$, $x^1 = a + \delta a$, $x^2 = b$, and $x^2 = b + \delta b$. A vector $\vec{V}$ defined at A is parallel-transported to B. From the parallel-transport law $\nabla_{\vec{e}_1} \vec{V} = 0$ we conclude

$$\frac{\partial V^\alpha}{\partial x^1} = -\Gamma^\alpha{}_{\mu 1} v^\mu. \tag{6.53}$$

So at B the vector has components

$$V^\alpha(B) = V^\alpha(A) + \int_A^B \frac{\partial V^\alpha}{\partial x^1} \, dx^1$$

$$= V^\alpha(A) - \int_{x^2 = b} \Gamma^\alpha{}_{\mu 1} V^\mu \, dx^1, \tag{6.54}$$

where the notation '$x^2 = b$' under the integral denotes the path AB. Similar transport from B to C to D gives

$$V^\alpha(C) = V^\alpha(B) - \int_{x^1 = a + \delta a} \Gamma^\alpha{}_{\mu 2} V^\mu \, dx^2, \tag{6.55}$$

$$V^\alpha(D) = V^\alpha(C) + \int_{x^2 = b + \delta b} \Gamma^\alpha{}_{\mu 1} V^\mu \, dx^1. \tag{6.56}$$

The integral in the last equation has a different sign because the direction of transport from C to D is in the negative x^1 direction. Similarly, the completion of the loop gives

$$V^\alpha(A_{\text{final}}) = V^\alpha(D) + \int_{x^1 = a} \Gamma^\alpha{}_{\mu 2} V^\mu \, dx^2. \tag{6.57}$$

The net change in $V^\alpha(A)$ is a vector δV^α, found by adding Eqs. (6.54)–(6.57):

$$\delta V^\alpha = V^\alpha(A_{\text{final}}) - V^\alpha(A_{\text{initial}})$$

$$= \int_{x^1 = a} \Gamma^\alpha{}_{\mu 2} V^\mu \, dx^2 - \int_{x^1 = a + \delta a} \Gamma^\alpha{}_{\mu 2} V^\mu \, dx^2$$

$$+ \int_{x^2 = b + \delta b} \Gamma^\alpha{}_{\mu 1} V^\mu \, dx^1 - \int_{x^2 = b} \Gamma^\alpha{}_{\mu 1} V^\mu \, dx^1. \tag{6.58}$$

Notice that these would cancel in pairs if $\Gamma^\alpha{}_{\mu\nu}$ and V^μ were constants on the loop, as they would be in flat space. But in curved space they are not, and we get to lowest order

$$\delta V^\alpha \approx - \int_b^{b + \delta b} \delta a \frac{\partial}{\partial x^1} (\Gamma^\alpha{}_{\mu 2} V^\mu) \, dx^2$$

$$+ \int_a^{a + \delta a} \delta b \frac{\partial}{\partial x^2} (\Gamma^\alpha{}_{\mu 1} V^\mu) \, dx^1 \tag{6.59}$$

$$\approx \delta a \, \delta b \left[-\frac{\partial}{\partial x^1} (\Gamma^\alpha{}_{\mu 2} V^\mu) + \frac{\partial}{\partial x^2} (\Gamma^\alpha{}_{\mu 1} V^\mu) \right]. \tag{6.60}$$

This involves derivatives of Christoffel symbols and of V^α. The derivatives V^α can be eliminated using Eq. (6.53) and its equivalent with 1 replaced by 2. Then Eq. (6.60) becomes

$$\delta V^\alpha = \delta a\, \delta b [\Gamma^\alpha{}_{\mu 1,2} - \Gamma^\alpha{}_{\mu 2,1} + \Gamma^\alpha{}_{\nu 2}\Gamma^\nu{}_{\mu 1} - \Gamma^\alpha{}_{\nu 1}\Gamma^\nu{}_{\mu 2}]V^\mu. \qquad (6.61)$$

(To obtain this, one needs to relabel dummy indices in the terms quadratic in Γs.) Notice that this turns out to be just a number times V^μ, summed on μ. Now, the indices 1 and 2 appear because the path was chosen to go along those coordinates. It is antisymmetric in 1 and 2 because the change δV^α would have to have the opposite sign if one went around the loop in the opposite direction (that is, interchanging the roles of 1 and 2). If one used general coordinate lines x^σ and x^λ, one would find

$$\delta V^\alpha = \text{change in } V^\alpha \text{ due to transport, first } \delta a\, \vec{e}_\sigma, \text{ then } \delta b\, \vec{e}_\lambda,$$
$$\text{then} - \delta a\, \vec{e}_\sigma, \text{ and finally} - \delta b\, \vec{e}_\lambda$$
$$= \delta a\, \delta b [\Gamma^\alpha{}_{\mu\sigma,\lambda} - \Gamma^\alpha{}_{\mu\lambda,\sigma} + \Gamma^\alpha{}_{\nu\lambda}\Gamma^\nu{}_{\mu\sigma} - \Gamma^\alpha{}_{\nu\sigma}\Gamma^\nu{}_{\mu\lambda}]V^\mu. \qquad (6.62)$$

Now, δV^α depends on $\delta a\, \delta b$, the coordinate 'area' of the loop. So it is clear that if the length of the loop in one direction is doubled, δV^α is doubled. This means that δV^α depends *linearly* on $\delta a\, \vec{e}_\sigma$ and $\delta b\, \vec{e}_\lambda$. Moreover, it certainly also depends linearly in Eq. (6.62) on V^α itself and on $\tilde{\omega}^\alpha$, which is the basis one-form that gives δV^α from the vector $\delta \vec{V}$. Hence we have the following result: if we define

$$\blacklozenge \qquad R^\alpha{}_{\beta\mu\nu} \equiv \Gamma^\alpha{}_{\beta\nu,\mu} - \Gamma^\alpha{}_{\beta\mu,\nu} + \Gamma^\alpha{}_{\sigma\mu}\Gamma^\sigma{}_{\beta\nu} - \Gamma^\alpha{}_{\sigma\nu}\Gamma^\sigma{}_{\beta\mu}, \qquad (6.63)$$

then $R^\alpha{}_{\beta\mu\nu}$ must be components of the $\binom{1}{3}$ tensor which, when supplied with arguments $\tilde{\omega}^\alpha$, $\vec{V}$, $\delta a\, \vec{e}_\mu$, $\delta b\, \vec{e}_\nu$, gives δV^α, the component of the change in $\vec{V}$ on parallel-transport around a loop given by $\delta a\, \vec{e}_\mu$ and $\delta b\, \vec{e}_\nu$. This tensor is called the *Riemann curvature tensor* **R**.[1]

It is useful to look at the components of **R** in a locally inertial frame at a point $\mathcal{P}$. We have $\Gamma^\alpha{}_{\mu\nu} = 0$ at $\mathcal{P}$ but from Eq. (6.32)

$$\Gamma^\alpha{}_{\mu\nu,\sigma} = \tfrac{1}{2}g^{\alpha\beta}(g_{\beta\mu,\nu\sigma} + g_{\beta\nu,\mu\sigma} - g_{\mu\nu,\beta\sigma}). \qquad (6.64)$$

Since second derivatives of $g_{\alpha\beta}$ don't vanish, we get at $\mathcal{P}$

$$R^\alpha{}_{\beta\mu\nu} = \tfrac{1}{2}g^{\alpha\sigma}(g_{\sigma\beta,\nu\mu} + g_{\sigma\nu,\beta\mu} - g_{\beta\nu,\sigma\mu}$$
$$- g_{\sigma\beta,\mu\nu} - g_{\sigma\mu,\beta\nu} + g_{\beta\mu,\sigma\nu}). \qquad (6.65)$$

Using the symmetry of $g_{\alpha\beta}$ and the fact that

$$g_{\alpha\beta,\mu\nu} = g_{\alpha\beta,\nu\mu}, \qquad (6.66)$$

because partial derivatives always commute, we find

$$R^\alpha{}_{\beta\mu\nu} = \tfrac{1}{2}g^{\alpha\sigma}(g_{\sigma\nu,\beta\mu} - g_{\sigma\mu,\beta\nu} + g_{\beta\mu,\sigma\nu} - g_{\beta\nu,\sigma\mu}). \qquad (6.67)$$

---

1   As with other definitions we have earlier introduced, there is no universal agreement about the overall sign of the Riemann tensor, or even on the placement of its indices. Always check the conventions of whatever book you read.

If we lower the index α we get (in this coordinate system at $\mathcal{P}$)

$$R_{\alpha\beta\mu\nu} \equiv g_{\alpha\lambda}R^{\lambda}{}_{\beta\mu\nu} = \tfrac{1}{2}(g_{\alpha\nu,\beta\mu} - g_{\alpha\mu,\beta\nu} + g_{\beta\mu,\alpha\nu} - g_{\beta\nu,\alpha\mu}). \qquad (6.68)$$

In this form it is easy to verify the following identities:

◆ $$R_{\alpha\beta\mu\nu} = -R_{\beta\alpha\mu\nu} = -R_{\alpha\beta\nu\mu} = R_{\mu\nu\alpha\beta}, \qquad (6.69)$$

◆ $$R_{\alpha\beta\mu\nu} + R_{\alpha\nu\beta\mu} + R_{\alpha\mu\nu\beta} = 0. \qquad (6.70)$$

Thus, $R_{\alpha\beta\mu\nu}$ is antisymmetric on the first pair and on the second pair of indices, and symmetric on exchange of the two pairs. Since Eqs. (6.69) and (6.70) are valid tensor equations true in one coordinate system, they are true in all bases. (Note that an equation like Eq. (6.67) is not a valid tensor equation, since it involves partial derivatives, not covariant ones. Therefore it is true only in the coordinate system in which it was derived.)

It can be shown (Exer. 18, § 6.9) that the various identities, Eqs. (6.69) and (6.70), reduce the number of independent components of $R_{\alpha\beta\mu\nu}$ (and hence of $R^{\alpha}{}_{\beta\mu\nu}$) to 20, in four dimensions. This is, *not* coincidentally, the same number of independent $g_{\alpha\beta,\mu\nu}$ that we found at the end of § 6.2 could not be made to vanish by a coordinate transformation. Thus $R^{\alpha}{}_{\beta\mu\nu}$ characterizes the curvature in a tensorial way.

A *flat* manifold is one which has a global definition of parallelism: a vector can be moved around parallel to itself on an arbitrary curve and will return to its starting point unchanged. This clearly means that

◆ $$R^{\alpha}{}_{\beta\mu\nu} = 0 \Leftrightarrow \text{flat manifold.} \qquad (6.71)$$

(Try showing that this is true in polar coordinates for the Euclidean plane.)

An important use of the curvature tensor comes when we examine the consequences of taking two covariant derivatives of a vector field $\vec{V}$. We found in § 6.3 that first derivatives were like flat-space ones, since we could find coordinates in which the metric was flat to first order. But second derivatives are a different story:

$$\nabla_{\alpha}\nabla_{\beta}V^{\mu} = \nabla_{\alpha}(V^{\mu}{}_{;\beta})$$
$$= (V^{\mu}{}_{;\beta})_{,\alpha} + \Gamma^{\mu}{}_{\sigma\alpha}V^{\sigma}{}_{;\beta} - \Gamma^{\sigma}{}_{\beta\alpha}V^{\mu}{}_{;\sigma}. \qquad (6.72)$$

In locally inertial coordinates at $\mathcal{P}$, all the Γs are zero, but their partial derivatives are not. Therefore we have at $\mathcal{P}$

$$\nabla_{\alpha}\nabla_{\beta}V^{\mu} = V^{\mu}{}_{,\beta\alpha} + \Gamma^{\mu}{}_{\nu\beta,\alpha}V^{\nu} \qquad (6.73)$$

in these coordinates only. Consider the same formula with α and β exchanged:

$$\nabla_{\beta}\nabla_{\alpha}V^{\mu} = V^{\mu}{}_{,\alpha\beta} + \Gamma^{\mu}{}_{\nu\alpha,\beta}V^{\nu}. \qquad (6.74)$$

If we subtract these we get the *commutator* of the covariant derivative perators ∇_{α} and ∇_{β}, written in the same notation as we would employ in quantum mechanics:

$$[\nabla_\alpha, \nabla_\beta] V^\mu \equiv \nabla_\alpha \nabla_\beta V^\mu - \nabla_\beta \nabla_\alpha V^\mu$$

$$= (\Gamma^\mu{}_{\nu\beta,\alpha} - \Gamma^\mu{}_{\nu\alpha,\beta}) V^\nu. \tag{6.75}$$

The terms involving the second derivatives of V^μ drop out here, since

$$V^\mu{}_{,\alpha\beta} = V^\mu{}_{,\beta\alpha}. \tag{6.76}$$

(Let us pause to recall that $V^\mu{}_{,\alpha}$ is the partial derivative of the component V^μ, so by the laws of partial differentiation the partial derivatives must commute. On the other hand, $\nabla_\alpha V^\mu$ is a component of the tensor $\nabla \vec{V}$, and $\nabla_\alpha \nabla_\beta V^\mu$ is a component of $\nabla \nabla \vec{V}$: there is no reason (from differential calculus) why it must be symmetric on α and β. We have proved it generally is not.) Now, in this frame (where $\Gamma^\mu{}_{\alpha\beta} = 0$), we can compare Eq. (6.75) with Eq. (6.63) and see that at $\mathscr{P}$

◆ $$[\nabla_\alpha, \nabla_\beta] V^\mu = R^\mu{}_{\nu\alpha\beta} V^\nu. \tag{6.77}$$

Since this is a valid tensor equation, it is true in *any* coordinate system. The Riemann tensor gives the commutator of covariant derivatives. This means that in curved spaces, one must be careful to know the order in which covariant derivatives are taken: they do not commute. This can be extended to tensors of higher rank. For example, a $\binom{1}{1}$ tensor has

$$[\nabla_\alpha, \nabla_\beta] F^\mu{}_\nu = R^\mu{}_{\sigma\alpha\beta} F^\sigma{}_\nu + R^\sigma{}_{\nu\alpha\beta} F^\mu{}_\sigma. \tag{6.78}$$

That is, *each* index gets a Riemann tensor on it, and each one comes in with a + sign. (They *must* all have the same sign because raising and lowering indices with **g** is unaffected by ∇_α, since $\nabla \mathbf{g} = 0$.)

Eq. (6.77) is closely related to our original derivation of the Riemann tensor from parallel-transport around loops, because the parallel-transport problem can be thought of as computing, first the change of $\vec{V}$ in one direction, and then in another, followed by subtracting changes in the reverse order: this is what commuting covariant derivatives also does.

Geodesic deviation. We have often mentioned that in a curved space, parallel lines when extended do not remain parallel. This can now be formulated mathematically in terms of the Riemann tensor. Consider two geodesics (with tangents $\vec{V}$ and $\vec{V}'$) that begin parallel and near each other, as in Fig. 6.6, at points A and A'. Let the affine parameter on the

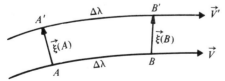

Fig. 6.6.   A connecting vector $\vec{\xi}$ between two geodesics connects points of the same parameter value.

geodesics be called λ. We define a 'connecting vector' $\vec{\xi}$ which 'reaches' from one geodesic to another, connecting points at equal intervals in λ (i.e. A to A', B to B', etc.). For simplicity, let us adopt a locally inertial coordinate system at A, in which the coordinate x^0 points along the geodesics. Thus at A we have $V^\alpha = \delta^\alpha_0$. The equation of the geodesic at A is

$$\left.\frac{d^2 x^\alpha}{d\lambda^2}\right|_A = 0, \tag{6.79}$$

since all Christoffel symbols vanish at A. The Christoffel symbols do not vanish at A', so the equation of the geodesic $\vec{V'}$ at A' is

$$\left.\frac{d^2 x^\alpha}{d\lambda^2}\right|_{A'} + \Gamma^\alpha{}_{00}(A') = 0, \tag{6.80}$$

where again at A' we have arranged the coordinates so that $V^\alpha = \delta^\alpha_0$. But, since A and A' are separated by $\vec{\xi}$, we have

$$\Gamma^\alpha{}_{00}(A') \cong \Gamma^\alpha{}_{00,\beta}\xi^\beta, \tag{6.81}$$

the right-hand side being evaluated at A. With Eq. (6.80) this gives

$$\left.\frac{d^2 x^\alpha}{d\lambda^2}\right|_{A'} = -\Gamma^\alpha{}_{00,\beta}\xi^\beta. \tag{6.82}$$

Now, the difference $x^\alpha(\lambda,\ \text{geodesic } \vec{V'}) - x^\alpha(\lambda,\ \text{geodesic } \vec{V})$ is just the component ξ^α of the vector $\vec{\xi}$. Therefore, at A, we have

$$\frac{d^2 \xi^\alpha}{d\lambda^2} = \left.\frac{d^2 x^\alpha}{d\lambda^2}\right|_{A'} - \left.\frac{d^2 x^\alpha}{d\lambda^2}\right|_A = -\Gamma^\alpha{}_{00,\beta}\xi^\beta. \tag{6.83}$$

This then gives how the components of $\vec{\xi}$ change. But since the coordinates are to some extent arbitrary, we want to have, not merely the second derivative of the component ξ^α, but the full second covariant derivative $\nabla_V \nabla_V \vec{\xi}$. We can use Eq. (6.48) to obtain

$$\nabla_V \nabla_V \xi^\alpha = \nabla_V(\nabla_V \xi^\alpha)$$
$$= \frac{d}{d\lambda}(\nabla_V \xi^\alpha) + \Gamma^\alpha{}_{\beta 0}(\nabla_V \xi^\beta). \tag{6.84}$$

Now, using $\Gamma^\alpha{}_{\beta\gamma} = 0$ at A, we have

$$\nabla_V \nabla_V \xi^\alpha = \frac{d}{d\lambda}\left(\frac{d}{d\lambda}\xi^\alpha + \Gamma^\alpha{}_{\beta 0}\xi^\beta\right) + 0$$
$$= \frac{d^2}{d\lambda^2}\xi^\alpha + \Gamma^\alpha{}_{\beta 0,0}\xi^\beta \tag{6.85}$$

at A. (We have also used $\xi^\beta{}_{,0} = 0$ at A, which is the condition that curves begin parallel.) So we get

$$\nabla_V \nabla_V \xi^\alpha = (\Gamma^\alpha{}_{\beta 0,0} - \Gamma^\alpha{}_{00,\beta})\xi^\beta$$
$$= R^\alpha{}_{00\beta}\xi^\beta = R^\alpha{}_{\mu\nu\beta} V^\mu V^\nu \xi^\beta, \qquad (6.86)$$

where the second equality follows from Eq. (6.63). The final expression is frame invariant, so we have, in *any* basis,

$$\blacklozenge \qquad \nabla_V \nabla_V \xi^\alpha = R^\alpha{}_{\mu\nu\beta} V^\mu V^\nu \xi^\beta. \qquad (6.87)$$

Geodesics in flat space maintain their separation; those in curved spaces don't. This is called the equation of geodesic deviation and shows mathematically that the tidal forces of a graviational field (which cause trajectories of neighboring particles to diverge) can be represented by curvature of a spacetime in which particles follow geodesics.

6.6 Bianchi identities; Ricci and Einstein tensors

Let us return to Eq. (6.63) for the Riemann tensor's components. If we differentiate it with respect to x^λ (just the partial derivative) and evaluate the result in locally inertial coordinates, we find

$$R_{\alpha\beta\mu\nu,\lambda} = \tfrac{1}{2}(g_{\alpha\nu,\beta\mu\lambda} - g_{\alpha\mu,\beta\nu\lambda} + g_{\beta\mu,\alpha\nu\lambda} - g_{\beta\nu,\alpha\mu\lambda}). \qquad (6.88)$$

From this equation, the symmetry $g_{\alpha\beta} = g_{\beta\alpha}$ and the fact that partial derivatives commute, one can show that

$$R_{\alpha\beta\mu\nu,\lambda} + R_{\alpha\beta\lambda\mu,\nu} + R_{\alpha\beta\nu\lambda,\mu} = 0. \qquad (6.89)$$

Since in our coordinates $\Gamma^\mu{}_{\alpha\beta} = 0$ at this point, this equation is *equivalent* to

$$\blacklozenge \qquad R_{\alpha\beta\mu\nu;\lambda} + R_{\alpha\beta\lambda\mu;\nu} + R_{\alpha\beta\nu\lambda;\mu} = 0. \qquad (6.90)$$

But this is a tensor equation, valid in *any* system. It is called the *Bianchi identities*, and will be very important for our work.

The Ricci tensor. Before pursuing the consequences of the Bianchi identities, we shall need to define the Ricci tensor $R_{\alpha\beta}$:

$$\blacklozenge \qquad R_{\alpha\beta} \equiv R^\mu{}_{\alpha\mu\beta} = R_{\beta\alpha}. \qquad (6.91)$$

It is the contraction of $R^\mu{}_{\alpha\nu\beta}$ on the first and third indices. Other contractions would in principle also be possible: on the first and second, the first and fourth, etc. But because $R_{\alpha\beta\mu\nu}$ is antisymmetric on α and β and on μ and ν, all these contractions either vanish identically or reduce to $\pm R_{\alpha\beta}$. Therefore the Ricci tensor is essentially the *only* contraction of the Riemann tensor. Note that Eq. (6.69) implies it is a *symmetric* tensor (Exer. 25, § 6.9).

Similarly, the *Ricci scalar* is defined as

◆
$$R \equiv g^{\mu\nu} R_{\mu\nu} = g^{\mu\nu} g^{\alpha\beta} R_{\alpha\mu\beta\nu}. \tag{6.92}$$

The Einstein tensor. Let us apply the Ricci contraction to the Bianchi identities, Eq. (6.90):

$$g^{\alpha\mu}[R_{\alpha\beta\mu\nu;\lambda} + R_{\alpha\beta\lambda\mu;\nu} + R_{\alpha\beta\nu\lambda;\mu}] = 0$$

or

$$R_{\beta\nu;\lambda} + (-R_{\beta\lambda;\nu}) + R^{\mu}{}_{\beta\nu\lambda;\mu} = 0. \tag{6.93}$$

To derive this result one needs two facts. First, by Eq. (6.31) we have

$$g_{\alpha\beta;\mu} = 0;$$

since $g^{\alpha\mu}$ is a function only of $g_{\alpha\beta}$ it follows that

$$g^{\alpha\beta}{}_{;\mu} = 0. \tag{6.94}$$

Therefore $g^{\alpha\mu}$ and $g_{\beta\nu}$ can be taken in and out of covariant derivatives at will: index-raising and -lowering commutes with covariant differentiation. The second fact is that

$$g^{\alpha\mu} R_{\alpha\beta\lambda\mu;\nu} = -g^{\alpha\mu} R_{\alpha\beta\mu\lambda;\nu} = -R_{\beta\lambda;\nu}, \tag{6.95}$$

accounting for the second term in Eq. (6.93). Eq. (6.93) is called the contracted Bianchi identities. A more useful equation is obtained by contracting again on the indices β and ν:

$$g^{\beta\nu}[R_{\beta\nu;\lambda} - R_{\beta\lambda;\nu} + R^{\mu}{}_{\beta\nu\lambda;\mu}] = 0$$

or

$$R_{;\lambda} - R^{\mu}{}_{\lambda;\mu} + (-R^{\mu}{}_{\lambda;\mu}) = 0. \tag{6.96}$$

Again the antisymmetry of **R** has been used to get the correct sign in the last term. Note that since R is a scalar, $R_{;\lambda} \equiv R_{,\lambda}$ in all coordinates. Now, Eq. (6.96) can be written in the form

$$(2R^{\mu}{}_{\lambda} - \delta^{\mu}{}_{\lambda} R)_{;\mu} = 0. \tag{6.97}$$

These are the twice-contracted Bianchi identities, often simply also called the Bianchi identities. If we define the symmetric tensor

◆
$$G^{\alpha\beta} \equiv R^{\alpha\beta} - \tfrac{1}{2} g^{\alpha\beta} R = G^{\beta\alpha}, \tag{6.98}$$

then we see that Eq. (6.97) is equivalent to

◆
$$G^{\alpha\beta}{}_{;\beta} = 0. \tag{6.99}$$

The tensor $G^{\alpha\beta}$ is constructed only from the Riemann tensor and the metric, and is automatically divergence free as an identity. It is called the Einstein tensor, since its importance for gravity was first understood by Einstein. (In fact we shall see that the Einstein field equations for GR are

$$G^{\alpha\beta} = 8\pi T^{\alpha\beta}$$

(where $T^{\alpha\beta}$ is the stress–energy tensor). The Bianchi identities then imply

$$T^{\alpha\beta}{}_{;\beta} \equiv 0,$$

which is the equation of local conservation of energy and momentum. But this is looking a bit far ahead.)

6.7 Curvature in perspective

The mathematical machinery for dealing with curvature is formidable. There are many important equations in this chapter, but few of them need to be memorized. It is far more important to understand their derivation and particularly their geometrical interpretation. This interpretation is something we will build up over the next few chapters, but the material already in hand should give the student some idea of what the mathematics means. Let us review the important features of curved spaces.

(1) We work on Riemannian manifolds, which are smooth spaces with a metric defined on them.

(2) The metric has signature +2, and there always exists a coordinate system in which, at a single point, one can have
$$g_{\alpha\beta} = \eta_{\alpha\beta},$$
$$g_{\alpha\beta,\gamma} = 0 \Rightarrow \Gamma^{\alpha}{}_{\beta\gamma} = 0.$$

(3) The element of proper volume is
$$|g|^{1/2}\, d^4x,$$
where g is the determinant of the matrix of components $g_{\alpha\beta}$.

(4) The covariant derivative is simply the ordinary derivative in locally inertial coordinates. Because of curvature ($\Gamma^{\alpha}{}_{\beta\gamma,\sigma} \neq 0$) these derivatives do not commute.

(5) The definition of parallel-transport is that the covariant derivative along the curve is zero. A geodesic parallel-transports its own tangent vector. Its affine parameter can be taken to be the proper distance itself.

(6) The Riemann tensor is the characterization of the curvature. Only if it vanishes identically is the manifold flat. It has 20 independent components (in four dimensions), and satisfies the Bianchi identities, which are differential equations. The Riemann tensor in a general coordinate system depends on $g_{\alpha\beta}$ and its first and second partial derivatives. The Ricci tensor, Ricci scalar, and Einstein tensor are contractions of the Riemann tensor. In particular, the Einstein tensor is symmetric and of second rank, so it has ten independent components. They satisfy the four differential identities, Eq. (6.99).

6.8   Bibliography

The theory of differentiable manifolds is introduced in a large number of books. The following are suitable for exploring the subject further with a view toward its physical applications, particularly outside of relativity: Abraham & Marsden (1978), Bishop & Goldberg (1968), Choquet-Bruhat *et al.* (1977), Dodson & Poston (1977), Hermann (1968), Lovelock & Rund (1975), Misner (1964), Schmidt (1973), and Schutz (1980*b*). Standard mathematical reference works include Kobayashi & Nomizu (1963, 1969), Schouten (1954), and Spivak (1979).

6.9   Exercises

1   Decide if the following sets are manifolds and say why. If there are exceptional points at which the sets are not manifolds, give them:

(a) phase space of Hamiltonian mechanics, the space of the canonical coordinates and momenta p_i and q^i;

(b) the interior of a circle of unit radius in two-dimensional Euclidean space;

(c) the set of permutations of n objects;

(d) the subset of Euclidean space of two dimensions (coordinates x and y) which is a solution to $xy\,(x^2 + y^2 - 1) = 0$.

2   Of the manifolds in Exer. 1, on which is it customary to use a metric, and what is that metric? On which would a metric not normally be defined, and why?

3   It is well known that for any symmetric matrix A (with real entries), there exists a matrix H for which the matrix $H^{\mathrm{T}} A H$ is a diagonal matrix whose entries are the eigenvalues of A.

(a) Show that there is a matrix R such that $R^{\mathrm{T}} H^{\mathrm{T}} A H R$ is the same matrix as $H^{\mathrm{T}} A H$ except with the eigenvalues rearranged in ascending order along the main diagonal from top to bottom.

(b) Show that there exists a third matrix N such that $N^{\mathrm{T}} R^{\mathrm{T}} H^{\mathrm{T}} A H R N$ is a diagonal matrix whose entries on the diagonal are -1, 0, or $+1$.

(c) Show that if A has an inverse, none of the diagonal elements in (b) is zero.

(d) Show from (a)–(c) that there exists a transformation matrix Λ which produces Eq. (6.2).

4   Prove the following results used in the proof of the local flatness theorem in § 6.2:

(a) The number of independent values of $\partial^2 x^\alpha / \partial x^{\gamma'} \partial x^{\mu'}|_0$ is 40.

(b) The corresponding number for $\partial^3 x^\alpha / \partial x^{\lambda'} \partial x^{\gamma'} \partial x^{\mu'}|_0$ is 80.

(c) The corresponding number for $g_{\alpha\beta,\gamma'\mu'}|_0$ is 100.

5(a) Prove that $\Gamma^\mu_{\ \alpha\beta} = \Gamma^\mu_{\ \beta\alpha}$ in any coordinate system in a curved Riemannian space.

(b) Use this to prove that Eq. (6.32) can be derived in the same manner as in flat space.

6 Prove that the first term in Eq. (6.37) vanishes.

7(a) Give the definition of the determinant of a matrix A in terms of cofactors of elements.

(b) Differentiate the determinant of an arbitrary 2×2 matrix and show that it satisfies Eq. (6.39).

(c) Generalize Eq. (6.39) (by induction or otherwise) to arbitrary $n \times n$ matrices.

8 Fill in the missing algebra leading to Eqs. (6.40) and (6.42).

9 Show that Eq. (6.42) leads to Eq. (5.55). Derive the divergence formula for the metric in Eq. (6.19).

10 A 'straight line' on a sphere is a great circle, and it is well known that the sum of the interior angles of any triangle on a sphere whose sides are arcs of great circles exceeds 180°. Show that the amount by which a vector is rotated by parallel transport around such a triangle (as in Fig. 6.3) equals the excess of the sum of the angles over 180°.

11 In this exercise we will determine the condition that a vector field $\vec{V}$ can be considered to be globally parallel on a manifold. More precisely, what guarantees that we can find a vector field $\vec{V}$ satisfying the equation
$$(\nabla \vec{V})^\alpha_{\ \beta} = V^\alpha_{\ ;\beta} = V^\alpha_{\ ,\beta} + \Gamma^\alpha_{\ \mu\beta} V^\mu = 0?$$

(a) A necessary condition, called the *integrability condition* for this equation, follows from the commuting of partial derivatives. Show that $V^\alpha_{\ ,\beta\nu} = V^\alpha_{\ ,\nu\beta}$ implies
$$(\Gamma^\alpha_{\ \mu\beta,\nu} - \Gamma^\alpha_{\ \mu\nu,\beta}) V^\mu = (\Gamma^\alpha_{\ \mu\beta}\Gamma^\mu_{\ \sigma\nu} - \Gamma^\alpha_{\ \mu\nu}\Gamma^\mu_{\ \sigma\beta}) V^\sigma.$$

(b) By relabeling indices, work this into the form
$$(\Gamma^\alpha_{\ \mu\beta,\nu} - \Gamma^\alpha_{\ \mu\nu,\beta} + \Gamma^\alpha_{\ \sigma\nu}\Gamma^\sigma_{\ \mu\beta} - \Gamma^\alpha_{\ \sigma\beta}\Gamma^\sigma_{\ \mu\nu}) V^\mu = 0.$$
This turns out to be *sufficient*, as well.

12 Prove that Eq. (6.52) defines a new affine parameter.

13(a) Show that if $\vec{A}$ and $\vec{B}$ are parallel-transported along a curve, then $g(\vec{A}, \vec{B}) = \vec{A} \cdot \vec{B}$ is constant on the curve.

(b) Conclude from this that if a *geodesic* is spacelike (or timelike or null) somewhere, it is spacelike (or timelike or null) everywhere.

14   The proper distance along a curve whose tangent is $\vec{V}$ is given by Eq. (6.8). Show that if the curve is a geodesic, then proper length is an affine parameter. (Use the result of Exer. 13.)

15   Use Exer. 13 and 14 to prove that the proper length of geodesic between two points is unchanged to first order by small changes in the curve that do not change its endpoints.

16(a)   Derive Eqs. (6.59) and (6.60) from Eq. (6.58).
 (b)   Fill in the algebra needed to justify Eq. (6.61).

17(a)   Prove that Eq. (6.5) implies $g^{\alpha\beta}{}_{,\mu}(P) = 0$.
 (b)   Use this to establish Eq. (6.64).
 (c)   Fill in the steps needed to establish Eq. (6.68).

18(a)   Derive Eqs. (6.69) and (6.70) from Eq. (6.68).
 (b)   Show that Eq. (6.69) reduces the number of independent components of $R_{\alpha\beta\mu\nu}$ from $4\times4\times4\times4 = 256$ to $6\times7/2 = 21$. (Hint: treat *pairs* of indices. Calculate how many independent choices of pairs there are for the first and the second pairs on $R_{\alpha\beta\mu\nu}$.)
 (c)   Show that Eq. (6.70) imposes only one further relation independent of Eq. (6.69) on the components, reducing the total of independent ones to 20.

19   Prove that $R^{\alpha}{}_{\beta\mu\nu} = 0$ for polar coordinates in the Euclidean plane. Use Eq. (5.44) or equivalent results.

20   Fill in the algebra necessary to establish Eq. (6.73).

21   Consider the sentences following Eq. (6.78). Why does the argument in parentheses *not* apply to the signs in
$$V^{\alpha}{}_{;\beta} = V^{\alpha}{}_{,\beta} + \Gamma^{\alpha}{}_{\mu\beta}V^{\mu} \quad \text{and} \quad V_{\alpha;\beta} = V_{\alpha,\beta} - \Gamma^{\mu}{}_{\alpha\beta}V_{\mu}?$$

22   Fill in the algebra necessary to establish Eqs. (6.84), (6.85), and (6.86).

23   Prove Eq. (6.88). (Be careful: one cannot simply differentiate Eq. (6.67) since it is valid only at P, not in the neighborhood of P.)

24   Establish Eq. (6.89) from Eq. (6.88).

25(a)   Prove that the Ricci tensor is the only independent contraction of $R^{\alpha}{}_{\beta\mu\nu}$: all others are multiples of it.
 (b)   Show that the Ricci tensor is symmetric.

26   Use Exer. 17(a) to prove Eq. (6.94).

27   Fill in the algebra necessary to establish Eqs. (6.95), (6.97) and (6.99).

28(a) Derive Eq. (6.19) by using the usual coordinate transformation from Cartesian to spherical polars.

(b) Deduce from Eq. (6.19) that the metric of the surface of a sphere of radius r has components $(g_{\theta\theta} = r^2, g_{\phi\phi} = r^2 \sin^2 \theta, g_{\theta\phi} = 0)$ in the usual spherical coordinates.

(c) Find the components $g^{\alpha\beta}$ for the sphere.

29 In polar coordinates, calculate the Riemann curvature tensor of the sphere of unit radius, whose metric is given in Exer. 28. (Note that in two dimensions there is only *one* independent component, by the same arguments as in Exer. 18(b). So calculate $R_{\theta\phi\theta\phi}$ and obtain all other components in terms of it.)

30 Calculate the Riemann curvature tensor of the cylinder. (Since the cylinder is flat, this should vanish. Use whatever coordinates you like, and make sure you write down the metric properly!)

31 Show that covariant differentiation obeys the usual product rule, e.g. $(V^{\alpha\rho} W_{\beta\gamma})_{;\mu} = V^{\alpha\beta}_{\ \ ;\mu} W_{\beta\gamma} + V^{\alpha\beta} W_{\beta\gamma;\mu}$. (Hint: use a locally inertial frame.)

32 A four-dimensional manifold has coordinates (u, v, w, p) in which the metric has components $g_{uv} = g_{ww} = g_{pp} = 1$, all other independent components vanishing.

(a) Show that the manifold is flat and the signature is +2.

(b) The result in (a) implies the manifold must be Minkowski spacetime. Find a coordinate transformation to the usual coordinates (t, x, y, z). (You may find it a useful hint to calculate $\vec{e}_v \cdot \vec{e}_v$ and $\vec{e}_u \cdot \vec{e}_u$.)

33 A 'three-sphere' is the three-dimensional surface in four-dimensional Euclidean space (coordinates x, y, z, w), given by the equation $x^2 + y^2 + z^2 + w^2 = r^2$, where r is the radius of the sphere.

(a) Define new coordinates (r, θ, ϕ, χ) by the equations $w = r \cos \chi$, $z = r \sin \chi \cos \theta$, $x = r \sin \chi \sin \theta \cos \phi$, $y = r \sin \chi \sin \theta \sin \phi$. Show that (θ, ϕ, χ) are coordinates for the sphere. These generalize the familiar polar coordinates.

(b) Show that the metric of the three-sphere of radius r has components in these coordinates $g_{\chi\chi} = r^2$, $g_{\theta\theta} = r^2 \sin^2 \chi$, $g_{\phi\phi} = r^2 \sin^2 \chi \sin^2 \theta$, all other components vanishing. (Use the same method as in Exer. 28.)

34 Establish the following identities for a general metric tensor in a general coordinate system. You may find Eqs. (6.39) and (6.40) useful.

(a) $\Gamma^{\mu}_{\ \mu\nu} = \frac{1}{2}(\ln |g|)_{,\nu}$;

(b) $g^{\mu\nu} \Gamma^{\alpha}_{\ \mu\nu} = -(g^{\alpha\beta} \sqrt{-g})_{,\beta} / \sqrt{-g}$;

(c) for an antisymmetric tensor $F^{\mu\nu}$, $F^{\mu\nu}_{\ \ ;\nu} = (\sqrt{-g} F^{\mu\nu})_{,\nu} / \sqrt{-g}$;

(d) $g^{\alpha\beta} g_{\beta\mu,\nu} = -g^{\alpha\beta}_{\ \ ,\nu} g_{\beta\mu}$ (hint: what is $g^{\alpha\beta} g_{\beta\mu}$?);

(e) $g^{\mu\nu}_{\ \ ,\alpha} = -\Gamma^{\mu}_{\ \beta\alpha} g^{\beta\nu} - \Gamma^{\nu}_{\ \beta\alpha} g^{\mu\beta}$ (hint: use Eq. (6.31)).

35   Compute 20 independent components of $R_{\alpha\beta\mu\nu}$ for a manifold wth line element $ds^2 = -e^{2\Phi} dt^2 + e^{2\Lambda} dr^2 + r^2(d\theta^2 + \sin^2\theta \, d\phi^2)$, where Φ and Λ are arbitrary functions of the coordinate r alone. (First, identify the coordinates and the components $g_{\alpha\beta}$; then compute $g^{\alpha\beta}$ and the Christoffel symbols. Then decide on the indices of the 20 components of $R_{\alpha\beta\mu\nu}$ you wish to calculate, and compute them. Remember that one can deduce the remaining 236 components from those 20.)

36   A four-dimensional manifold has coordinates (t, x, y, z) and line element $ds^2 = -(1+2\phi) dt^2 + (1-2\phi)(dx^2 + dy^2 + dz^2)$,

where $|\phi(t, x, y, z)| \ll 1$ everywhere. At any point P with coordinates (t_0, x_0, y_0, z_0), find a coordinate transformation to a locally inertial coordinate system, to first order in ϕ. At what rate does such a frame accelerate with respect to the original coordinates, again to first order in ϕ?

37(a) 'Proper volume' of a two-dimensional manifold is usually called 'proper area'. Using the metric in Exer. 28, integrate Eq. (6.18) to find the proper area of a sphere of radius r.

(b)   Do the analogous calculation for the three-sphere of Exer. 33.

38   Integrate Eq. (6.8) to find the length of a circle of constant coordinate θ on a sphere of radius r.

39(a) For any two vector fields $\vec{U}$ and $\vec{V}$, their *Lie bracket* is defined to be the vector field $[\vec{U}, \vec{V}]$ with components

$$[\vec{U}, \vec{V}]^\alpha = U^\beta \nabla_\beta V^\alpha - V^\beta \nabla_\beta U^\alpha. \tag{6.100}$$

Show that

$$[\vec{U}, \vec{V}] = -[\vec{V}, \vec{U}],$$
$$[\vec{U}, \vec{V}]^\alpha = U^\beta \, \partial V^\alpha / \partial x^\beta - V^\beta \, \partial U^\alpha / \partial x^\beta.$$

This is one tensor field in which partial derivatives need not be accompanied by Christoffel symbols!

(b)   Show that $[\vec{U}, \vec{V}]$ is a derivative operator on $\vec{V}$ along $\vec{U}$, i.e. show that for any scalar f,

$$[\vec{U}, f\vec{V}] = f[\vec{U}, \vec{V}] + \vec{V}(\vec{U} \cdot \nabla f). \tag{6.101}$$

This is sometimes called the *Lie derivative* with respect to $\vec{U}$ and denoted by

$$[\vec{U}, \vec{V}] \equiv \pounds_{\vec{U}} \vec{V}, \qquad \vec{U} \cdot \nabla f \equiv \pounds_{\vec{U}} f. \tag{6.102}$$

Then Eq. (6.101) would be written in the more conventional form of the Leibnitz rule for the derivative operator $\pounds_{\vec{U}}$:

$$\pounds_{\vec{U}}(f\vec{V}) = f\pounds_{\vec{U}}\vec{V} + \vec{V}\pounds_{\vec{U}} f. \tag{6.103}$$

The result of (a) shows that this derivative operator may be defined without a connection or metric, and is therefore very fundamental. See Schutz (1980b) for an introduction.

(c) Calculate the components of the Lie derivative of a one-form field $\tilde{\omega}$ from the knowledge that, for any vector field $\vec{V}$, $\tilde{\omega}(\vec{V})$ is a scalar like f above, and from the definition that $\pounds_{\vec{U}}\tilde{\omega}$ is a one-form field:

$$\pounds_{\vec{U}}[\tilde{\omega}(\vec{V})] = (\pounds_{\vec{U}}\tilde{\omega})(\vec{V}) + \tilde{\omega}(\pounds_{\vec{U}}\vec{V}).$$

This is the analogue of Eq. (6.103).

7

Physics in a curved spacetime

7.1 The transition from differential geometry to gravity

The essence of a physical theory expressed in mathematical form is the identification of the mathematical concepts with certain physically measurable quantities. This must be our first concern when we look at the relation of the concepts of geometry we have developed to the effects of gravity in the physical world. We have already discussed this to some extent. In particular, we have assumed that spacetime is a differentiable manifold, and we have shown that there do not exist global inertial frames in the presence of nonuniform gravitational fields. Behind these statements are the two identifications:

◆   (i)   Spacetime (the set of all events) is a four-dimensional manifold with a metric.

◆   (ii)   The metric is measurable by rods and clocks. The distance along a rod between two nearby points is $|\mathrm{d}\vec{x} \cdot \mathrm{d}\vec{x}|^{1/2}$ and the time measured by a clock that experiences two events closely separated in time is $|-\mathrm{d}\vec{x} \cdot \mathrm{d}\vec{x}|^{1/2}$.

So there do not generally exist coordinates in which $\mathrm{d}\vec{x} \cdot \mathrm{d}\vec{x} = -(\mathrm{d}x^0)^2 + (\mathrm{d}x^1)^2 + (\mathrm{d}x^2)^2 + (\mathrm{d}x^3)^2$ everywhere. On the other hand, we have also argued that such frames *do* exist locally. This clearly suggests a curved manifold, in which coordinates can be found which make the dot product at a particular point look like it does in a Minkowski spacetime.

Therefore we make a further requirement:

◆ (III) The metric of spacetime can be put in the Lorentz form $\eta_{\alpha\beta}$ at any particular event by an appropriate choice of coordinates.

Having chosen this way of representing spacetime, we must do two more things to get a complete theory. First, we must specify how physical objects (particles, electric fields, fluids) behave in a curved spacetime and, second, we need to say how the curvature is generated or determined by the objects in the spacetime.

Let us consider Newtonian gravity as an example of a physical theory. For Newton, spacetime consisted of three-dimensional Euclidean space, repeated endlessly in time. (Mathematically, this is called $R^3 \times R$.) There was no metric on spacetime as a whole manifold, but the Euclidean space had its usual metric and time was measured by a universal clock. Thus, 'distance' had meaning *only* at a fixed point in R^3 (when it meant 'time elapsed') *or* at a fixed time (when it meant 'distance between objects'). There was no invariant measure of the length of a general curve that changed position and time as it went along. Newton gave a law for the behavior of objects in spacetime: $F = ma$, where $F = -m\nabla\phi$ for a given gravitational field ϕ. And he also gave a law determining how ϕ is generated: $\nabla^2\phi = 4\pi G\rho$. These two laws are the ones we must now find analogues for in our relativistic point of view on spacetime. The second one will be dealt with in the next chapter. In this chapter, we ask only how a given metric affects bodies in spacetime.

We have already discussed this for the simple case of particle motion. Since we know that the 'acceleration' of a particle in a gravitational field is independent of its mass, we can go to a freely falling frame in which nearby particles have no acceleration. This is what we have identified as a locally inertial frame. Since freely falling particles have no acceleration in that frame, they follow straight lines, at least locally. But straight lines in a local inertial frame are, of course, the *definition* of geodesics in the full curved manifold. So we have our first postulate on the way particles are affected by the metric:

◆ (IV) Freely falling particles move on timelike geodesics of the spacetime.

By 'freely falling' we mean particles unaffected by other forces, such as electric fields, etc. All other known forces in physics are distinguished from gravity by the fact that there *are* particles unaffected by them. Postulate (IV) is a very strong prediction, capable of experimental test. But it refers only to particles. How are, say, electromagnetic fields and fluids affected by a nonflat metric? We need a generalization of (IV) which is called the *strong equivalence principle*:

♦ (IV') Any physical law which can be expressed in tensor notation in SR has exactly the same form in a locally inertial frame of a curved spacetime.

This principle is often called the 'comma-goes-to-semicolon rule', because if a law contains derivatives in its special-relativistic form ('commas'), then it has these same derivatives in the local inertial frame. To convert the law into an expression valid in *any* coordinate frame, one simply makes the derivatives covariant ('semicolons'). It is an extremely simple way to generalize the physical laws. In particular, it forbids 'curvature coupling': it is conceivable that the correct form of, say, Maxwell's equations in a curved spacetime would involve the Riemann tensor somehow, which would vanish in SR. Postulate (IV') would not put any Riemann-tensor terms into the equations.

As an example of (IV'), we discuss fluid dynamics, which will be our main interest in this course. The law of conservation of particles in SR was expressed as

$$(nU^{\alpha})_{,\alpha} = 0, \tag{7.1}$$

where n was the density of particles in the momentarily comoving reference frame (MCRF), and where U^{α} was the four-velocity of a fluid element. In a curved spacetime, one can find a locally inertial frame comoving momentarily with the fluid element and define n in exactly the same way. Similarly one can define $\vec{U}$ to be the time basis vector of that frame, just as in SR. Then, in the locally inertial frame, conservation of particles is *exactly* Eq. (7.1). But this, because the Christoffel symbols are zero, is equivalent to

$$(nU^{\alpha})_{;\alpha} = 0. \tag{7.2}$$

This is valid in *all* frames and so serves as the generalization of the law to a curved spacetime, according to the strong equivalence principle. Notice that, in principle, the correct equation in a curved spacetime *might* be

$$(nU^{\alpha})_{;\alpha} = R, \tag{7.3}$$

where R is the Ricci scalar defined in Eq. (6.92). This would also reduce to Eq. (7.1) in SR, since in a flat spacetime the Riemann tensor vanishes. The strong equivalence principle asserts that we should generalize Eq. (7.1) in the simplest possible manner, that is Eq. (7.2). It is of course a matter for experiment, or astronomical observation, to decide whether

Eq. (7.2) or Eq. (7.3) is correct. In this book we shall simply make the assumption that is nearly universally made, that the strong equivalence principle is correct. There is no observational evidence to the contrary.

Similarly, the law of conservation of entropy in SR was

$$U^\alpha S_{,\alpha} = 0. \tag{7.4}$$

Since there are no Christoffel symbols in the covariant derivative of a scalar like S, this law is *unchanged* in a curved spacetime. Finally, conservation of four-momentum was

$$T^{\mu\nu}{}_{,\nu} = 0. \tag{7.5}$$

The generalization is

$$T^{\mu\nu}{}_{;\nu} = 0, \tag{7.6}$$

with the definition

$$T^{\mu\nu} = (\rho + p) U^\mu U^\nu + p g^{\mu\nu}, \tag{7.7}$$

exactly as before. (Notice that $g^{\mu\nu}$ replaces $\eta^{\mu\nu}$ in Eq. (7.7), since we need the metric in a general coordinate system.)

7.2 Physics in slightly curved spacetimes

To see the implications of (IV′) for the motion of a particle or fluid, one must know the metric on the manifold. Since we have not yet studied the way a metric is generated, we will at this stage have to be content with assuming a form for the metric which we shall derive later. It turns out that for *weak* gravitational fields (where, in Newtonian language, the gravitational potential energy of a particle is much less than its rest-mass energy) the ordinary Newtonian potential ϕ completely determines the metric, which has the form

$$\blacklozenge \qquad ds^2 \stackrel{.}{=} -(1 + 2\phi)\, dt^2 + (1 - 2\phi)\, (dx^2 + dy^2 + dz^2). \tag{7.8}$$

(The sign of ϕ is chosen negative, so that, far from a source of mass M, $\phi = -GM/r$.) Now, the condition above that the field be weak means the $|m\phi| \ll m$, so that $|\phi| \ll 1$. The metric, Eq. (7.8), is really only correct to first order in ϕ, so we shall work to this order from now on.

Let us compute the motion of a freely falling particle. Let us denote its four-momentum by $\vec{p}$. For all except massless particles, this is $m\vec{U}$, where $\vec{U} = d\vec{x}/d\tau$. Now, by (IV), the particle's path is a geodesic, and we know that proper time is an affine parameter on such a path. Therefore $\vec{U}$ must satisfy the geodesic equation,

$$\nabla_{\vec{U}} \vec{U} = 0. \tag{7.9}$$

For convenience later, however, we note that any constant times proper time is an affine parameter, in particular τ/m. Then $d\vec{x}/d(\tau/m)$ is also a vector satisfying the geodesic equation. This vector is just $md\vec{x}/d\tau = \vec{p}$. So we can also write the equation of motion of the particle as

$$\nabla_{\vec{p}}\vec{p} = 0. \tag{7.10}$$

This equation can also be used for photons, which have a well-defined $\vec{p}$ but no $\vec{U}$ since $m = 0$.

If the particle has a nonrelativistic velocity in the coordinates of Eq. (7.8), we can find an approximate form for Eq. (7.10). First let us consider the 0 component of the equation,

$$p^{\alpha}p^{0}{}_{,\alpha} + \Gamma^{0}{}_{\alpha\beta}p^{\alpha}p^{\beta} = 0. \tag{7.11}$$

Because the particle has a nonrelativistic velocity we have $p^{0} \gg p^{1}$. Moreover, $p^{\alpha}\partial_{\alpha} = mU^{\alpha}\partial_{\alpha} = m\,d/d\tau$, so Eq. (7.11) is approximately

$$m\frac{d}{d\tau}p^{0} + \Gamma^{0}{}_{00}(p^{0})^{2} = 0. \tag{7.12}$$

We need to compute $\Gamma^{0}{}_{00}$:

$$\Gamma^{0}{}_{00} = \tfrac{1}{2}g^{0\alpha}(g_{\alpha0,0} + g_{\alpha0,0} - g_{00,\alpha}). \tag{7.13}$$

Now because $[g_{\alpha\beta}]$ is diagonal, $[g^{\alpha\beta}]$ is also diagonal and its elements are the reciprocals of those of $[g_{\alpha\beta}]$. Therefore $g^{0\alpha}$ is nonzero only when $\alpha = 0$, so Eq. (7.13) becomes

$$\Gamma^{0}{}_{00} = \tfrac{1}{2}g^{00}g_{00,0} = \frac{1}{2}\frac{1}{-(1+2\phi)}(-2\phi)_{,0}$$
$$= \phi_{,0} + 0(\phi^{2}). \tag{7.14}$$

To lowest order in the velocity of the particle and in ϕ, we can replace $(p^{0})^{2}$ in the second term of Eq. (7.12) by m^{2}, obtaining

$$\frac{d}{d\tau}p^{0} = -m\frac{\partial\phi}{\partial\tau}. \tag{7.15}$$

Since p^{0} is the energy of the particle in this frame, this means the energy is conserved unless the gravitational field depends on time. This result is true also in Newtonian theory. Here, however, we must note that p^{0} is the energy of the particle with respect to this frame only.

The spatial components of the geodesic equation give the counterpart of the Newtonian $F = ma$. They are

$$p^{\alpha}p^{i}{}_{,\alpha} + \Gamma^{i}{}_{\alpha\beta}p^{\alpha}p^{\beta} = 0, \tag{7.16}$$

or to, lowest order in the velocity,

$$m\frac{dp^{i}}{d\tau} + \Gamma^{i}{}_{00}(p^{0})^{2} = 0. \tag{7.17}$$

Again we have neglected p^i compared to p^0 in the Γ summation. Consistent with this we can again put $(p^0)^2 = m^2$ to a first approximation and get

$$\frac{dp^i}{d\tau} = -m\Gamma^i{}_{00}. \tag{7.18}$$

We calculate the Christoffel symbol:

$$\Gamma^i{}_{00} = \tfrac{1}{2}g^{i\alpha}(g_{\alpha 0,0} + g_{\alpha 0,0} - g_{00,\alpha}). \tag{7.19}$$

Now, since $[g^{\alpha\beta}]$ is diagonal, we can write

$$g^{i\alpha} = (1 - 2\phi)^{-1}\delta^{i\alpha} \tag{7.20}$$

and get

$$\Gamma^i{}_{00} = \tfrac{1}{2}(1 - 2\phi)^{-1}\delta^{ij}(2g_{j0,0} - g_{00,j}), \tag{7.21}$$

where we have changed α to j because δ^{i0} is zero. Now we notice that $g_{j0} \equiv 0$ and so we get

$$\Gamma^i{}_{00} = -\tfrac{1}{2}g_{00,j}\delta^{ij} + 0(\phi^2) \tag{7.22}$$

$$= -\tfrac{1}{2}(-2\phi)_{,j}\delta^{ij}. \tag{7.23}$$

With this the equation of motion, Eq. (7.17), becomes

$$dp^i/d\tau = -m\phi_{,j}\delta^{ij}. \tag{7.24}$$

This is the usual equation in Newtonian theory, since the force of a gravitational field is $-m\nabla\phi$.

Both the energy-conservation equation and the equation of motion were derived as approximations based on two things: the metric was nearly the Minkowski metric ($|\phi| \ll 1$), and the particle's velocity was nonrelativistic ($p^0 \gg p^i$). These two limits are just the circumstances under which Newtonian gravity is verified, so it is reassuring – indeed, essential – that we have recovered the Newtonian equations. However, there is no magic here. It almost *had* to work, given that we know that particles fall on straight lines in freely falling frames.

One can do the same sort of calculation to verify that the Newtonian equations hold for other systems in the appropriate limit. For instance, the student has an opportunity to do this for the perfect fluid in Exer. 5, § 7.6. Note that the condition that the fluid be nonrelativistic means not only that its velocity is small but also that the random velocities of its particles be nonrelativistic, which means $p \ll \rho$.

This correspondence of our relativistic point of view with the older, Newtonian theory in the appropriate limit is very important. *Any* new theory must make the same predictions as the old theory in the regime in which the old theory was known to be correct. The equivalence principle plus the form of the metric, Eq. (7.8), does this.

7.3   Curved intuition

Although in the appropriate limit our curved-spacetime picture of gravity predicts the same things as Newtonian theory predicts, it is very different from Newton's theory in concept. One must therefore work gradually toward an understanding of its new point of view.

The first difference is the absence of a preferred frame. In Newtonian physics *and* in SR, inertial frames are preferred. Since 'velocity' cannot be measured locally but 'acceleration' can be, both theories single out special classes of coordinate systems for spacetime in which particles which have no physical acceleration (i.e. $d\vec{U}/d\tau = 0$) also have no coordinate acceleration ($d^2x^i/dt^2 = 0$). In our new picture, there is no coordinate system which is inertial everywhere, i.e. in which $d^2x^i/dt^2 = 0$ for every particle for which $d\vec{U}/d\tau = 0$. Therefore we have to allow all coordinates on an equal footing. By using the Christoffel symbols we correct coordinate-dependent quantities like d^2x^i/dt^2 to obtain coordinate-independent quantities like $d\vec{U}/d\tau$. Therefore, one need not, and in fact one *should not*, develop coordinate-dependent ways of thinking.

A second difference concerns energy and momentum. In Newtonian physics, SR, and our geometrical gravity theory, each particle has a definite energy and momentum, whose values depend on the frame they are evaluated in. In the latter two theories, energy and momentum are components of a single four-vector $\vec{p}$. In SR, the total four-momentum of a system is the sum of the four-momenta of all the particles, $\sum_i \vec{p}_{(i)}$. But in a curved spacetime, one *cannot* add up vectors that are defined at different points, because one does not know how: two vectors can only be said to be parallel if they are compared at the same point, and the value of a vector at a point to which it has been parallel-transported depends on the curve along which it was moved. So there is *no* invariant way of adding up all the $\vec{p}$s, and so if a system has definable four-momentum, it is not just the simple thing it was in SR.

It turns out that for any system whose spatial extent is bounded (i.e. an isolated system), a total energy and momentum *can* be defined, in a manner which we will discuss later. One way to see that the *total* mass energy of a system should not be the sum of the energies of the particles is that this neglects what in Newtonian language is called its gravitational self-energy, a negative quantity which is the work one gains by assembling the system from isolated particles at infinity. This energy, if it is to be included, cannot be assigned to any particular particle but resides in the geometry itself. The notion of gravitational potential energy, however, is itself not well defined in the new picture: it must in some sense represent the difference between the sum of the energies of the particles and the total mass of the system, but since the sum of the energies of the particles is not well defined, neither is the gravitational potential energy. Only the *total* energy–momentum of a system is, in general, definable, in addition to the four-momentum of individual particles.

7.4 Conserved quantities

The previous discussion of energy may make one wonder what one can say about conserved quantities associated with a particle or system. For a particle, one must realize that gravity, in the old viewpoint, is a 'force', so that a particle's kinetic energy and momentum need not be conserved under its action. In our new viewpoint, then, one cannot expect to find a coordinate system in which the components of $\vec{p}$ are constants along the trajectory of a particle. There is one notable exception to this, and it is important enough to look at in detail.

The geodesic equation can be written for the 'lowered' components of $\vec{p}$ as follows:

$$p^{\alpha}p_{\beta:\alpha} = 0, \tag{7.25}$$

or

$$p^{\alpha}p_{\beta,\alpha} - \Gamma^{\gamma}{}_{\beta\alpha}p^{\alpha}p_{\gamma} = 0,$$

or

$$m\frac{\mathrm{d}p_{\beta}}{\mathrm{d}\tau} = \Gamma^{\gamma}{}_{\beta\alpha}p^{\alpha}p_{\gamma}. \tag{7.26}$$

Now, the right-hand side turns out to be simple:

$$\begin{aligned}\Gamma^{\gamma}{}_{\alpha\beta}p^{\alpha}p_{\gamma} &= \tfrac{1}{2}g^{\gamma\nu}(g_{\nu\beta,\alpha} + g_{\nu\alpha,\beta} - g_{\alpha\beta,\nu})p^{\alpha}p_{\gamma}\\ &= \tfrac{1}{2}(g_{\nu\beta,\alpha} + g_{\nu\alpha,\beta} - g_{\alpha\beta,\nu})g^{\gamma\nu}p_{\gamma}p^{\alpha}\\ &= \tfrac{1}{2}(g_{\nu\beta,\alpha} + g_{\nu\alpha,\beta} - g_{\alpha\beta,\nu})p^{\nu}p^{\alpha}. \end{aligned} \tag{7.27}$$

The product $p^{\nu}p^{\alpha}$ is symmetric on ν and α, while the first and third terms inside parentheses are, together, antisymmetric on ν and α. Therefore they cancel, leaving only the middle term:

$$\Gamma^{\gamma}{}_{\beta\alpha}p^{\alpha}p_{\gamma} = \tfrac{1}{2}g_{\nu\alpha,\beta}p^{\nu}p^{\alpha}. \tag{7.28}$$

The geodesic equation can thus, in complete generality, be written

$$\blacklozenge \qquad m\frac{\mathrm{d}p_{\beta}}{\mathrm{d}\tau} = \tfrac{1}{2}g_{\nu\alpha,\beta}p^{\nu}p^{\alpha}. \tag{7.29}$$

We therefore have the following important result: *if all the components $g_{\alpha\nu}$ are independent of x^{β} for some fixed index β, then p_{β} is a constant along any particle's trajectory.*

For instance, suppose we have a stationary gravitational field. Then a coordinate system can be found in which the metric components are time independent, and in that system p_0 is conserved. Therefore p_0 (or, really, $-p_0$) is usually called the 'energy' of the particle, *without* qualifying it with 'in this frame'. Notice that coordinates can also be found in which the same metric has time-dependent components: any time-dependent coordinate transformation from the 'nice' system will do this. In fact, most freely falling locally inertial systems are like this, since a freely falling particle sees a gravitational field that varies with its position, and therefore with time in its coordinate system. The frame in which the metric components are stationary is special, and is the usual 'laboratory frame' on Earth. Therefore p_0 in this frame is related to the usual energy defined in the lab, and includes the particle's gravitational potential energy, as we shall now show. Consider the equation

$$\vec{p} \cdot \vec{p} = -m^2 = g_{\alpha\beta} p^\alpha p^\beta$$
$$= -(1+2\phi)(p^0)^2 + (1-2\phi)[(p^x)^2 + (p^y)^2 + (p^z)^2], \quad (7.30)$$

where we have used the metric, Eq. (7.8). This can be solved to give

$$(p^0)^2 = [m^2 + (1-2\phi)(p^2)](1+2\phi)^{-1}, \quad (7.31)$$

where, for shorthand, we denote by p^2 the sum $(p^x)^2 + (p^y)^2 + (p^z)^2$. Keeping within the approximation $\phi \ll 1$, $|p| \ll m$, we can simplify this to

$$(p^0)^2 \approx m^2(1 - 2\phi + p^2/m^2)$$

or

$$p^0 \approx m(1 - \phi + p^2/2m^2). \quad (7.32)$$

Now we lower the index and get

$$p_0 = g_{0\alpha}p^\alpha = g_{00}p^0 = -(1+2\phi)p^0, \quad (7.33)$$

◆ $\quad -p_0 \approx m(1 + \phi + p^2/2m^2) = m + m\phi + p^2/2m. \quad (7.34)$

The first term is the rest mass of the particle. The second and third are the Newtonian pieces of its energy: gravitational potential energy and kinetic energy. This means that the constancy of p_0 along a particle's trajectory generalizes the Newtonian concept of a conserved energy.

Notice that a *general* gravitational field will not be stationary in *any* frame,[1] so no conserved energy can be defined.

---

1   It is easy to see that there is generally no coordinate system which makes a given metric time independent. The metric has ten independent components (same as a 4×4 symmetric matrix), while a change of coordinates enables one to introduce only four degrees of freedom to change the components (these are the four functions $x^\alpha(x^\mu)$). It is a special metric indeed if all ten components can be made time independent this way.

In a similar manner, if a metric is axially symmetric, then coordinates can be found in which $g_{\alpha\beta}$ is independent of the angle ψ around the axis. Then p_ψ will be conserved. This is the particle's angular momentum. In the nonrelativistic limit we have

$$p_\psi = g_{\psi\psi}p^\psi = g_{\psi\psi}m\, d\psi/dt = mg_{\psi\psi}\Omega, \tag{7.35}$$

where Ω is the angular velocity of the particle. Now, for a nearly flat metric we have

$$g_{\psi\psi} = \vec{e}_\psi \cdot \vec{e}_\psi \approx r^2 \tag{7.36}$$

in cylindrical coordinates (r, ψ, z) so that the conserved quantity is

$$p_\psi = mr^2\Omega. \tag{7.37}$$

This is the usual Newtonian definition of angular momentum.

So much for conservation laws of particle motion. Similar considerations apply to fluids, since they are just large collections of particles. But the situation with regard to the total mass and momentum of a self-gravitating system is more complicated. It turns out that an isolated system's mass and momentum *are* conserved, but we must postpone any discussion of this until we see how they are defined.

7.5 Bibliography

The question of how curvature and physics fit together is discussed in more detail by Buchdahl (1981) and Geroch (1978). Conserved quantities are discussed in detail in any of the advanced texts, in Trautman (1962), and in Goldberg (1980). The material in this chapter is preparation for one of the most active research areas in GR, the theory of quantum fields in a fixed curved spacetime. See Davies (1980) and Gibbons (1979).

7.6 Exercises

1 If Eq. (7.3) were the correct generalization of Eq. (7.1) to a curved spacetime, how would you interpret it?

2 To first order in ϕ, compute $g^{\alpha\beta}$ for Eq. (7.8).

3 Calculate all the Christoffel symbols for the metric given by Eq. (7.8), to first order in ϕ. Assume ϕ is a general function of t, x, y, and z.

4 Verify that the results, Eqs. (7.15) and (7.24), depended only on g_{00}: the form of g_{xx} doesn't affect them, as long as it is $1 + 0(\phi)$.

5(a) For a perfect fluid, verify that the spatial components of Eq. (7.6) in the Newtonian limit reduce to

$$v_{,t} + (v \cdot \nabla)v + \nabla p/\rho + \nabla\phi = 0 \tag{7.38}$$

for the metric, Eq. (7.8). This is known as Euler's equation for nonrelativistic fluid flow in a gravitational field.

(b)   Examine the time-component of Eq. (7.6) under the same assumptions, and interpret each term.

(c)   Eq. (7.38) implies that a static fluid ($v = 0$) in a static Newtonian gravitational field obeys the equation of hydrostatic equilibrium:

$$\nabla p + \rho \nabla \phi = 0. \tag{7.39}$$

A metric tensor is said to be static if there exist coordinates in which $\vec{e}_0$ is timelike, $g_{i0} = 0$, and $g_{\alpha\beta,0} = 0$. Deduce from Eq. (7.6) that a static fluid ($U^i = 0$, $p_{,0} = 0$, etc.) obeys the relativistic equation of hydrostatic equilibrium:

$$p_{,i} + (\rho + p)[\tfrac{1}{2}\ln(-g_{00})]_{,i} = 0. \tag{7.40}$$

(d)   This suggests that, at least for static situations, there is a close relation between g_{00} and $-\exp(2\phi)$, where ϕ is the Newtonian potential for a similar physical situation. Show that Eq. (7.8) and Exer. 4 are consistent with this.

6   Deduce Eq. (7.25) from Eq. (7.10).

7   Consider the following four different metrics, as given by their line elements:

(i) $ds^2 = -dt^2 + dx^2 + dy^2 + dz^2$;

(ii) $ds^2 = -(1 - 2M/r)\,dt^2 + (1 - 2M/r)^{-1}\,dr^2 + r^2\,(d\theta^2 + \sin^2\theta\,d\phi^2)$,

where M is a constant;

(iii) $ds^2 = -\dfrac{\Delta - a^2\sin^2\theta}{\rho^2}\,dt^2 - 2a\,\dfrac{2Mr\sin^2\theta}{\rho^2}\,dt\,d\phi$

$\qquad + \dfrac{(r^2 + a^2)^2 - a^2\Delta\sin^2\theta}{\rho^2}\sin^2\theta\,d\phi^2 + \dfrac{\rho^2}{\Delta}\,dr^2 + \rho^2\,d\theta^2$,

where M and a are constants and we have introduced the short-hand notation $\Delta = r^2 - 2Mr + a^2$, $\rho^2 = r^2 + a^2\cos^2\theta$;

(iv) $ds^2 = -dt^2 + R^2(t)\left[\dfrac{dr^2}{1 - kr^2} + r^2(d\theta^2 + \sin^2\theta\,d\phi^2)\right]$,

where k is a constant and $R(t)$ is an arbitrary function of t alone.

The first one should be familiar by now. We shall encounter the other three in later chapters. Their names are, respectively, the Schwarzschild, Kerr, and Robertson–Walker metrics.

(a)   For each metric find as many conserved components p_α of a freely falling particle's four momentum as possible.

(b)   Use the result of Exer. 28, § 6.9 to put (i) in the form

(i′) $ds^2 = -dt^2 + dr^2 + r^2(d\theta^2 + \sin^2\theta\,d\phi^2)$.

From this, argue that (ii) and (iv) are spherically symmetric. Does this increase the number of conserved components p_α?

(c) It can be shown that for (i') and (ii)–(iv), a geodesic which begins with $\theta = \pi/2$ and $p^\theta = 0$ – i.e. one which begins tangent to the equatorial plane – always has $\theta = \pi/2$ and $p^\theta = 0$. For cases (i'), (ii) and (iii), use the equation $\vec{p} \cdot \vec{p} = m^2$ to solve for p^r in terms of m, other conserved quantities, and known functions of position.

(d) For (iv), spherical symmetry implies that if a geodesic begins with $p^\theta = p^\phi = 0$, these remain zero. Use this to show from Eq. (7.29) that when $k = 0$, p_r is a conserved quantity.

8 Suppose that in some coordinate system the components of the metric $g_{\alpha\beta}$ are independent of some coordinate x^μ.

(a) Show that the conservation law $T^\nu{}_{\mu\,;\nu} = 0$ for *any* stress–energy tensor becomes

$$\frac{1}{\sqrt{-g}} (\sqrt{-g}\, T^\nu{}_\mu)_{,\nu} = 0. \tag{7.41}$$

(b) Suppose that in these coordinates $T^{\alpha\beta} \neq 0$ only in some bounded region of each spacelike hypersurface $x^0 = \text{const}$. Show that Eq. (7.41) implies

$$\int_{x^0 = \text{const.}} T^\nu{}_\mu \sqrt{-g}\, n_\nu\, \mathrm{d}^3 x$$

is independent of x^0, if n_ν is the unit normal to the hypersurface. This is the generalization to continua of the conservation law stated after Eq. (7.29).

(c) Consider flat Minkowski space in a global inertial frame with spherical polar coordinates (t, r, θ, ϕ). Show from (b) that

$$J = \int_{t = \text{const.}} T^0{}_\phi r^2 \sin \theta\, \mathrm{d}r\, \mathrm{d}\theta\, \mathrm{d}\phi \tag{7.42}$$

is independent of t. This is the total angular momentum of the system.

(d) Express the integral in (c) in terms of the components of $T^{\alpha\beta}$ on the Cartesian basis (t, x, y, z), showing that

$$J = \int (x T^{y0} - y T^{x0})\, \mathrm{d}x\, \mathrm{d}y\, \mathrm{d}z. \tag{7.43}$$

This is the continuum version of the nonrelativistic expression $(\mathbf{r} \times \mathbf{p})_z$ for a particle's angular momentum about the z axis.

9(a) Find the components of the Riemann tensor $R_{\alpha\beta\mu\nu}$ for the metric, Eq. (7.8), to first order in ϕ.

(b) Show that the equation of geodesic deviation, Eq. (6.87), implies (to lowest order in ϕ and velocities)

$$\frac{\mathrm{d}^2 \xi^i}{\mathrm{d}t^2} = -\phi_{,ij} \xi^j. \tag{7.44}$$

(c) Interpret this equation when the geodesics are world lines of freely falling particles which begin from rest at nearby points in a Newtonian gravitational field.

10(a)   Show that if a vector field ξ^{α} satisfies *Killing's equation*,

$$\nabla_{\alpha}\xi_{\beta} + \nabla_{\beta}\xi_{\alpha} = 0, \tag{7.45}$$

then along a geodesic, $p^{\alpha}\xi_{\alpha} = \text{const}$. This is a coordinate-invariant way of characterizing the conservation law we deduced from Eq. (7.29). One only has to know whether a metric admits Killing fields.

(b)   Find ten Killing fields of Minkowski spacetime.

(c)   Show that if $\vec{\xi}$ and $\vec{\eta}$ are Killing fields, then so is $\alpha\vec{\xi} + \beta\vec{\eta}$ for *constant* α and β.

(d)   Show that Lorentz transformations of the fields in (b) simply produce linear combinations as in (c).

(e)   If you did Exer. 7, use the results of Exer. 7a to find Killing vectors of metrics (ii)–(iv).

8

The Einstein field equations

8.1 Purpose and justification of the field equations

Having decided upon a description of gravity and its action on matter that is based on the idea of a curved manifold with a metric, we must now complete the theory by postulating a law which shows how the sources of the gravitational field determine the metric. The Newtonian analogue is

$$\nabla^2 \phi = 4\pi G \rho, \tag{8.1}$$

where ρ is the density of mass. Its solution for a point particle of mass m is (see Exer. 1, § 8.6).

$$\phi = -\frac{Gm}{r}, \tag{8.2}$$

which is dimensionless in units where $c = 1$. The source of the gravitational field in Newton's theory is the mass density. In our relativistic theory of gravity the source must be related to this, but it must be a relativistically meaningful concept, which 'mass' alone is not. An obvious relativistic generalization is the total energy, including rest mass. In the MCRF of a fluid element, we have denoted the density of total energy by ρ in Ch. 4. So we might be tempted to use this ρ as the source of the relativistic gravitational field. This would not be very satisfactory, however, because ρ is the energy density as measured by only one observer, the MCRF. Other observers measure the energy density to be the component T^{00} in their own reference frames. If we were to use ρ as the source of the field, we would be saying that one class of observers is preferred above all others, namely those for whom ρ is the energy density. This point of view is at variance with the approach we adopted in the previous chapter, where we stressed that one must allow *all* coordinate systems on an equal footing. So we shall reject ρ as the source and instead insist that the generalization of Newton's mass density should be T^{00}. But again, if T^{00} alone were the source, one would have to specify a frame in which T^{00} was evaluated. An invariant theory can avoid introducing preferred coordinate systems by using the *whole* of the stress–energy tensor **T** as the source of the gravitational field. The generalization of Eq. (8.1) to relativity would then have the form

$$\mathbf{O}(\mathbf{g}) = k\mathbf{T}, \tag{8.3}$$

where k is a constant (as yet undetermined) and **O** is a differential operator on the metric tensor **g**, which we have already decided is the generalization of ϕ. There will thus be 16 differential equations (one for each component of Eq. (8.3)) in place of the single one, Eq. (8.1).

By analogy with Eq. (8.1), we should look for a second-order differential operator **O** that produces a tensor of rank $\binom{2}{0}$, since in Eq. (8.3) it is equated to the $\binom{2}{0}$ tensor **T**. In other words, $\{O^{\alpha\beta}\}$ must be the components of a $\binom{2}{0}$ tensor and must be combinations of $g_{\mu\nu,\lambda\sigma}$, $g_{\mu\nu,\lambda}$, and $g_{\mu\nu}$. It is clear from Ch. 6 that the Ricci tensor $R^{\alpha\beta}$ satisfies these conditions. In fact, *any* tensor of the form

$$O^{\alpha\beta} = R^{\alpha\beta} + \mu g^{\alpha\beta} R + \Lambda g^{\alpha\beta} \tag{8.4}$$

satisfies these conditions, if μ and Λ are constants. To determine μ we use a property of $T^{\alpha\beta}$ which we have not yet used, namely that the strong equivalence principle demands local conservation of energy and momentum (Eq. (7.6)):

$$T^{\alpha\beta}{}_{;\beta} = 0.$$

This equation must be true for *any* metric tensor. Then Eq. (8.3) implies that

$$O^{\alpha\beta}{}_{;\beta} = 0, \tag{8.5}$$

which again must be true for any metric tensor. Since $g^{\alpha\beta}{}_{;\mu} = 0$, we now find, from Eq. (8.4),

$$(R^{\alpha\beta} + \mu g^{\alpha\beta} R)_{;\beta} = 0. \tag{8.6}$$

By comparing this with Eq. (6.98), we see that we must have $\mu = -\frac{1}{2}$ if Eq. (8.6) is to be an identity for arbitrary $g_{\alpha\beta}$. So we are led by this chain of argument to the equation

$$G^{\alpha\beta} + \Lambda g^{\alpha\beta} = kT^{\alpha\beta}, \tag{8.7}$$

with undetermined constants Λ and k. In index-free form this is

$$\mathbf{G} + \Lambda\mathbf{g} = k\mathbf{T}. \tag{8.8}$$

These are called the field equations of GR, or Einstein's field equations.

In summary, we have been led to Eq. (8.7) by asking for equations that (i) resemble but generalize Eq. (8.1); (ii) introduce no preferred coordinate system; and (iii) guarantee local conservation of energy–momentum for any metric tensor. Eq. (8.7) is not the only equation which satisfies (i)–(iii). Many alternatives have been proposed, beginning even before Einstein arrived at equations like Eq. (8.7). In recent years, when technology has made it possible to test Einstein's equations fairly precisely, even in the weak gravity of the solar system, many new alternative theories have been proposed. Some have even been designed to agree with Einstein's predictions at the precision of foreseeable solar-system experiments, differing only for much stronger fields. GR's competitors are, however, invariably more complicated than Einstein's equations themselves, and on simply aesthetic grounds are unlikely to attract much attention from physicists unless Einstein's equations are eventually found to conflict with some experiment. A number of the competing theories and the experimental tests which have been used to eliminate them in recent years are discussed in Misner *et al.* (1973) and Will (1981). (We will study two classical tests in Ch. 11.) Einstein's equations have stood up well to these tests, so we will not discuss any alternative theories in this book. In this we are in the good company of the distinguished physicist S. Chandrasekhar (1980):

> The element of controversy and doubt, that have continued to shroud the general theory of relativity to this day, derives precisely from this fact, namely that in the formulation of his theory Einstein incorporates aesthetic criteria; and every critic feels that he is entitled to his own differing aesthetic and philosophic criteria. Let me simpy say that I do not share these doubts; and I shall leave it at that.

As remarked above, it is customary to refer to Eq. (8.7) as Einstein's equations, a custom which we follow in this book, but it might be fairer to call them the Einstein–Hilbert equations, because the mathematician D. Hilbert derived them independently of Einstein in the same year (see Mehra 1974). Hilbert was motivated by Einstein's own physical arguments, which were known to him, but his mathematical approach to deriving the equations was far more elegant than Einstein's (or than ours above). It was Einstein, however, who motivated the endeavor and Einstein who deduced the solar-system implications of the new theory – the perihelion precession of Mercury and the deflection of light by the Sun – that led to the theory's acceptance, so it is Einstein whose name remains attached to the theory in the minds of physicists and the public alike.

Geometrized units. We have not yet said anything about the constant k in Eq. (8.7), which plays the same role as $4\pi G$ in Eq. (8.1). Before discussing it below we will establish a more convenient set of units, namely those in which $G = 1$. Just as in SR we found it convenient to choose units in which the fundamental constant c was set to unity, so in studies of gravity it is more natural to work in units where G has the value unity. A convenient conversion factor from SI units to these geometrized units (where $c = G = 1$) is

$$1 = G/c^2 = 7.425 \times 10^{-28} \text{ m kg}^{-1}. \tag{8.9}$$

We shall use this to eliminate kg as a unit, measuring mass in meters. We list in Table 8.1 the values of certain useful constants in SI and geometrized units. Exer. 2, § 8.6, should help the student to become accustomed to these units.

An illustration of the fundamental nature of geometrized units in gravitational problems is provided by the uncertainties in the two values given for $M_\oplus$. Earth's mass is measured by examining satellite orbits and

Table 8.1. *Comparison of SI and geometrized values of fundamental constants*

Constant	SI value	Geometrized value
c	$2.998 \times 10^8 \text{ ms}^{-1}$	1
G	$6.673 \times 10^{-11} \text{ m}^3 \text{ kg}^{-1} \text{ s}^{-2}$	1
$\hbar$	$1.055 \times 10^{-34} \text{ kg m}^2 \text{ s}^{-1}$	$2.612 \times 10^{-70} \text{ m}^2$
m_e	$9.110 \times 10^{-31} \text{ kg}$	$6.764 \times 10^{-58} \text{ m}$
m_p	$1.673 \times 10^{-27} \text{ kg}$	$1.242 \times 10^{-54} \text{ m}$
$M_\odot$	$1.989 \times 10^{30} \text{ kg}$	$1.477 \times 10^3 \text{ m}$
$M_\oplus$	$5.973 \times 10^{24} \text{ kg}$	$4.435 \times 10^{-3} \text{ m}$
$L_\odot$	$3.90 \times 10^{26} \text{ kg m}^2 \text{ s}^{-3}$	1.07×10^{-26}

Notes: The symbols m_e and m_p stand respectively for the rest masses of the electron and proton; $M_\odot$ and $M_\oplus$ denote, respectively, the masses of the Sun and Earth; and $L_\odot$ is the Sun's luminosity (the SI unit is equivalent to joules per second).

using Kepler's laws. This measures the Newtonian potential, which involves the product $GM_\oplus$, c^2 times the *geometrized* value of the mass. With the advent of satellite tracking by laser ranging (Lerch 1978), this number can be measured very accurately. Moreover, the speed of light c is known to great accuracy. Thus, the geometrized value of $M_\oplus$ is known to an accuracy of five parts in 10^8. The value of G, however, is measured in laboratory experiments, where the weakness of gravity introduces large uncertainty. The conversion factor G/c^2 is uncertain by five parts in 10^4, so that is also the accuracy of the SI value of $M_\oplus$. Similarly, the Sun's geometrized mass is known very accurately by precise radar tracking of the planets; the uncertainty is only one part in 10^8. Again, its mass in kilograms is far more uncertain.

8.2 Einstein's equations

In component notation, Einstein's equations, Eq. (8.7), take the form

$$G^{\alpha\beta} = 8\pi T^{\alpha\beta}. \tag{8.10}$$

These are a specialization of Eq. (8.7), with $\Lambda = 0$ and $k = 8\pi$. The constant Λ is called the cosmological constant, and was originally not present in Einstein's equations; he inserted it many years later in order to obtain static cosmological solutions – solutions for the large-scale behavior of the universe – that he felt at the time were desirable. Observations of the expansion of the universe subsequently made him reject the term and regret he had ever invented it. We shall return to the discussion of Λ in the chapter on cosmology, but until then we shall set $\Lambda = 0$. The justification for doing this, and the possible danger of it, are discussed in Exer. 18, § 8.6.

The value of $k = 8\pi$ is obtained by demanding that Einstein's equations predict the correct behavior of planets in the solar system. This is the Newtonian limit, in which we must demand that the predictions of GR agree with those of Newton's theory when the latter are well tested by observation. We saw in the last chapter that the Newtonian motions are produced when the metric has the form Eq. (7.8). One of our tasks in this chapter is to show that Einstein's equations, Eq. (8.10), do indeed have Eq. (7.8) as a solution when we assume that gravity is weak (see Exer. 3, § 8.6). We could, of course, keep k arbitrary until then, adjusting its value to whatever is required to obtain the solution, Eq. (7.8). It is more convenient, however, for our subsequent use of the equations of this chapter if we simply set k to 8π at the outset and verify that this value is correct.

Eq. (8.10) should be regarded as a system of ten coupled differential equations (not 16, since $T^{\alpha\beta}$ and $G^{\alpha\beta}$ are symmetric). They are to be solved for the ten components $g_{\alpha\beta}$ when the source $T^{\alpha\beta}$ is given. The equations are nonlinear, but they have a well-posed initial-value structure – that is, they determine future values of $g_{\alpha\beta}$ from given initial data. However, one point must be made: since $\{g_{\alpha\beta}\}$ are the components of a tensor in some coordinate system, a change in coordinates induces a change in them. In particular, there are four coordinates, so there are four arbitrary functional degrees of freedom among the ten $g_{\alpha\beta}$. It should be impossible, therefore, to determine all ten $g_{\alpha\beta}$ from any initial data, since the coordinates to the future of the initial moment can be changed arbitrarily. In fact, Einstein's equations have exactly this property: the Bianchi identities:

$$G^{\alpha\beta}{}_{;\beta} = 0 \tag{8.11}$$

mean that there are four differential identities (one for each value of α above) among the ten $G^{\alpha\beta}$. These ten, then, are not independent, and the ten Einstein equations are really only six independent differential equations for the six functions among the ten $g_{\alpha\beta}$ that characterize the geometry independently of the coordinates.

The exact mathematical representation of these six functions is a difficult subject, which we will not have much to say about, since we will rarely be solving the initial-value problem in GR. There now exists a well-defined approach to the problem of separating the coordinate freedom in $g_{\alpha\beta}$ from the true geometric and dynamical freedom. This is described in more advanced texts, for instance Misner *et al.* (1973), or Hawking & Ellis (1973). See also Choquet-Bruhat & York (1980). It will suffice here simply to note that there are really only six equations for six quantities among the $g_{\alpha\beta}$, and that Einstein's equations permit complete freedom in choosing the coordinate system.

8.3 Einstein's equations for weak gravitational fields

Nearly Lorentz coordinate systems. Since the absence of gravity leaves spacetime flat, a weak gravitational field is one in which spacetime is 'nearly' flat. This is defined as a manifold on which coordinates exist in which the metric has components

$$\blacklozenge \qquad g_{\alpha\beta} = \eta_{\alpha\beta} + h_{\alpha\beta}, \tag{8.12}$$

where

$$|h_{\alpha\beta}| \ll 1, \tag{8.13}$$

everywhere in spacetime. Such coordinates are called nearly Lorentz coordinates. It is important to say 'there exist coordinates' rather than 'for all coordinates', since one can find coordinates even in Minkowski space in which $g_{\alpha\beta}$ is not close to the simple diagonal $(-1, +1, +1, +1)$ form of $\eta_{\alpha\beta}$. On the other hand, if one coordinate system exists in which Eq. (8.12) and (8.13) are true, then there are many such coordinate systems. Two fundamental types of coordinate transformations that take one nearly Lorentz coordinate system into another will be discussed below: background Lorentz transformations and gauge transformations.

But why should we specialize to nearly Lorentz coordinates at all? Haven't we just said that Einstein's equations allow complete coordinate freedom, so shouldn't their physical predictions be the same in any coordinates? Of course the answer is yes, the physical predictions will be the same. On the other hand, the amount of work we would have to do to arrive at the physical predictions could be enormous in a poorly chosen coordinate system. (For example, try to solve Newton's equation of motion for a particle free of all forces in spherical polar coordinates, or try to solve Poisson's equation in a coordinate system in which it does not separate!) Perhaps even more serious is the possibility that in a crazy coordinate system we may not have sufficient creativity and insight into the physics to know what calculations to make in order to arrive at interesting physical predictions. Therefore it is extremely important that the first step in the solution of any problem in GR must be an attempt to construct coordinates which will make the calculation simplest. Precisely because Einstein's equations have complete coordinate freedom, we should use this freedom intelligently. The construction of helpful coordinate systems is an art, and it is often rather difficult. In the present problem, however, it should be clear that $\eta_{\alpha\beta}$ is the simplest form for the flat-space metric, so that Eqs. (8.12) and (8.13) give the simplest and most natural 'nearly flat' metric components.

Background Lorentz transformations. The matrix of a Lorentz transformation in SR is

$$(\Lambda^{\bar{\alpha}}{}_{\beta}) = \begin{pmatrix} \gamma & -v\gamma & 0 & 0 \\ -v\gamma & \gamma & 0 & 0 \\ 0 & 0 & 1 & 0 \\ 0 & 0 & 0 & 1 \end{pmatrix}, \qquad \gamma = (1-v^2)^{-1/2} \qquad (8.14)$$

(for a boost of velocity v in the x direction). For weak gravitational fields we define a 'background Lorentz transformation' to be one which has the form

$$x^{\bar{\alpha}} = \Lambda^{\bar{\alpha}}{}_{\beta} x^{\beta}, \qquad (8.15)$$

in which $\Lambda^{\bar{\alpha}}{}_\beta$ is identical to a Lorentz transformation in SR, whose matrix elements are constant everywhere. Of course, we are not in SR, so this is only one class of transformations out of all possible ones. But it has a particularly nice feature, which we discover by transforming the metric tensor:

$$g_{\bar{\alpha}\bar{\beta}} = \Lambda^{\mu}{}_{\bar{\alpha}}\Lambda^{\nu}{}_{\bar{\beta}}g_{\mu\nu} = \Lambda^{\mu}{}_{\bar{\alpha}}\Lambda^{\nu}{}_{\bar{\beta}}\eta_{\mu\nu} + \Lambda^{\mu}{}_{\bar{\alpha}}\Lambda^{\nu}{}_{\bar{\beta}}h_{\mu\nu}. \tag{8.16}$$

But the Lorentz transformation is designed so that

$$\Lambda^{\mu}{}_{\bar{\alpha}}\Lambda^{\nu}{}_{\bar{\beta}}\eta_{\mu\nu} = \eta_{\bar{\alpha}\bar{\beta}}, \tag{8.17}$$

so we get

$$g_{\bar{\alpha}\bar{\beta}} = \eta_{\bar{\alpha}\bar{\beta}} + h_{\bar{\alpha}\bar{\beta}} \tag{8.18}$$

with

◆ $$h_{\bar{\alpha}\bar{\beta}} \equiv \Lambda^{\mu}{}_{\bar{\alpha}}\Lambda^{\nu}{}_{\bar{\beta}}h_{\mu\nu}. \tag{8.19}$$

We see that, under a background Lorentz transformation, $h_{\mu\nu}$ transforms *as if* it were a tensor in SR all by itself! It is, of course, not a tensor, but just a piece of $g_{\alpha\beta}$. But this restricted transformation property leads to a convenient fiction: we can think of a slightly curved spacetime as a *flat* spacetime with a 'tensor' $h_{\mu\nu}$ defined on it. Then all physical fields – like $R_{\mu\nu\alpha\beta}$ – will be defined in terms of $h_{\mu\nu}$, and they will 'look like' fields on a flat background spacetime. It is important to bear in mind, however, that spacetime is really curved, that this fiction results from considering only one type of coordinate transformation. We shall find this fiction to be useful, however, in our calculations below.

Gauge transformations. There is another very important kind of coordinate change which leaves Eqs. (8.12) and (8.13) unchanged: a very small change in coordinates, of the form

$$x^{\alpha'} = x^\alpha + \xi^\alpha(x^\beta),$$

generated by a 'vector' ξ^α, whose components are functions of position. If we demand that ξ^α be small in the sense that $|\xi^\alpha{}_{,\beta}| \ll 1$, then we have

$$\Lambda^{\alpha'}{}_\beta = \frac{\partial x^{\alpha'}}{\partial x^\beta} = \delta^\alpha{}_\beta + \xi^\alpha{}_{,\beta}, \tag{8.20}$$

$$\Lambda^\alpha{}_{\beta'} = \delta^\alpha{}_\beta - \xi^\alpha{}_{,\beta} + 0(|\xi^\alpha{}_{,\beta}|^2). \tag{8.21}$$

One can easily verify that, to first order in small quantities,

$$g_{\alpha'\beta'} = \eta_{\alpha\beta} + h_{\alpha\beta} - \xi_{\alpha,\beta} - \xi_{\beta,\alpha}, \tag{8.22}$$

where we define

$$\xi_\alpha = \eta_{\alpha\beta}\xi^\beta. \tag{8.23}$$

This means that the effect of the coordinate change is to re-define $h_{\alpha\beta}$:

♦ $\qquad h_{\alpha\beta} \to h_{\alpha\beta} - \xi_{\alpha,\beta} - \xi_{\beta,\alpha}.$ $\qquad\qquad$ (8.24)

If all $|\xi^{\alpha}{}_{,\beta}|$ are small, then the new $h_{\alpha\beta}$ is still small, and we are still in an acceptable coordinate system. This change is called a *gauge transformation*, which is a term used because of strong analogies between Eq. (8.24) and gauge transformations of electromagnetism. This analogy is explored in Exer. 11, § 8.6. The coordinate freedom of Einstein's equations means that we are free to choose an arbitrary (small) 'vector' ξ^{α} in Eq. (8.24). We will use this freedom below to simplify our equations enormously.

A word about the role of indices like α' and β' in Eqs. (8.21) and (8.22) may be helpful here, as beginning students are often uncertain on this point. A prime or bar on an index is an indication that it refers to a particular coordinate system, e.g. that $g_{\alpha'\beta'}$ is a component of $\mathbf{g}$ in the $\{x^{\nu'}\}$ coordinates. But the index still takes the same values $(0, 1, 2, 3)$. On the right-hand side of Eq. (8.22) there are no primes because all quantities are defined in the unprimed system. Thus, if $\alpha = \beta = 0$ we read Eq. (8.22) as: 'The 0–0 component of $\mathbf{g}$ in the primed coordinate system is a function whose value at any point is the value of the 0–0 component of $\boldsymbol{\eta}$ plus the value of the 0–0 'component' of $h_{\alpha\beta}$ in the unprimed coordinates at that point minus twice the derivative of the function ξ_0 – defined by Eq. (8.23) – with respect to the unprimed coordinate x^0 there.' Eq. (8.22) may look strange because – unlike, say, Eq. (8.15) – its indices do not 'match up'. But that is acceptable, since Eq. (8.22) is *not* what we have called a valid tensor equation. It expresses the relation between components of a tensor in two specific coordinates; it is not intended to be a general coordinate-invariant expression.

Riemann tensor. Using Eq. (8.12) it is easy to show that, to first order in $h_{\mu\nu}$,

♦ $\qquad R_{\alpha\beta\mu\nu} = \tfrac{1}{2}(h_{\alpha\nu,\beta\mu} + h_{\beta\mu,\alpha\nu} - h_{\alpha\mu,\beta\nu} - h_{\beta\nu,\alpha\mu}).$ $\qquad$ (8.25)

As demonstrated in Exer. 5, § 8.6, these components are *independent* of the gauge, unaffected by Eq. (8.24). The reason for this is that a coordinate transformation transforms the components of $\mathbf{R}$ into linear combinations of one another. A small coordinate transformation – a gauge transformation – changes the components by a small amount; but since they are already small, this change is of second order, and so the first-order expression, Eq. (8.25), remains unchanged.

Weak-field Einstein equations. We shall now consistently adopt the point of view mentioned earlier, the fiction that $h_{\alpha\beta}$ is a tensor on a 'background' Minkowski spacetime, i.e. a tensor in SR. Then all our equations will be expected to be valid tensor equations when interpreted in SR, but not necessarily valid under more general coordinate transformations. Gauge transformations will be allowed, of course, but we will not regard them as coordinate transformations. Rather, they define equivalence classes among all symmetric tensors $h_{\alpha\beta}$: any two related by Eq. (8.24) for some ξ_α will produce equivalent physical effects. Consistent with this point of view we can define index-raised quantities

$$h^\mu{}_\beta = \eta^{\mu\alpha} h_{\alpha\beta}, \tag{8.26}$$
$$h^{\mu\nu} = \eta^{\nu\beta} h^\mu{}_\beta, \tag{8.27}$$

the trace

$$h \equiv h^\alpha{}_\alpha, \tag{8.28}$$

and a 'tensor' called the 'trace reverse' of $h_{\alpha\beta}$

$$\bar{h}^{\alpha\beta} = h^{\alpha\beta} - \tfrac{1}{2} \eta^{\alpha\beta} h. \tag{8.29}$$

It has this name because

$$\bar{h} \equiv \bar{h}^\alpha{}_\alpha = -h. \tag{8.30}$$

Moreover, one can show that the inverse of Eq. (8.29) is the same:

$$h^{\alpha\beta} = \bar{h}^{\alpha\beta} - \tfrac{1}{2} \eta^{\alpha\beta} \bar{h}. \tag{8.31}$$

With these definitions it is not difficult to show, beginning with Eq. (8.25), that the Einstein tensor is

$$G_{\alpha\beta} = -\tfrac{1}{2}[\bar{h}_{\alpha\beta,\mu}{}^{,\mu} + \eta_{\alpha\beta} \bar{h}_{\mu\nu}{}^{,\mu\nu} - \bar{h}_{\alpha\mu,\beta}{}^{,\mu}$$
$$-\bar{h}_{\beta\mu,\alpha}{}^{,\mu} + 0(h_{\alpha\beta}^2)]. \tag{8.32}$$

(Recall that for any function f,

$$f^{,\mu} \equiv \eta^{\mu\nu} f_{,\nu})$$

It is clear that Eq. (8.32) would simplify considerably if we could require

◆ $$\bar{h}^{\mu\nu}{}_{,\nu} = 0. \tag{8.33}$$

These are four equations, and since we have four free gauge functions ξ^α, we might expect to be able to find a gauge in which Eq. (8.33) is true. We shall show that this expectation is correct: it is always possible to choose a gauge to satisfy Eq. (8.33). Thus, we refer to it as a gauge condition and, specifically, as the *Lorentz gauge* condition. If we have an $h_{\mu\nu}$ satisfying this, we say we are working in the Lorentz gauge. The gauge has this name, again by analogy with electromagnetism (see Exer. 11, § 8.6). Other names one encounters in the literature for the same gauge include the harmonic gauge and the de Donder gauge.

That this gauge exists can be shown as follows. Suppose we have some arbitrary $\bar{h}^{(old)}_{\mu\nu}$ for which $\bar{h}^{(old)\mu\nu}{}_{,\nu} \neq 0$. Then under a gauge change Eq. (8.24), one can show (Exer. 12, § 8.6) that $\bar{h}_{\mu\nu}$ changes to

$$\bar{h}^{(new)}_{\mu\nu} = \bar{h}^{(old)}_{\mu\nu} - \xi_{\mu,\nu} - \xi_{\nu,\mu} + \eta_{\mu\nu}\xi^{\alpha}{}_{,\alpha} \tag{8.34}$$

Then the divergence is

$$\bar{h}^{(new)\mu\nu}{}_{,\nu} = \bar{h}^{(old)\mu\nu}{}_{,\nu} - \xi^{\mu,\nu}{}_{,\nu}. \tag{8.35}$$

If we want a gauge in which $\bar{h}^{(new)\mu\nu}{}_{,\nu} = 0$ then ξ^{μ} is determined by the equation

$$\Box \xi^{\mu} = \xi^{\mu,\nu}{}_{,\nu} = \bar{h}^{(old)\mu\nu}{}_{,\nu} \tag{8.36}$$

where the symbol $\Box$ is used for the four-dimensional Laplacian:

$$\Box f = f^{,\mu}{}_{,\mu} = \eta^{\mu\nu}f_{,\mu\nu} = \left(-\frac{\partial^2}{\partial t^2} + \nabla^2\right)f. \tag{8.37}$$

This operator is also called the *D'Alembertian* or *wave operator*, and is sometimes denoted by Δ. The equation

$$\Box f = g \tag{8.38}$$

is the three-dimensional inhomogeneous wave equation, and it always has a solution for any (sufficiently well-behaved) g (see Choquet–Bruhat *et al.*, 1977), so there always exists some ξ^{μ} which will transform from an arbitrary $h_{\mu\nu}$ to the Lorentz gauge. In fact, this ξ^{μ} is not unique, since any vector η^{μ} satisfying the homogeneous wave equation

$$\Box \eta^{\mu} = 0 \tag{8.39}$$

can be added to ξ^{μ} and the result will still obey

$$\Box(\xi + \eta^{\mu}) = \bar{h}^{(old)\mu\nu}{}_{,\nu} \tag{8.40}$$

and so will still give a Lorentz gauge. Thus, the Lorentz gauge is really a class of gauges.

In this gauge, Eq. (8.32) becomes (see Exer. 10, § 8.6)

$$G^{\alpha\beta} = -\tfrac{1}{2}\Box \bar{h}^{\alpha\beta}. \tag{8.4!}$$

Then the weak-field Einstein equations are

◆ $$\Box \bar{h}^{\mu\nu} = -16\pi\, T^{\mu\nu}. \tag{8.42}$$

These are called the field equations of '*linearized theory*', since they result from keeping terms linear in $h_{\alpha\beta}$.

8.4 Newtonian gravitational fields

Newtonian limit. Newtonian gravity is known to be valid when gravitational fields are too weak to produce velocities near the speed of light: $|\phi| \ll 1$, $|v| \ll 1$. In such situations, GR must make the same predictions as Newtonian gravity. The fact that velocities are small means that the components $T^{\alpha\beta}$ obey the inequalities $|T^{00}| \gg |T^{0i}| \gg |T^{ij}|$. Because of Eq. (8.42), these inequalities transfer to $\bar{h}_{\alpha\beta}$: $|\bar{h}^{00}| \gg |\bar{h}^{0i}| \gg |\bar{h}^{ij}|$. Thus, the dominant 'Newtonian' field comes from the equation

$$\Box \bar{h}^{00} = -16\pi\rho, \tag{8.43}$$

where we use the fact that $T^{00} = \rho + 0(\rho v^2)$. For fields that change only because the sources move with velocity v, we have that $\partial/\partial t$ is of the same order as $v\,\partial/\partial x$, so that

$$\Box = \nabla^2 + 0(v^2\nabla^2). \tag{8.44}$$

Thus, our equation is

$$\nabla^2 \bar{h}^{00} = -16\pi\rho. \tag{8.45}$$

Comparing this with the Newtonian equation, Eq. (8.1),

$$\nabla^2 \phi = 4\pi\rho$$

(with $G = 1$), we see that we must identify

$$\bar{h}^{00} = -4\phi. \tag{8.46}$$

Since all other components of $\bar{h}^{\alpha\beta}$ are negligible at this order, we have

$$h = h^\alpha{}_\alpha = -\bar{h}^\alpha{}_\alpha = \bar{h}^{00}, \tag{8.47}$$

and this implies

$$h^{00} = -2\phi, \tag{8.48}$$

$$h^{xx} = h^{yy} = h^{zz} = -2\phi, \tag{8.49}$$

or

$$ds^2 = -(1+2\phi)\,dt^2 + (1-2\phi)(dx^2 + dy^2 + dz^2). \tag{8.50}$$

This is identical to the metric given in Eq. (7.8). We saw there that this metric gives the correct Newtonian laws of motion, so the demonstration here that it follows from Einstein's equations completes the proof that Newtonian gravity is a limiting case of GR. (Incidentally it confirms that the constant 8π in Einstein's equations is the correct one.)

The far field of stationary relativistic sources. For any source of the full Einstein equations which is confined within a limited region of space (a 'localized' source), one can always go far enough away from it that its gravitational field becomes weak enough that linearized theory applies in that region. One might be tempted, then, to carry the discussion we have just gone through over to this case and say that Eq. (8.50) describes the far field of the source, with ϕ the Newtonian potential. This method would be wrong, for two related reasons. First, the derivation of Eq. (8.50) assumed that gravity was weak everywhere, including inside the source, because a crucial step was the identification of Eq. (8.45) with

Eq. (8.1) inside the source. In the present discussion we wish to make no assumptions about the weakness of gravity in the source. The second reason the method would be wrong is that we do not know how to define the Newtonian potential ϕ of a highly relativistic source anyway, so Eq. (8.50) would not make sense.

So we shall work from the linearized field equations directly. Since at first we assume the source of the field $T^{\mu\nu}$ is stationary (i.e. independent of time), we can assume that far away from it $h_{\mu\nu}$ is independent of time. (Later we will relax this assumption.) Then Eq. (8.42) becomes

$$\nabla^2 \bar{h}^{\mu\nu} = 0, \tag{8.51}$$

far from the source. This has the solution

$$\bar{h}^{\mu\nu} = A^{\mu\nu}/r + 0(r^{-2}), \tag{8.52}$$

where $A^{\mu\nu}$ is constant. In addition, we must demand that the gauge condition, Eq (8.33), be satisfied:

$$0 = \bar{h}^{\mu\nu}{}_{,\nu} = \bar{h}^{\mu j}{}_{,j} = -A^{\mu j} n_j/r^2 + 0(r^{-3}), \tag{8.53}$$

where the sum on ν collapses to a sum on the spatial index j because $\bar{h}^{\mu\nu}$ is time independent, and where n_j is the unit radial normal,

$$n_j = x_j/r. \tag{8.54}$$

The consequence of Eq. (8.53) for all x^i is

$$A^{\mu j} = 0, \tag{8.55}$$

for all μ and j. This means that *only* $\bar{h}^{00}$ survives or, in other words, that, far from the source

$$|\bar{h}^{00}| \gg |\bar{h}^{ij}|, |\bar{h}^{00}| \gg |h^{0j}|. \tag{8.56}$$

These conditions guarantee that the gravitational field does indeed behave like a Newtonian field out there, so we can reverse the identification that led to Eq. (8.46) and *define* the 'Newtonian potential' for the far field of any stationary source to be

$$(\phi)_{\text{relativistic far field}} = -\tfrac{1}{4}(\bar{h}^{00})_{\text{far field}}. \tag{8.57}$$

With this identification, Eq. (8.50) now does make sense for our problem, and it describes the far field of our source. Now, far from a Newtonian source the potential is

$$(\phi)_{\text{Newtonian far field}} = -M/r + 0(r^{-2}), \tag{8.58}$$

where M is the mass of the source (with $G = 1$). Thus, if in Eq. (8.52) we rename the constant A^{00} to be $4M$, the identification, Eq. (8.57), says that

$$(\phi)_{\text{relativistic far field}} = -M/r. \tag{8.59}$$

Any small body, for example a planet, that falls freely in the relativistic source's gravitational field but stays far away from it will follow the geodesics of the metric, Eq. (8.49), with ϕ given by Eq. (8.59). In Ch. 7 we saw that these geodesics obey Kepler's laws for the gravitational field of a body of mass M. We therefore *define* this constant M to be the *total mass* of the relativistic source.

Notice that this definition is not an integral over the source: we do not add up the masses of its constituent particles. Instead, we simply measure its mass – 'weigh it' – by the orbits it produces in test bodies far away. This definition enables us to write Eq. (8.50) in its form far from *any* stationary source:

$$ds^2 = -[1 - 2M/r + 0(r^{-2})]\, dt^2$$
$$+[1 + 2M/r + 0(r^{-2})](dx^2 + dy^2 + dz^2). \tag{8.60}$$

The assumption that the source was stationary was necessary to reduce the wave equation, Eq. (8.42), to Laplace's equation, Eq. (8.51). A source which changes with time can emit gravitational waves, and these, as we shall see in the next chapter, travel out from it at the speed of light and do not obey the inequalities, Eq. (8.56), so they cannot be regarded as Newtonian fields. Nevertheless, there are situations in which the definition of the mass we have just given may be used with confidence: the waves may be very weak, so that the stationary part of $\bar{h}^{00}$ dominates the wave part; or the source may have been stationary in the distant past, so that one can choose r large enough that any waves have not yet had time to reach that large an r. The definition of the mass of a time-dependent source is discussed in greater detail in more-advanced texts, such as Weinberg (1972) or Misner *et al.* (1973).

8.5   Bibliography

There are a wide variety of ways to 'derive' (really, to justify) Einstein's field equations, and a selection of them may be found in the texts listed below. The weak-field or linearized equations are useful for many investigations where the full equations are too difficult to solve. We shall use them frequently in subsequent chapters, and most texts discuss them. Our extraction of the Newtonian limit is very heuristic, but there are more rigorous approaches which reveal the geometric nature of Newton's equations (Misner *et al.* 1973, Cartan 1923) and the asymptotic nature of the limit (Futamase & Schutz 1983).

It seems appropriate here to list a sampling of widely available textbooks on GR. They differ in the background and sophistication they assume of the reader. There are those which expect little–Foster & Nightingale (1979), Buchdahl (1981), Rindler (1969), Frankel (1979), Bowler (1976), Clarke (1979); those which might be classed as first-year graduate texts–Adler *et al.* (1975), Anderson (1967), Burke (1980), McVittie (1965), parts of Misner *et al.* (1973), Møller (1972), Robertson & Noonan (1968), Stephani (1982), and Weinberg (1972); and those which make heavy demands of the student–Fock (1964), Hawking & Ellis (1973), Landau & Lifshitz (1962), much of Misner *et al.* (1973), Papapetrou (1974), Sachs & Wu (1977), Synge (1960), and Thirring (1978). The material in the present text ought, in most cases, to be sufficient preparation for supplementary reading in even the most advanced texts.

Solving problems is an essential ingredient of learning a theory, and the book of problems by Lightman *et al.* (1975), is an excellent supplement to those in this book.

The centenary of Einstein's birth in 1979 saw the appearance of two comprehensive collections of articles surveying research in most aspects of GR, one edited by Hawking & Israel (1979) and the other by Held (1980*a,b*). The student may find many of their articles helpful in supplementing later chapters, and several of them will be referred to explicitly in later bibliographies.

We shall discuss various strong-field metrics in later chapters, but few of them will be time dependent. For discussions of how to formulate and solve Einstein's equations as an initial-value system, see Misner *et al.* (1973), Fischer & Marsden (1979), Choquet-Bruhat & York (1980), and Isenberg & Nester (1980).

The measurement of the mass and angular momentum of a source by looking at its distant gravitational field is discussed in Misner *et al.* (1973), Weinberg (1972), Ashtekar (1980), and Winicour (1980).

8.6 Exercises

1 Show that Eq. (8.2) is a solution of Eq. (8.1) by the following method. Assume the point particle to be at the origin, $r = 0$, and to produce a spherically symmetric field. Then use Gauss' law on a sphere of radius r to conclude

$$\frac{d\phi}{dr} = Gm/r^2.$$

Deduce Eq. (8.2) from this. (Consider the behavior at infinity.)

2(a) Derive the following useful conversion factors from the SI values of G and c:

$$G/c^2 = 7.425 \times 10^{-28} \text{ m kg}^{-1} = 1,$$
$$c^5/G = 3.629 \times 10^{52} \text{ J s}^{-1} = 1.$$

(b) Derive the values in geometrized units of the constants in Table 8.1 from their given values in SI units.

(c) Express the following quantities in geometrized units:
 (i) a density (typical of neutron stars) $\rho = 10^{17} \text{ kg m}^{-3}$;
 (ii) a pressure (also typical of neutron stars) $p = 10^{33} \text{ kg s}^{-2} \text{ m}^{-1}$;
 (iii) the acceleration of gravity on Earth's surface $g = 9.80 \text{ m s}^{-2}$;
 (iv) the luminosity of a supernova $L = 10^{41} \text{ J s}^{-1}$.

(d) Three dimensioned constants in nature are regarded as fundamental: c, G, and $\hbar$. With $c = G = 1$, $\hbar$ has units m^2, so $\hbar^{1/2}$ defines a fundamental unit of length, called the Planck length. From Table 8.1, we calculate $\hbar^{1/2} = 1.616 \times 10^{-35}$ m. Since this number involves the fundamental constants of relativity, gravitation, and quantum theory, many physicists feel that this length will play an important role in quantum gravity. Express this length in terms of the SI values of c, G, and $\hbar$. Similarly, use the conversion factors to calculate the Planck mass and Planck time, fundamental numbers formed from c, G, and $\hbar$ that have the units of mass and time respectively. Compare these fundamental numbers with characteristic masses, lengths, and timescales that are known from elementary particle theory.

3 Calculate in geometrized units:
(a) the Newtonian potential ϕ of the Sun at the Sun's surface, radius 6.960×10^8 m;
(b) the Newtonian potential ϕ of the Sun at the radius of Earth's orbit, $r = 1 \text{ AU} = 1.496 \times 10^{11}$ m;
(c) the Newtonian potential ϕ of Earth at its surface, radius $= 6.371 \times 10^6$ m;
(d) the velocity of Earth in its orbit around the Sun.
(e) You should have found that your answer to (b) was larger than to (c). Why, then, do we on Earth feel Earth's gravitational pull much more than the Sun's?
(f) Show that a circular orbit around a body of mass M has an orbital velocity, in Newtonian theory, of $v^2 = -\phi$, where ϕ is the Newtonian potential.

4(a) Let A be an $n \times n$ matrix whose entries are all very small, $|A_{ij}| \ll 1/n$, and let I be the unit matrix. Show that

$$(I + A)^{-1} = I - A + A^2 - A^3 + A^4 - + \cdots$$

by proving that (i) the series on the right-hand side converges absolutely for each of the n^2 entries, and (ii) $(I + A)$ times the right-hand side equals I.

(b) Use (a) to establish Eq. (8.21) from Eq. (8.20).

5(a) Show that if $h_{\alpha\beta} = \xi_{\alpha,\beta} + \xi_{\beta,\alpha}$, then Eq. (8.25) vanishes.
 (b) Argue from this that Eq. (8.25) is gauge invariant.
 (c) Relate this to Exer. 10, § 7.6.

6 Weak-field theory assumes $g_{\mu\nu} = \eta_{\mu\nu} + h_{\mu\nu}$, with $|h_{\mu\nu}| \ll 1$. Similarly, $g^{\mu\nu}$ must be close to $\eta^{\mu\nu}$, say $g^{\mu\nu} = \eta^{\mu\nu} + \delta g^{\mu\nu}$. Show from Exer. 4a that $\delta g^{\mu\nu} = -h^{\mu\nu} + 0(h^2)$. Thus, $\eta^{\mu\alpha}\eta^{\nu\beta}h_{\alpha\beta}$ is *not* the deviation of $g^{\mu\nu}$ from flatness.

7(a) Prove that $\bar{h}^{\alpha}{}_{\alpha} = -h^{\alpha}{}_{\alpha}$.
 (b) Prove Eq. (8.31).

8 Derive Eq. (8.32) in the following manner.
(a) Show that $R^{\alpha}{}_{\beta\mu\nu} = \eta^{\alpha\sigma}R_{\sigma\beta\mu\nu} + 0(h^2_{\alpha\beta})$.
(b) From this calculate $R_{\alpha\beta}$ to first order in $h_{\mu\nu}$.
(c) Show that $g_{\alpha\beta}R = \eta_{\alpha\beta}\eta^{\mu\nu}R_{\mu\nu} + 0(h^2_{\alpha\beta})$.
(d) From this conclude that
$$G_{\alpha\beta} = R_{\alpha\beta} - \tfrac{1}{2}\eta_{\alpha\beta}R,$$
i.e. that the linearized $G_{\alpha\beta}$ is the *trace reverse* of the linearized $R_{\alpha\beta}$, in the sense of Eq. (8.29).
(e) Use this to simplify somewhat the calculation of Eq. (8.32).

9(a) Show from Eq. (8.32) that G_{00} and G_{0i} do not contain second time derivatives of any $\bar{h}_{\alpha\beta}$. Thus only the *six* equations, $G_{ij} = 8\pi T_{ij}$, are true dynamical equations. Relate this to the discussion at the end of § 8.2. The equations $G_{0\alpha} = 8\pi T_{0\alpha}$ are called *constraint* equations because they are relations among the initial data for the other six equations, which prevent one choosing all these data freely.
 (b) Eq. (8.42) contains second time derivatives even when μ or ν is zero. Does this contradict (a)? Why?

10 Use the Lorentz gauge condition, Eq. (8.33), to simplify $G_{\alpha\beta}$ to Eq. (8.41).

11 When one writes Maxwell's equations in special-relativistic form, one identifies the scalar potential ϕ and three-vector potential A_i (signs defined by $E_i = -\phi_{,i} - A_{i,0}$) as components of a one-form $A_0 = -\phi$, A_i (one-form) = A_i (three-vector). A gauge transformation is the replacement $\phi \rightarrow \phi - \partial f/\partial t$, $A_i \rightarrow A_i + f_{,i}$. This leaves the electric and magnetic fields unchanged. The Lorentz gauge is a gauge in which $\partial\phi/\partial t + \nabla_i A^i = 0$. Write both the gauge transformation and the Lorentz gauge condition in four-tensor notation. Draw the analogy with similar equations in linearized gravity.

12 Prove Eq. (8.34).

13   The inequalities $|T^{00}| \gg |T^{0i}| \gg |T^{ij}|$ for a Newtonian system are illustrated in Exer. 2c and 3d–e. Devise physical arguments to justify them in general.

14   From Eq. (8.46) and the inequalities among the components $h_{\alpha\beta}$, derive Eqs. (8.47)–(8.50).

15   We have argued that we should use convenient coordinates to solve the weak-field problem (or any other!), but that any physical results should be expressible in coordinate-free language. From this point of view our demonstration of the Newtonian limit is as yet incomplete, since in Ch. 7 we merely showed that the metric Eq. (7.50), led to Newton's law $\mathrm{d}\boldsymbol{p}/\mathrm{d}t = -m\boldsymbol{\nabla}\phi$.

But surely this is a coordinate-dependent equation, involving coordinate time and position. It is certainly not a valid four-dimensional tensor equation. Fill in this gap in our reasoning by showing that one can make physical measurements to verify that the relativistic predictions match the Newtonian ones. (For example, what is the relation between the proper time one orbit takes and its proper circumference?)

16   Re-do the derivation of the Newtonian limit by replacing 8π in Eq. (8.10) by k and following through the changes this makes in subsequent equations. Verify that one recovers Eq. (8.50) only if $k = 8\pi$.

17(a)   A small planet orbits a static neutron star in a circular orbit whose proper circumference is 6×10^{11} m. The orbital period takes 200 days of the planet's proper time. Estimate the mass M of the star.

  (b)   Five satellites are placed into circular orbits around a static black hole. The proper circumferences and proper periods of their orbits are given in the table below. Use the method of (a) to estimate the hole's mass. Explain the trend of the results you get for the satellites.

Proper circumference	2.5×10^6 m	6.3×10^6 m	6.3×10^7 m	3.1×10^8 m	6.3×10^9 m
Proper period	8.4×10^{-3} s	0.055 s	2.1 s	23 s	2.1×10^3 s

18   Consider the field equations with cosmological constant, Eq. (8.7), with Λ arbitrary and $k = 8\pi$.

  (a)   Find the Newtonian limit and show that one recovers the motion of the planets only if $|\Lambda|$ is very small. Given that the radius of Pluto's orbit is 5.9×10^{12} m, set an upper bound on $|\Lambda|$ from solar-system measurements.

  (b)   By bringing Λ over to the right-hand side of Eq. (8.7), one can regard $-\Lambda g^{\mu\nu}/8\pi$ as the stress–energy tensor of 'empty space'. Given that the observed mass of the region of the universe near our Galaxy would have a density of about 10^{-27} kg m^{-3} if it were uniformly distributed, do you think that a value of $|\Lambda|$ near the limit you established in (a) could have observable consequences for cosmology?

19 In this exercise we shall compute the first correction to the Newtonian solution caused by a source that rotates. In Newtonian gravity, the angular momentum of the source does not affect the field: two sources with the same $\rho(x^i)$ but different angular momenta have the same field. Not so in relativity, since *all* components of $T^{\mu\nu}$ generate the field.

(a) Suppose a spherical body of uniform density ρ and radius R rotates rigidly about the x^3 axis with constant angular velocity Ω. Write down the components $T^{0\nu}$ in a Lorentz frame at rest with respect to the center of mass of the body, assuming ρ, Ω and R are independent of time. For each component, work to lowest nonvanishing order in ΩR.

(b) The general solution to the equation $\nabla^2 f = g$, which vanishes at infinity, is the generalization of Eq. (8.2),

$$f(x) = -\frac{1}{4\pi} \int \frac{g(y)}{|x-y|} \, d^3 y,$$

which reduces to Eq. (8.2) when g is nonzero in a very small region. Use this to solve Eq. (8.42) for $\bar{h}^{00}$ and $\bar{h}^{0j}$ for the source described in (a). Obtain the solutions only outside the body, and only to lowest nonvanishing order in r^{-1}, where r is the distance from the body's center. Express the result for $\bar{h}^{0j}$ in terms of the body's angular momentum. Find the metric tensor within this approximation, and transform it to spherical coordinates.

(c) Because the metric is independent of t and the azimuthal angle ϕ, particles orbiting this body will have p_0 and p_ϕ constant along their trajectories (see § 7.4). Consider a particle of nonzero rest mass in a circular orbit of radius r in the equatorial plane. To lowest order, calculate the difference between its orbital period in the positive sense (i.e., rotating in the sense of the central body's rotation) and in the negative sense. (Define the period to be the coordinate time taken for one orbit of $\Delta\phi = 2\pi$.)

(d) From this devise an experiment to measure the angular momentum J of the central body. We take the central body to be the Sun ($M = 2 \times 10^{30}$ kg, $R = 7 \times 10^8$ m, $\Omega = 3 \times 10^{-6}$ s^{-1}) and the orbiting particle Earth ($r = 1.5 \times 10^{11}$ m). What would be the difference in the year between positive and negative orbits?

9

Gravitational radiation

9.1 The propagation of gravitational waves

It may happen that in a region of spacetime the gravitational field is weak but not stationary. This can happen far from a fully relativistic source undergoing rapid changes that took place long enough ago for the disturbances produced by the changes to reach the distant region under consideration. We shall study this problem by using the weak-field equations developed in the last chapter. The Einstein equations Eq. (8.42), in vacuum ($T^{\mu\nu} = 0$) are

$$\left(-\frac{\partial^2}{\partial t^2} + \nabla^2 \right) \bar{h}^{\alpha\beta} = 0. \tag{9.1}$$

In this chapter we will not neglect $\partial^2/\partial t^2$. Eq. (9.1) is called the three-dimensional wave equation. We shall show that it has a solution of the form

$$\bar{h}^{\alpha\beta} = A^{\alpha\beta} \exp(\mathrm{i}k_\alpha x^\alpha), \tag{9.2}$$

where $\{k_\alpha\}$ are the constant components of some one-form and $\{A^{\alpha\beta}\}$ the constant components of some tensor. Eq. (9.1) can be written as

$$\eta^{\mu\nu} \bar{h}^{\alpha\beta}{}_{,\mu\nu} = 0, \tag{9.3}$$

and, from Eq. (9.2), we have

$$\bar{h}^{\alpha\beta}{}_{,\mu} = k_\mu \bar{h}^{\alpha\beta}. \tag{9.4}$$

Therefore Eq. (9.3) becomes

$$\eta^{\mu\nu}\bar{h}^{\alpha\beta}{}_{,\mu\nu} = \eta^{\mu\nu}k_{\mu}k_{\nu}\bar{h}^{\alpha\beta} = 0.$$

This can vanish only if

$$\eta^{\mu\nu}k_{\mu}k_{\nu} = k^{\nu}k_{\nu} = 0. \tag{9.5}$$

So Eq. (9.2) gives a solution to Eq. (9.1) if k_{α} is a *null* one-form or, equivalently, if the associated four-vector k^{α} is null, i.e. tangent to the world line of a photon. (Recall that we raise and lower indices with the flat-space metric tensor $\eta^{\mu\nu}$, so k^{α} is a Minkowski null vector.) Eq. (9.2) describes a wavelike solution. The value of $\bar{h}^{\alpha\beta}$ is constant on a hypersurface on which $k_{\alpha}x^{\alpha}$ is constant:

$$k_{\alpha}x^{\alpha} = k_0 t + \mathbf{k} \cdot \mathbf{x} = \text{const.}, \tag{9.6}$$

where $\mathbf{k}$ refers to $\{k^i\}$. It is conventional to refer to k^0 as ω, which is called the frequency of the wave:

$$\vec{k} \to (\omega, \mathbf{k}) \tag{9.7}$$

is the time–space decomposition of $\vec{k}$. Imagine a photon moving in the direction of the null vector $\vec{k}$. It travels on a curve

$$x^{\mu}(\lambda) = k^{\mu}\lambda + l^{\mu}, \tag{9.8}$$

where λ is a parameter and l^{μ} is a constant vector (the photon's position at $\lambda = 0$). From Eqs. (9.8) and (9.5) we find

$$k_{\mu}x^{\mu}(\lambda) = k_{\mu}l^{\mu} = \text{const.} \tag{9.9}$$

Comparing this with Eq. (9.6), we see that the photon travels with the gravitational wave, staying forever at the same phase. We express this by saying that the wave itself travels at the speed of light, and $\vec{k}$ is its direction of travel. The nullity of $\vec{k}$ implies

$$\omega^2 = |\mathbf{k}|^2, \tag{9.10}$$

which is referred to as the dispersion relation for the wave. Readers familiar with wave theory will immediately see from Eq. (9.10) that the wave's phase velocity is 1, as is its group velocity.

The Einstein equations only assume the simple form, Eq. (9.1), if we impose the gauge condition

$$\bar{h}^{\alpha\beta}{}_{,\beta} = 0, \tag{9.11}$$

whose consequences we must therefore consider. From Eq. (9.4) we find

$$A^{\alpha\beta}k_{\beta} = 0, \tag{9.12}$$

which is a restriction on $A^{\alpha\beta}$: it must be orthogonal to $\vec{k}$.

The solution $A^{\alpha\beta}\exp(ik_{\mu}\mu)$ is called a *plane wave*. (Of course, in physical applications, one uses only the real part of this expression, allowing $A^{\alpha\beta}$ to be complex.) By the theorems of Fourier analysis, *any* solution of Eqs. (9.1) and (9.11) is a superposition of plane wave solutions. (See Exer. 3, § 9.6.)

The transverse–traceless gauge. We so far have only one constraint, Eq. (9.12), on the amplitude $A^{\alpha\beta}$, but we can use our gauge freedom to restrict it further. Recall from Eq. (8.38) that we can change the gauge while remaining within the Lorentz class of gauges using any vector solving

$$\left(-\frac{\partial^2}{\partial t^2}+\nabla^2\right)\xi_\alpha = 0. \tag{9.13}$$

Let us choose a solution

$$\xi_\alpha = B_\alpha \exp(\mathrm{i}k_\mu x^\mu), \tag{9.14}$$

where B_α is a constant and k^μ is the same null vector as for our wave solution. This produces a change in $h^{\alpha\beta}$, given by Eq. (8.24),

$$h^{(\mathrm{NEW})}_{\alpha\beta} = h^{(\mathrm{OLD})}_{\alpha\beta} - \xi_{\alpha,\beta} - \xi_{\beta,\alpha} \tag{9.15}$$

and a consequent change in $\bar{h}_{\alpha\beta}$, given by Eq. (8.34),

$$\bar{h}^{(\mathrm{NEW})}_{\alpha\beta} = \bar{h}^{(\mathrm{OLD})}_{\alpha\beta} - \xi_{\alpha,\beta} - \xi_{\beta,\alpha} + \eta_{\alpha\beta}\xi^\mu_{,\mu}. \tag{9.16}$$

Using Eq. (9.14) and dividing out the exponential factor common to all terms gives

$$A^{(\mathrm{NEW})}_{\alpha\beta} = A^{(\mathrm{OLD})}_{\alpha\beta} - \mathrm{i}B_\alpha k_\beta - \mathrm{i}B_\beta k_\alpha + \mathrm{i}\eta_{\alpha\beta}B^\mu k_\mu. \tag{9.17}$$

In Exer. 5, § 9.6 it is shown that B_α can be chosen to impose two further restrictions on $A^{(\mathrm{NEW})}_{\alpha\beta}$:

$$A^\alpha{}_\alpha = 0 \tag{9.18}$$

and

$$A_{\alpha\beta}U^\beta = 0, \tag{9.19}$$

where $\vec{U}$ is some fixed four-velocity, i.e. any constant timelike unit vector we wish to choose. Eqs. (9.12), (9.18), and (9.19) together are called the *transverse–traceless* (TT) gauge conditions. (The word 'traceless' refers to Eq. (9.18); 'transverse' will be explained below.) We have now used up all our gauge freedom, so any remaining independent components of $A_{\alpha\beta}$ must be physically important. Notice, by the way, that the trace condition, Eq. (9.18), implies (see Eq. (8.29))

$$\bar{h}^{\mathrm{TT}}_{\alpha\beta} = h^{\mathrm{TT}}_{\alpha\beta}, \tag{9.20}$$

Let us go to a Lorentz frame for the background Minkowski spacetime (i.e. make a background Lorentz transformation), in which the vector $\vec{U}$ upon which we have based the TT gauge is the time basis vector $U^\beta = \delta^\beta{}_0$. Then Eq. (9.19) implies $A_{\alpha 0} = 0$ for all α. In this frame, let us orient our spatial coordinate axes so that the wave is travelling in the z direction,

$\vec{k} \rightarrow (\omega, 0, 0, \omega)$. Then, with Eq. (9.19), Eq. (9.12) implies $A_{\alpha z} = 0$ for all α. (This is the origin of the adjective 'transverse' for the gauge: $A_{\mu\nu}$ is 'across' the direction of propagation $\vec{e}_z$.) These two restrictions mean that only A_{xx}, A_{yy}, and $A_{xy} = A_{yx}$ are nonzero. Moreover, the trace condition, Eq. (9.18), implies $A_{xx} = -A_{yy}$. In matrix form, we therefore have in this specially chosen frame

$$(A_{\alpha\beta}^{TT}) = \begin{pmatrix} 0 & 0 & 0 & 0 \\ 0 & A_{xx} & A_{xy} & 0 \\ 0 & A_{xy} & -A_{xx} & 0 \\ 0 & 0 & 0 & 0 \end{pmatrix}. \tag{9.21}$$

There are only *two* independent constants, A_{xx}^{TT} and A_{xy}^{TT}. What is their physical significance?

The effect of waves on free particles. As we remarked earlier, any wave is a superposition of plane waves; if the wave travels in the z direction we can put all the plane waves in the form of Eq. (9.21), so that any wave has only the two independent components h_{xx}^{TT} and h_{xy}^{TT}. Consider a situation in which a particle initially in a wave-free region of spacetime encounters a gravitational wave. Choose a background Lorentz frame in which the particle is initially at rest, and choose the TT gauge referred to this frame (i.e. the four-velocity U^α in Eq. (9.19) is the initial four-velocity of the particle). A free particle obeys the geodesic equation, Eq. (7.9),

$$\frac{d}{d\tau} U^\alpha + \Gamma^\alpha{}_{\mu\nu} U^\mu U^\nu = 0. \tag{9.22}$$

Since the particle is initially at rest, the initial value of its acceleration is

$$(dU^\alpha/d\tau)_0 = -\Gamma^\alpha{}_{00} = -\tfrac{1}{2}\eta^{\alpha\beta}(h_{\beta 0,0} + h_{0\beta,0} - h_{00,\beta}). \tag{9.23}$$

But by Eq. (9.21), $h_{\beta 0}^{TT}$ vanishes, so initially the acceleration vanishes. This means the particle will still be at rest a moment later, and then, by the same argument, the acceleration will still be zero a moment later. The result is that the particle remains at rest forever, regardless of the wave! However, being 'at rest' simply means remaining at constant coordinate position, so we should not be too hasty in its interpretation. All we have discovered is that by choosing the TT gauge – which means making a particular adjustment in the 'wiggles' of our coordinates – we have found a coordinate system that stays attached to individual particles. This in itself has no invariant geometrical meaning.

To get a better measure of the effect of the wave, let us consider two nearby particles, one at the origin and another at $x = \varepsilon$, $y = z = 0$, both beginning at rest. Both then remain at these coordinate positions, and the proper distance between them is

$$\Delta l \equiv \int |ds^2|^{1/2} = \int |g_{\alpha\beta} \, dx^\alpha \, dx^\beta|^{1/2}$$

$$= \int_0^\varepsilon |g_{xx}|^{1/2} \, dx \approx |g_{xx}(x=0)|^{1/2} \varepsilon$$

$$\approx [1 + \tfrac{1}{2} h_{xx}^{TT}(x=0)]\varepsilon. \tag{9.24}$$

Now, since h_{xx}^{TT} is not generally zero, the *proper* distance (as opposed to the coordinate distance) does change with time. This is an illustration of the difference between computing a coordinate-dependent number (the position of a particle) and a coordinate-independent number (the proper distance between two particles). The effect of the wave is unambiguously seen in the coordinate-independent number.

Another approach to the same question involves the equation of geodesic deviation, Eq. (6.87). Between the two particles set up the connecting vector ξ^α. It obeys the equation

$$\frac{d^2}{d\tau^2} \xi^\alpha = R^\alpha{}_{\mu\nu\beta} U^\mu U^\nu \xi^\beta, \tag{9.25}$$

where $\vec{U} = d\vec{x}/d\tau$ is the four-velocity of the two particles. In these coordinates the components of $\vec{U}$ are needed only to lowest (i.e. flat-space) order, since any corrections to U^α that depend on $h_{\mu\nu}$ will give terms second order in $h_{\mu\nu}$ in the above equation (because $R^\alpha{}_{\mu\nu\beta}$ is already first order in $h_{\mu\nu}$). Therefore $\vec{U} \to (1, 0, 0, 0)$ and, initially, $\vec{\xi} \to (0, \varepsilon, 0, 0)$. Then, to first order in $h_{\mu\nu}$, Eq. (9.25) reduces to

$$\frac{d^2}{d\tau^2} \xi^\alpha = \frac{\partial^2}{\partial t^2} \xi^\alpha = \varepsilon R^\alpha{}_{00x} = -\varepsilon R^\alpha{}_{0x0}. \tag{9.26}$$

Now, it is a simple matter to use Eq. (8.25) to show that, in the TT gauge,

$$\left. \begin{aligned} R^x{}_{0x0} &= R_{x0x0} = -\tfrac{1}{2} h_{xx,00}^{TT}, \\ R^y{}_{0x0} &= R_{y0x0} = -\tfrac{1}{2} h_{xy,00}^{TT}, \\ R^y{}_{0y0} &= R_{y0y0} = -\tfrac{1}{2} h_{yy,00}^{TT} = -R^x{}_{0x0}, \end{aligned} \right\} \tag{9.27}$$

and all other independent components vanish. This means that two particles initially separated in the x direction have a separation vector which obeys

$$\frac{\partial^2}{\partial t^2} \xi^x = \tfrac{1}{2}\varepsilon \frac{\partial^2}{\partial t^2} h_{xx}^{TT}, \qquad \frac{\partial^2}{\partial t^2} \xi^y = \tfrac{1}{2}\varepsilon \frac{\partial^2}{\partial t^2} h_{xy}^{TT}, \tag{9.28a}$$

which is clearly consistent with Eq. (9.24), Similarly, two particles initially separated by ε in the y direction obey

$$\frac{\partial^2}{\partial t^2} \xi^y = \tfrac{1}{2}\varepsilon \frac{\partial^2}{\partial t^2} h_{yy}^{TT} = -\tfrac{1}{2}\varepsilon \frac{\partial^2}{\partial t^2} h_{xx}^{TT},$$

$$\frac{\partial^2}{\partial t^2} \xi^x = \tfrac{1}{2}\varepsilon \frac{\partial^2}{\partial t^2} h_{xy}^{TT}. \tag{9.28b}$$

Polarization of gravitational waves. These equations help us describe the polarization of the wave. Consider a ring of particles initially at rest in the x–y plane, as in Fig. 9.1(a). Suppose a wave has $h_{xx}^{TT} \neq 0, h_{xy}^{TT} = 0$. Then the particles will be moved (in terms of proper distance relative to the one in the center) in the way shown in Fig. 9.1(b), first in (say), then

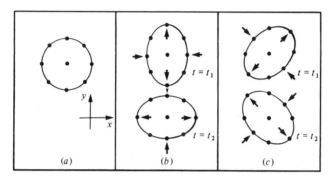

Fig. 9.1 (a) A circle of free particles before a wave travelling in the z direction reaches them. (b) Distortions of the circle produced by a wave with the '+' polarization. The two pictures represent the same wave at phases separated by 180°. Particles are positioned according to their proper distances from one another. (c) As (b) for the '×' polarization.

out, as the wave oscillates and h_{xx}^{TT} changes sign. If, instead, the wave had $h_{xy}^{TT} \neq 0$ but $h_{xx}^{TT} = h_{yy}^{TT} = 0$, then the picture would distort as in Fig. 9.1(c). Since h_{xy}^{TT} and h_{xx}^{TT} are independent, (b) and (c) provide a pictorial representation for two different linear polarizations. Notice that the two states are simply rotated 45° relative to one another. This constrasts with the two polarization states of an electromagnetic wave, which are 90° to each other. As Exer. 15, § 9.6, shows, this pattern of polarization is due to the fact that gravity is represented by the second-rank symmetric tensor $h_{\mu\nu}$. (By contrast, electromagnetism is represented by the vector potential A^μ of Exer. 11, § 8.6.)

An exact plane wave. Although all waves that we can expect to detect on Earth are so weak that linearized theory ought to describe them very accurately, it is interesting to see if the linear plane wave coresponds to some exact solution of the nonlinear equations that has similar properties. We shall briefly derive such a solution.

Suppose the wave is to travel in the z direction. By analogy with Eq. (9.2) we might hope to find a solution that depends only on

$$u \equiv t - z.$$

This suggests using u as a coordinate, as we did in Exer. 34, § 3.10 (with x replaced by z). In flat space it is natural, then, to define a complementary null coordinate

$$v = t + z,$$

so that the line element of flat spacetime becomes

$$ds^2 = -du\,dv + dx^2 + dy^2.$$

Now, we have seen that the linear wave affects only proper distances perpendicular to its motion, i.e. in the x–y coordinate plane. So let us look for a nonlinear generalization of this, i.e. for a solution with the metric

$$ds^2 = -du\,dv + f^2(u)\,dx^2 + g^2(u)\,dy^2,$$

where f and g are functions to be determined by Einstein's equations. It is a straightforward calculation to discover that the only nonvanishing Christoffel symbols and Riemann tensor components are

$$\Gamma^x{}_{xu} = \dot{f}/f, \qquad \Gamma^y{}_{yu} = \dot{g}/g,$$
$$\Gamma^v{}_{xx} = 2\dot{f}/f, \qquad \Gamma^v{}_{yy} = 2\dot{g}/g,$$
$$R^x{}_{uxu} = -\ddot{f}/f, \qquad R^y{}_{uyu} = -\ddot{g}/g,$$

and others obtainable by symmetries. Here, dots denote derivatives with respect to u. The only vacuum field equation then becomes

$$\ddot{f}/f + \ddot{g}/g = 0. \tag{9.29}$$

We can therefore prescribe an arbitrary function $g(u)$ and solve this equation for $f(u)$. This is the same freedom as we had in the linear case, where Eq. (9.2) can be multiplied by an arbitrary $f(k_z)$ and integrated over k_z to give the Fourier representation of an arbitrary function of $(z - t)$. In fact, if g is nearly 1,

$$g \approx 1 + \varepsilon(u),$$

so we are near the linear case, then Eq. (9.29) has a solution

$$f \approx 1 - \varepsilon(a).$$

This is just the linear wave in Eq. (9.21), with plane polarization with $h_{xy} = 0$, i.e. the polarization shown in Fig. 9.1(b). Moreover, it is easy to see that the geodesic equation implies, in the nonlinear case, that a particle initially at rest on our coordinates remains at rest. We have, therefore, a simple nonlinear solution corresponding to our approximate linear one.

9.2 The detection of gravitational waves

General considerations. The great progress that astronomy has made since about 1960 is due largely to the fact that technology has permitted astronomers to begin to observe in many different parts of the electromagnetic spectrum. Because they were restricted to observing visible light, the astronomers of the 1940s could have had no inkling of such diverse and exciting phenomena as giant radio galaxies, quasars, pulsars, compact X-ray binaries, molecular-line masers in dense clouds, and the cosmic microwave background radiation. As technology has progressed, each new wavelength region has revealed unexpected and important information. There are still regions of the electromagnetic spectrum that are largely unexplored, but there is another spectrum which is as yet completely untouched: the gravitational wave spectrum. As we shall see in § 9.3 below, nearly all astrophysical phenomena emit gravitational waves, and the most violent ones (which are of course among the most interesting ones!) give off radiation in copious amounts. In some situations, gravitational radiation carries information that no electromagnetic radiation can give us. For example, gravitational waves come to us direct from the heart of supernova explosions; the electromagnetic radiation from the same region is scattered countless times by the dense material surrounding the explosion, taking days to eventually make its way out, and in the process losing most of the detailed information it might carry about the explosion. We can also expect long-wavelength gravitational waves to tell us about the formation of very large black holes, should they exist in the centers of quasars and active galaxies. Beyond this, we can be virtually certain the gravitational-wave spectrum has surprises for us, clues to phenomena we never suspected. It is not surprising, therefore, that considerable effort is now being devoted to the development of sufficiently sensitive gravitational-wave antennas.

We have already encountered one type of 'antenna' in Exer. 9, § 9.6: by monitoring electromagnetic signals traveling between two freely falling particles, one can detect the passage of a wave. In principle, one could do this by putting several spacecraft carrying accurate atomic clocks in orbit about the Sun, and having them continually exchanging signals among one another. A less expensive alternative is precise tracking of existing interplanetary space probes, but this is less sensitive (see Hellings

1983). In the laboratory, one can use the same principle to monitor the changes in separation between two heavy masses, suspended from supports that isolate the masses from outside vibrations. This approach is being actively pursued in a number of laboratories, generally using lasers for the electromagnetic radiation and interferometers to monitor changes in the separation of the masses. These and other devices are described in some detail in Smarr (1979*a*) and Bertotti (1974, 1977).

The technical difficulties involved in the detection of gravitational radiation are enormous, because the amplitudes of the metric perturbations $h_{\mu\nu}$ that can be expected from distant sources are so small (see § 9.3 below). To get some feeling for what is involved, we shall look at the oldest type of detector, the resonant oscillator pioneered by J. Weber (1961). A deeper discussion of detection theory may be found in Misner *et al.* (1973).

A resonant detector. Let us consider the following idealized detector, depicted in Fig. 9.2. Two point particles, each of mass *m*, are connected

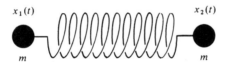

$x_1(t)$ $x_2(t)$

m m

Fig. 9.2   A spring with two identical masses as a detector of gravitational waves.

by a massless spring with spring constant k, damping constant ν, and unstretched length l_0. The system lies on the x axis of our TT coordinate system, with the masses at coordinate positions x_1 and x_2. In flat spacetime, the masses would obey the equations

$$mx_{1,00} = -k(x_1 - x_2 + l_0) - \nu(x_1 - x_2)_{,0} \tag{9.30}$$

and

$$mx_{2,00} = -k(x_2 - x_1 - l_0) - \nu(x_2 - x_1)_{,0}. \tag{9.31}$$

If we define

$$\xi = x_2 - x_1 - l_0, \qquad \omega_0^2 = 2k/m, \qquad \gamma = \nu/m, \tag{9.32}$$

then we can combine Eqs. (9.30) and (9.31) to give

$$\xi_{,00} + 2\gamma\xi_{,0} + \omega_0^2\xi = 0, \tag{9.33}$$

the usual damped harmonic-oscillator equation.

What is the situation as a gravitational wave passes? We shall analyze the problem in three steps.

(1) A *free* particle remains at rest in the TT coordinates. This means that a local inertial frame at rest at, say, x_1, before the wave arrives remains at rest there after the wave hits. Let its coordinates be $\{x^{\alpha'}\}$. Suppose that the only motions in the system are those produced by the wave, i.e. that $\xi = 0(l_0|h_{\mu\nu}|) \ll l_0$. Then the masses' velocities will be small as well, and Newton's equations for the masses will apply in the local inertial frame:

$$mx^{j'}_{,0'0'} = F^{j'}, \tag{9.34}$$

where $\{F^{j'}\}$ are the components of any nongravitational forces on the masses. Because $\{x^{\alpha'}\}$ can differ from our TT coordinates $\{x^{\alpha}\}$ only by terms of order $h_{\mu\nu}$, and because x_1, $x_{1,0}$, and $x_{1,00}$ are all of order $h_{\mu\nu}$, we can use the TT coordinates in Eq. (9.34) with negligible error:

$$mx^{j}_{,00} = F^{j} + 0(|h_{\mu\nu}|^2). \tag{9.35}$$

(2) The only nongravitational force on each mass is that due to the spring. Since all the motions are slow, the spring will exert a force proportional to its instantaneous *proper* extension, as measured using the metric. If the proper length of the spring is l, and if the gravitational wave travels in the z direction, then

$$l(t) = \int_{x_1(t)}^{x_2(t)} [1 + h^{TT}_{xx}(t)]^{1/2} \, dt, \tag{9.36}$$

and Eq. (9.35) for our system gives

$$mx_{1,00} = -k(l_0 - l) - \nu(l_0 - l)_{,0}, \tag{9.37}$$

$$mx_{2,00} = -k(l - l_0) - \nu(l - l_0)_{,0}. \tag{9.38}$$

(3) Let us define ω_0 and γ as before, and

$$\xi = l - l_0, \tag{9.39}$$

$$= x_2 - x_1 - l_0 + \tfrac{1}{2}h^{TT}_{xx}(x_2 - x_1) + 0(|h_{\mu\nu}|^2). \tag{9.40}$$

This can be solved to give

$$x_2 - x_1 = l_0 + \xi - \tfrac{1}{2}h^{TT}_{xx}l_0 + 0(|h_{\mu\nu}|^2). \tag{9.41}$$

If we use this in the difference between Eqs. (9.38) and (9.37), we obtain

◆ $$\xi_{,00} + 2\gamma\xi_{,0} + \omega_0^2\xi = \tfrac{1}{2}l_0 h^{TT}_{xx,00}, \tag{9.42}$$

correct to first order in h^{TT}_{xx}. This is the fundamental equation governing the response of the detector to the gravitational wave. It has the simple form of a forced, damped harmonic oscillator. The generalization of this to waves incident from other directions is dealt with in Exer. 20, § 9.6, and an alternative derivation using the equation of geodesic deviation may be found in Exer. 21, § 9.6.

One might use a detector of this sort as a resonant detector for sources of gravitational radiation of a fixed frequency (e.g. pulsars or close binary stars). (It can also be used to detect bursts – short wave packets of radiation – but we will not discuss those.) Suppose that

$$h_{xx}^{TT} = A \cos \Omega t,$$ (9.43)

then the steady solution for ξ is

$$\xi = R \cos (\Omega t + \phi),$$ (9.44)

with

$$R = \tfrac{1}{2} l_0 \Omega^2 A / [(\omega_0 - \Omega)^2 + 4\Omega^2 \gamma^2]^{1/2},$$ (9.45)

$$\tan \phi = 2\gamma \Omega / (\omega_0^2 - \Omega^2).$$ (9.46)

(Of course, the general initial-value solution for ξ will also contain transients, which damp away on a timescale $1/\gamma$.) The energy of oscillation of the detector is, to lowest order in h_{xx}^{TT},

$$E = \tfrac{1}{2} m(x_{1,0})^2 + \tfrac{1}{2} m(x_{2,0})^2 + \tfrac{1}{2} k \xi^2.$$ (9.47)

For a detector which was at rest before the wave arrived, we have $x_{1,0} = -x_{2,0} = -\xi_0/2$ (see Exer. 22, § 9.6), so that

$$E = \tfrac{1}{4} m[(\xi_0)^2 + \omega_0^2 \xi^2]$$ (9.48)

$$= \tfrac{1}{4} m R^2 [\Omega^2 \sin^2 (\Omega t + \phi) + \omega_0^2 \cos^2 (\Omega t + \phi)].$$ (9.49)

The mean value of this is its average over one period, $2\pi/\Omega$:

$$\langle E \rangle = \tfrac{1}{8} m R^2 (\omega_0^2 + \Omega^2).$$ (9.50)

We shall always use angle brackets $\langle \ \rangle$ to denote time averages.

If we wish to detect a specific source whose frequency Ω is known, then we should adjust ω_0 to equal Ω for maximum response (resonance), as we see from Eq. (9.45). In this case the amplitude of the response will be

$$R_{\text{resonant}} = \tfrac{1}{4} l_0 A(\Omega/\gamma)$$ (9.51)

and the energy of vibration is

$$E_{\text{resonant}} = \tfrac{1}{64} m l_0^2 \Omega^2 A^2 (\Omega/\gamma)^2.$$ (9.52)

The ratio Ω/γ is related to what is usually called the quality factor Q of an oscillator, where $1/Q$ is defined as the average fraction of the energy of the undriven oscillator, which it loses (to friction) in one radian of oscillation (see Exer. 24, § 9.6):

$$Q = \omega_0/2\gamma.$$ (9.53)

In the resonant case we have

$$E_{\text{resonant}} = \tfrac{1}{16} m l_0^2 \Omega^2 A^2 Q^2.$$ (9.54)

What numbers are realistic for laboratory detectors? Most such detectors are massive cylindrical bars, in which the 'spring' is the elasticity of the bar when it is stretched along its axis. When waves hit the bar broadside, they excite its longitudinal modes of vibration. The first detectors, built by J. Weber of the University of Maryland in the 1960s, were aluminum bars of mass 1.4×10^3 kg, length $l_0 = 1.5$ m, resonant frequency $\omega_0 = 10^4 \, s^{-1}$, and Q about 10^5. This means that a strong resonant gravitational wave of $A = 10^{-20}$ (see § 9.3 below) will excite the bar to an energy of the order of 10^{-20} J. The resonant amplitude given by Eq. (9.51) is only about 10^{-15} m, roughly the diameter of an atomic nucleus! Many realistic gravitational waves will have amplitudes many orders of magnitude smaller than this, and will last for much too short a time to bring the bar to its full resonant amplitude.

Clearly, the detection of such small levels of excitation will be hampered by random 'noise' in the oscillator. For example, thermal noise in any oscillator induces random vibrations with a mean energy of kT, where T is the absolute temperature and k is Boltzmann's constant,

$$k = 1.38 \times 10^{-23} \, \text{J/K}.$$

In our example, this will be comparable to the energy of excitation if T is room temperature (~300 K). Other sources of noise, such as vibrations from passing vehicles and everyday seismic disturbances, could be considerably larger than this, so the apparatus has to be very carefully isolated.

We have confined our discussion to on-resonance detection of a continuous wave, in the case when there are no motions in the detector, other than those produced by the wave. If the wave comes in as a burst with a wide range of frequencies, or if the excitation amplitude is well below noise levels, then rather different considerations apply. These are discussed in the references quoted earlier.

As of 1983, no gravitational waves have yet unambiguously been detected. This is not surprising because detector sensitivities are still smaller than required for most theoretical predictions of the gravitational wave flux incident on Earth. The main problem is, of course, noise, and various methods are being employed to overcome the problem: the growth of very large single crystals for use as high-Q detectors; the cooling of the massive aluminum bars to liquid-helium temperatures; the development of very stable high-power lasers for the laser-interferometer detectors; and the design of electronics which can detect excitations of the bars when there are only a few quanta in the excitation, without changing the number of quanta (i.e. when the excitation energy equals a few times $\hbar \omega_0$). These all stretch the limits of modern technology, with the result that many of the advances made by the builders of these antennae have much wider technological applications. Astronomers, meanwhile, wait and hope that these advances will also soon achieve their main purpose!

9.3 The generation of gravitational waves

Simple estimates. It is easy to see that the amplitude of any gravitational waves incident on Earth should be small. A 'strong' gravitational wave would have $h_{\mu\nu} = 0(1)$, and we should expect amplitudes like this only near the source, where the Newtonian potential would be of order 1. For a source of mass M, this would be at distances of order M from it. As with all radiation fields, the amplitude of the gravitational waves falls off as r^{-1} far from the source. (Readers who are not familiar with solutions of the wave equation will find demonstrations of this in the next sections.) So if Earth is a distance R from a source of mass M, the largest amplitude waves we should expect are of order M/R. For the formation of a $10M_\odot$ black hole in a supernova explosion in a nearby galaxy 10^{23} m away, this is about 10^{-17}. This is in fact an upper limit in this case, and less-violent events will lead to very much smaller amplitudes.

An approximate calculation of wave generation. Our object is to solve Eq. (8.42):

$$\left(-\frac{\partial^2}{\partial t^2} + \nabla^2\right) \bar{h}_{\mu\nu} = -16\pi T_{\mu\nu}. \tag{9.55}$$

We will find the exact solution in a later section. Here we will make some simplifying – but realistic – assumptions. We assume that the time-dependent part of $T_{\mu\nu}$ is in sinusoidal oscillation with frequency Ω, i.e. that it is the real part of

$$T_{\mu\nu} = S_{\mu\nu}(x^i)\,e^{-i\Omega t}, \tag{9.56}$$

and that the region of space in which $S_{\mu\nu} \neq 0$ is small compared with $2\pi/\Omega$, the wavelength of a gravitational wave of frequency Ω. The first assumption is not much of a restriction, since a general time dependence can be reduced to a sum over sinusoidal motions by Fourier analysis. Besides, many interesting astrophysical sources *are* roughly periodic: pulsating stars, pulsars, binary systems. The second assumption is called the slow-motion assumption, since it implies that the typical velocity inside the source region, which is Ω times the size of that region, should be much less than 1. All but the most powerful sources of gravitational waves probably satisfy this condition.

Let us look for a solution for $\bar{h}_{\mu\nu}$ of the form

$$\bar{h}_{\mu\nu} = B_{\mu\nu}(x^i)\, e^{-i\Omega t}. \tag{9.57}$$

(In the end we must take the real part of this for our answer.) Putting this and Eq. (9.56) into Eq. (9.55) gives

$$(\nabla^2 + \Omega^2)B_{\mu\nu} = -16\pi S_{\mu\nu}. \tag{9.58}$$

It is important to bear in mind as we proceed that the indices on $\bar{h}_{\mu\nu}$ in Eq. (9.55) play almost no role. We shall regard each component $\bar{h}_{\mu\nu}$ as simply a function on Minkowski space, satisfying the wave equation. All our steps would be the same if we were solving the scalar equation $(-\partial^2/\partial t^2 + \nabla^2)f = g$, until we come to Eq. (9.66).

Outside the source (i.e. where $S_{\mu\nu} = 0$) we want a solution $B_{\mu\nu}$ of Eq. (9.58), which represents outgoing radiation far away; and of all such possible solutions we want the one which dominates in the slow-motion limit. Let us define r to be the spherical polar radial coordinate whose origin is chosen inside the source. We show in Exer. 26, § 9.6, that the solution we seek is the simplest of all the solutions of Eq. (9.58) outside the source,

$$B_{\mu\nu} = \frac{A_{\mu\nu}}{r}\, e^{i\Omega r} + \frac{Z_{\mu\nu}}{r}\, e^{-i\Omega r}, \tag{9.59}$$

where $A_{\mu\nu}$ and $Z_{\mu\nu}$ are constants. The term in $e^{-i\Omega r}$ represents a wave traveling toward the origin $r = 0$ (called an ingoing wave), while the other term is outgoing (see Exer. 25, § 9.6). We want waves emitted by the source, so we choose $Z_{\mu\nu} = 0$.

Our problem is to determine $A_{\mu\nu}$ in terms of the source. Here we make our approximation that the source is nonzero only inside a sphere of radius $\varepsilon \ll 2\pi/\Omega$. Let us integrate Eq. (9.58) over the interior of this sphere. One term we get is

$$\int \Omega^2 B_{\mu\nu}\, d^3x \leqslant \Omega^2 |B_{\mu\nu}|_{\max} 4\pi\varepsilon^3/3, \tag{9.60}$$

where $|B_{\mu\nu}|_{\max}$ is the maximum value $B_{\mu\nu}$ takes inside the source. We will see that this term is negligible. The other term from integrating the left-hand side of Eq. (9.58) is

$$\int \nabla^2 B_{\mu\nu}\, d^3x = \oint \boldsymbol{n} \cdot \nabla B_{\mu\nu}\, dS, \tag{9.61}$$

by Gauss' theorem. But the surface integral is outside the source, where $B_{\mu\nu}$ is given by Eq. (9.59), which is spherically symmetric:

$$\oint \boldsymbol{n} \cdot \nabla B_{\mu\nu}\, dS = 4\pi\varepsilon^2 \left(\frac{d}{dr} B_{\mu\nu}\right)_{r=\varepsilon} \approx -4\pi A_{\mu\nu}, \tag{9.62}$$

again with the approximation $\varepsilon \ll 2\pi/\Omega$. Finally, we define the integral of the right-hand side of Eq. (9.58) to be

$$J_{\mu\nu} = \int S_{\mu\nu}\, d^3x. \tag{9.63}$$

Combining these results in the limit $\varepsilon \to 0$ gives

$$A_{\mu\nu} = 4 J_{\mu\nu}, \tag{9.64}$$

$$\bar{h}_{\mu\nu} = 4 J_{\mu\nu}\, e^{i\Omega(r-t)}/r. \tag{9.65}$$

These are the expressions for the gravitational waves generated by the source, neglecting terms of order r^{-2} *and* any r^{-1} terms that are higher order in $\varepsilon\Omega$.

It is possible to simplify these considerably. Here we begin to use the fact that $\{\bar{h}_{\mu\nu}\}$ are components of a single tensor, not the unrelated functions we have solved for in Eq. (9.65). From Eq. (9.63) we learn

$$J_{\mu\nu}\, e^{-i\Omega t} = \int T_{\mu\nu}\, d^3x, \tag{9.66}$$

which has as one consequence:

$$-i\Omega J^{\mu 0}\, e^{-i\Omega t} = \int T^{\mu 0}{}_{,0}\, d^3x. \tag{9.67}$$

Now, from the conservation law for $T^{\mu\nu}$,

$$T^{\mu\nu}{}_{,\nu} = 0, \tag{9.68}$$

we conclude that

$$T^{\mu 0}{}_{,0} = - T^{\mu k}{}_{,k} \tag{9.69}$$

and hence that

$$i\Omega J^{\mu 0}\, e^{-i\Omega t} = \int T^{\mu k}{}_{,k}\, d^3x = \oint T^{\mu k} n_k\, dS, \tag{9.70}$$

the last step being the application of Gauss' theorem to any volume completely containing the source. This means that $T^{\mu\nu} = 0$ on the surface bounding this volume, so that the right-hand side of Eq. (9.70) vanishes. This means that if $\Omega \neq 0$ we have

$$J^{\mu 0} = 0, \qquad \bar{h}^{\mu 0} = 0. \tag{9.71}$$

The expression for J_{ij} can also be rewritten in an instructive way, by using the result of Exer. 23, § 4.10.

$$\frac{d^2}{dt^2} \int T^{00} x^l x^m\, d^3x = 2 \int T^{lm}\, d^3x. \tag{9.72}$$

For a source in slow motion, we have seen in Ch. 7 that $T^{00} \approx \rho$, the Newtonian mass density. It follows that the integral on the left-hand side of Eq. (9.72) is what is often referred to as the quadrupole moment tensor of the mass distribution,

$$I^{lm} \equiv \int T^{00} x^l x^m \, d^3 x \tag{9.73a}$$

$$= D^{lm} e^{-i\Omega t} \tag{9.73b}$$

(Conventions for defining the quadrupole moment vary from one text to another. We follow Misner *et al.* (1973).) In terms of this we have

$$\bar{h}_{jk} = -2\Omega^2 D_{jk} e^{i\Omega(r-t)} / r. \tag{9.74}$$

It is important to remember that Eq. (9.74) is an approximation which neglects not merely all terms of order r^{-2} but also r^{-1} terms that are not dominant in the slow-motion approximation. In particular, $\bar{h}_{jk}{}^{,k}$ is of higher order, and this guarantees that the gauge condition $\bar{h}^{\mu\nu}{}_{,\nu} = 0$ is satisfied by Eqs. (9.74) and (9.71) at the lowest order in r^{-1} and Ω. Because of Eq. (9.74), this approximation is often called the *quadrupole* approximation for gravitational radiation.

As for the plane waves we studied earlier, we have here the freedom to make a further restriction of the gauge. The obvious choice is to try to find a TT gauge, transverse to the direction of motion of the wave (the radial direction), whose unit vector is $n^j = x^j / r$. Exer. 29, § 9.6, shows that this is possible, so that in the TT gauge we have the simplest form of the wave. If we choose our axes so that at the point where we measure the wave it is traveling in the z direction, then we can supplement Eq. (9.71) by

$$\bar{h}_{zi}^{TT} = 0, \tag{9.75}$$

$$\bar{h}_{xx}^{TT} = -\bar{h}_{yy}^{TT} = -\Omega^2 (\not{I}_{xx} - \not{I}_{yy}) e^{i\Omega r} / r, \tag{9.76}$$

$$\bar{h}_{xy}^{TT} = -2\Omega^2 \not{I}_{xy} e^{i\Omega r} / r, \tag{9.77}$$

where

$$\not{I}_{jk} = I_{jk} - \tfrac{1}{3} \delta_{jk} I^l_l \tag{9.78}$$

is called the trace-free or reduced quadrupole moment tensor.

Examples. Let us consider the waves emitted by a simple oscillator like the one we used as a detector in § 9.2. If both masses oscillate with angular frequency ω and amplitude A about mean equilibrium positions a distance l_0, apart, then, by Exer. 27, § 9.6, the quadrupole tensor has only one nonzero component,

$$I_{xx} = m[(x_1)^2 + (x_2)^2]$$
$$= [(-\tfrac{1}{2}l_0 - A \cos \omega t)^2 + (\tfrac{1}{2}l_0 + A \cos \omega t)^2]$$
$$= \text{const.} + mA^2 \cos 2\omega t + 2ml_0 A \cos \omega t. \tag{9.79}$$

Recall that only the sinusoidal part of I_{xx} should be used in the formulae developed in the previous paragraph. In this case there are two such pieces, with frequencies ω and 2ω. Since the wave equation, Eq. (9.55) is linear, we shall treat each term separately and simply add the results later. The ω term in I_{xx} is the real part of $2ml_0 A \exp(-i\omega t)$. The trace-free quadrupole tensor has components

$$\left.\begin{aligned} \mathscr{I}_{xx} &= I_{xx} - \tfrac{1}{3}I_j^j = \tfrac{2}{3}I_{xx} = \tfrac{4}{3}ml_0 A\, e^{-i\omega t}, \\ \mathscr{I}_{yy} &= \mathscr{I}_{zz} = -\tfrac{1}{3}I_{xx} = -\tfrac{2}{3}ml_0 A\, e^{-i\omega t}, \end{aligned}\right\}$$
(9.80)

all off-diagonal components vanishing. If we consider the radiation traveling in the z direction, we get, from Eqs. (9.75)–(9.77),

$$\bar{h}_{xx}^{TT} = -\bar{h}_{yy}^{TT} = -2m\omega^2 l_0 A\, e^{i\omega(r-t)}/r, \qquad \bar{h}_{xy}^{TT} = 0.$$
(9.81)

The radiation is linearly polarized, with an orientation such that the ellipse in Fig. 9.1 is aligned with the line joining the two masses. The same is true for the radiation going in the y direction, by symmetry. But for the radiation traveling in the x direction (i.e. along the line joining the masses), we need to make the substitutions $z \to x$, $x \to y$, $y \to z$ in Eqs. (9.76)–(9.77), and we find

$$\bar{h}_{ij}^{TT} = 0.$$
(9.82)

There is *no* radiation in the x direction. In Exer. 33, § 9.6, we will fill in this radiation pattern by calculating the amount of radiation and its polarization in arbitrary directions.

A similar calculation for the 2ω piece of I_{xx} gives the same radiation pattern, replacing Eq. (9.81) by

$$\bar{h}_{xx}^{TT} = -\bar{h}_{yy}^{TT} = -4m\omega^2 A^2\, e^{2i\omega(r-t)}/r, \qquad \bar{h}_{xy}^{TT} = 0.$$
(9.83)

The total radiation field is the real part of the sum of Eqs. (9.81) and (9.83), e.g.

$$\bar{h}_{xx}^{TT} = -[2m\omega^2 l_0 A \cos \omega(r-t) + 4m\omega^2 A^2 \cos 2\omega(r-t)]/r.$$
(9.84)

Let us estimate the radiation from a laboratory-sized generator of this type. If we take $m = 10^3$ kg $= 7 \times 10^{-24}$ m, $l_0 = 1$ m, $A = 10^{-4}$ m, and $\omega = 10^4\,\mathrm{s}^{-1} = 3 \times 10^{-4}\,\mathrm{m}^{-1}$, then the 2ω contribution is negligible and we find that the amplitude is about $10^{-34}/r$, where r is measured in meters. This shows that laboratory generators are unlikely to produce useful gravitational waves in the near future!

A more interesting example of a gravitational wave source is a binary star system. Strictly speaking, our derivation applies only to sources whose motions result from nongravitational forces (this is the content of Eq. (9.68)), but our final result, Eqs. (9.75)–(9.78), makes use only of the motions produced, not of the forces. It is perhaps not so surprising, then, that one can show that Eqs. (9.75)–(9.78) are a good first approximation for systems dominated by Newtonian gravitational forces. (See bibliography for references.) Let us suppose, then, that we have two stars of mass m, idealized as points in circular orbit about one another, separated a distance l_0 (i.e. moving on a circle of radius $\frac{1}{2}l_0$). Their orbit equation (gravitational force = 'centrifugal force') is

$$\frac{m^2}{l_0^2} = m\omega^2 \left(\frac{l_0}{2}\right) \Rightarrow \omega = (2m/l_0^3)^{1/2}, \tag{9.85}$$

where ω is the angular velocity of the orbit. Then, with an appropriate choice of coordinates, the masses move on the curves

$$\left.\begin{array}{ll} x_1(t) = \frac{1}{2}l_0 \cos \omega t, & y_1(t) = \frac{1}{2}l_0 \sin \omega t, \\ x_2(t) = -x_1(t), & y_2(t) = -y_1(t), \end{array}\right\} \tag{9.86}$$

where the subscripts 1 and 2 refer to the respective stars. These equations give

$$\left.\begin{array}{l} I_{xx} = \frac{1}{4}ml_0^2 \cos 2\omega t + \text{const.}, \\ I_{yy} = -\frac{1}{4}ml_0^2 \cos 2\omega t + \text{const.}, \\ I_{xy} = \frac{1}{4}ml_0^2 \sin 2\omega t. \end{array}\right\} \tag{9.87}$$

The reduced quadrupole tensor is, in complex notation and omitting time-independent terms,

$$\left.\begin{array}{l} \mathbf{I}_{xx} = -\mathbf{I}_{yy} = \frac{1}{4}ml_0^2 \, e^{-2i\omega t}, \\ \mathbf{I}_{xy} = \frac{1}{4}i ml_0^2 \, e^{-2i\omega t}. \end{array}\right\} \tag{9.88}$$

All the radiation comes out with frequency $\Omega = 2\omega$. The radiation along the z direction (perpendicular to the plane of the orbit) is, by Eqs. (9.75)–(9.77),

$$\left.\begin{array}{l} \bar{h}_{xx} = -\bar{h}_{yy} = -2ml_0^2\omega^2 \, e^{2i\omega(r-t)}/r, \\ \bar{h}_{xy} = -2i ml_0^2\omega^2 \, e^{2i\omega(r-t)}/r. \end{array}\right\} \tag{9.89}$$

This is *circularly polarized* radiation (see Exer. 14, § 9.6). The radiation in the plane of the orbit, say in the x direction, is found in the same manner used to derive Eq. (9.82). This gives

$$\bar{h}_{yy}^{TT} = -\bar{h}_{zz}^{TT} = \frac{1}{2}ml_0^2\omega^2 \, e^{2i\omega(r-t)}/r, \tag{9.90}$$

all others vanishing. This shows linear polarization aligned with the orbital plane. The antenna pattern and polarization are examined in greater detail in Exer. 35, § 9.6, and the calculation is generalized to unequal masses in elliptical orbits in Exer. 36.

The amplitude of the radiation is of order $ml_0^2\omega^2/r$, which, by Eq. (9.85), is $\sim (m\omega)^{2/3}m/r$. One particularly important system is the one containing the pulsar PSR 1913 + 16; this apparently consists of two very compact stars orbiting each other closely. The orbital period, inferred from the Doppler shift of the pulsar's period, is 7 h 45 min 7 s (27907 s, or 8.3721×10^{12} m), and both stars have masses approximately equal to $1.4 \, M_\odot$ (2.07 km) (Taylor & Weisberg 1982). If the system is 5 kpc = 1.5×10^{17} m away, then its radiation will have the approximate amplitude 10^{-20} at Earth. We will calculate the effect of this radiation on the binary orbit itself later in this chapter. In Ch. 10 we will discuss the dynamics of the system, including how the masses are measured.

Order-of-magnitude estimates. Although our simple approach does not enable us to write down solutions for $\bar{h}_{\mu\nu}$ generated by more complicated, nonperiodic motions, we can use Eq. (9.74) to obtain some order-of-magnitude estimates. Since D_{jk} is of order MR^2, for a system of mass M and size R, the radiation will have amplitude about $M(\Omega R)^2/r \approx v^2(M/r)$, where v is a typical internal velocity in the source. This applies directly to Eq. (9.90); note that in Eq. (9.84) the first term uses, instead of R^2, the product $l_0 A$ of the two characteristic lengths in the problem. If we are dealing with, say, a collapsing mass moving under its own gravitational forces, then by the virial theorem $v^2 \sim \phi_0$, the typical Newtonian potential in the source, while $M/r \sim \phi_r$, the Newtonian potential of the source at the observer's distance r. Then we get the simple estimate

$$\blacklozenge \qquad h \sim \phi_0 \phi_r. \qquad (9.91)$$

So the wave amplitude is always less than, or of the order of, the Newtonian potential ϕ_r. Why then can we detect h but not ϕ_r itself; why can we hope to find waves from a supernova in a distant galaxy, without being able to detect its presence gravitationally before the explosion? The answer lies in the forces involved. The Newtonian tidal gravitational force on a detector of size l_0 at a distance r is about $\phi_r l_0/r^2$, while the wave force is $h l_0 \omega^2$ (see Eq. (9.42)). The wave force is thus a factor $\phi_0(\omega r)^2 \sim (\phi_0 r/R)^2$ larger than the Newtonian force. For a relativistic system ($\phi_0 \sim 0.1$) of size 1 AU ($\sim 10^{11}$ m), observed by a detector a distance 10^{23} m away, this factor is 10^{22}. This estimate, incidentally, gives the largest distance r at which we may still approximate the gravitational field of a dynamical system as Newtonian (i.e. neglecting wave effects): $r = R/\phi_0$, where R is the size of the system.

The estimate in Eq. (9.91) is really an optimistic upper limit, because it assumed that all the mass motions contributed to D_{jk}. In realistic situations this could be a serious overestimate because of the following fundamental fact: *spherically symmetric motions do not radiate.* The rigorous proof of this is discussed in Ch. 10, but in Exer. 37, § 9.6, we derive it from linearized theory, Eq. (9.92) below. It also seems to follow from Eq. (9.73a): if T^{00} is spherically symmetric, then I^{lm} is proportional to δ^{lm} and I^{lm} vanishes. But this argument has to be treated with care, since Eq. (9.73a) is part of an approximation designed to give only the dominant radiation. One would have to show that spherically symmetric motions would not contribute to terms of higher order in the approximation if they were present. This is in fact true, and it is interesting to ask what eliminates them. The answer is Eq. (9.68): conservation of energy eliminates 'monopole' radiation in linearized theory, just as conservation of charge eliminates monopole radiation in electromagnetism.

The danger of using Eq. (9.73a) too glibly is illustrated by Exer. 28e, § 9.6: four equal masses at the corners of a rotating square give no time-dependent I^{lm} and hence no radiation in this approximation. But they *would* contribute radiation at the next higher order of approximation, octupole radiation.

Exact solution of the wave equation. Readers who have studied the wave equation, Eq. (9.55), will know that its outgoing-wave solution for arbitrary $T_{\mu\nu}$ is given by the retarded integral

$$\bar{h}_{\mu\nu}(t, x^i) = 4 \int \frac{T_{\mu\nu}(t - R, y^i)}{R} d^3y,$$ (9.92)

$$R = |x^i - y^i|,$$

where the integral is over the past light cone of the event (t, x^i) at which $\bar{h}_{\mu\nu}$ is evaluated. We let the origin be inside the source and we suppose that the field point x^i is far away,

$$|x^i| \equiv r \gg |y^i| \equiv y,$$ (9.93)

and that time derivatives of $T_{\mu\nu}$ are small. Then, inside the integral, Eq. (9.92), the dominant contribution comes from replacing R by r:

$$\bar{h}_{\mu\nu}(t, x^i) \approx \frac{4}{r} \int T_{\mu\nu}(t - r, y^i) \, d^3y.$$ (9.94)

This is the generalization of Eq. (9.65). Now, by virtue of the conservation laws

$$T^{\mu\nu}{}_{,\nu} = 0,$$ (9.68)

we have

$$\int T_{0\mu} \, d^3y = \text{const.},$$ (9.95)

i.e. the total energy and momentum are conserved. It follows that the $1/r$ part of $\bar{h}_{0\mu}$ is time independent, so it will not contribute to any wave field. This generalizes Eq. (9.71). Then, using Eq. (9.72), we get the generalization of Eq. (9.74):

$$\bar{h}_{jk}(t, x^i) = \frac{2}{r} I_{jk,00}(t - r). \tag{9.96}$$

As before, we can adopt the TT gauge to get

$$\bar{h}_{xx}^{TT} = \frac{1}{r} [I_{xx,00}(t - r) - I_{yy,00}(t - r)],$$

$$\bar{h}_{xy}^{TT} = \frac{2}{r} I_{xy,00}(t - r). \tag{9.97}$$

9.4 The energy carried away by gravitational waves

Preview. We have seen that gravitational waves can put energy into things they pass through. This is how detectors work. It stands to reason, then, that they also carry energy away from their sources. This is a very important aspect of gravitational wave theory because, as we shall see, there are some circumstances in which the effects of this loss of energy on a source can be observed, even when the gravitational waves themselves cannot be detected. There are a number of different methods of deriving the formula for this energy loss (see Misner *et al.* 1973) and, indeed, some of the mathematical questions raised by these derivations have not yet been settled (Ehlers *et al.* 1976; Schutz 1980*a*; Futamase 1983). Our approach here will make the maximum use of what we already know about the waves.

In our discussion of the harmonic oscillator as a detector of waves in § 9.2, we implicitly assumed that the detector was a kind of 'test body', whose influence on the gravitational wave field is negligible. But this is, strictly speaking, inconsistent. If the detector extracts energy from the waves, then surely the waves must be weaker after passing through the detector. That is, 'downstream' of the detector they should have slightly lower amplitude than 'upstream'. It is easy to see how this comes about once we realize that in § 9.2 we ignored the fact that the oscillator, once set in motion by the waves, will radiate waves itself. We have solved this in § 9.3 and found, in Eq. (9.79), that waves of two frequencies will be emitted. Consider the emitted waves with frequency Ω, the same as the incident wave. The part which is emitted exactly downstream has the same frequency as the incident wave, so the *total* downstream wave field has an amplitude which is the sum of the two. We will see below that the two interfere destructively, producing a net decrease in the down-stream amplitude (see Fig. 9.3). (In other directions, there is no net interference: the waves simply pass through each other.) By assuming that this amplitude change signals a change in the energy actually carried by the waves, and by equating this energy change to the energy extracted from the waves by the detector, we shall arrive at a simple expression for the energy carried by the waves as a function of their amplitude. We will then be able to calculate the energy lost by bodies which radiate arbitrarily, since we know from § 9.3 what waves they produce.

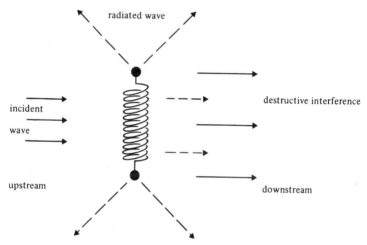

Fig. 9.3   When the detector of Fig. 9.2 is excited by a wave, it re-radiates some waves itself.

The energy flux of a gravitational wave. What we are after is the energy flux, the energy carried by a wave across a surface per unit area per unit time. It is more convenient, therefore, to consider not just one oscillator but an array of them filling the plane $z = 0$. We suppose they are very close together, so we may regard them as a nearly continuous distribution of oscillators, σ oscillators per unit area (Fig. 9.4). If the incident wave is, in the TT gauge,

$$\bar{h}^{TT}_{xx} = A \cos \Omega(z - t),$$
$$\bar{h}^{TT}_{yy} = - \bar{h}^{TT}_{xx}, \tag{9.98}$$

all other components vanishing, then in § 9.2 we have seen that each oscillator responds with a steady oscillation (after transients have died out) of the form

$$\xi = R \cos (\Omega t + \phi), \tag{9.99}$$

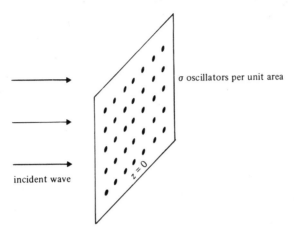

Fig. 9.4   The situation when detectors of Fig. 9.3 are arranged in a plane at a density of σ per unit area.

where R and ϕ are given by Eqs. (9.45) and (9.46) respectively. This motion is steady because the energy dissipated by friction in the oscillators is compensated by the work done on the spring by the tidal gravitational forces of the wave. It follows that the wave supplies an energy to each oscillator equal to

$$dE/dt = \nu\,(d\xi/dt)^2 = m\gamma\,(d\xi/dt)^2. \tag{9.100}$$

Averaging this over one period of oscillation, $2\pi/\Omega$, in order to get a steady energy loss, gives (angle brackets denote the average)

$$\langle dE/dt \rangle = \frac{1}{2\pi/\Omega} \int_0^{2\pi/\Omega} m\gamma\Omega^2 R^2 \sin^2\,(\Omega t + \phi)\,dt$$

$$= \tfrac{1}{2}m\gamma\Omega^2 R^2. \tag{9.101}$$

This is the energy supplied to each oscillator per unit time. With σ oscillators per unit area, the net energy flux F of the wave must decrease on passing through the $z = 0$ plane by

$$\delta F = -\tfrac{1}{2}\sigma m\gamma\Omega^2 R^2. \tag{9.102}$$

We calculate the change in the amplitude downstream independently of the calculation that led to Eq. (9.102). Each oscillator has a quadrupole tensor given by Eq. (9.79), with ωt replaced by $\Omega t + \phi$ and A replaced by $R/2$. (Each mass moves an amplitude A, one-half of the total stretching of the spring R.) Since in our case R is tiny compared to $l_0(R = 0(h_{xx}^{TT}l_0))$, the 2Ω term in Eq. (9.79) is negligible compared to the Ω term. So each oscillator has

$$I_{xx} = m l_0 R \cos\,(\Omega t + \phi). \tag{9.103}$$

By Eq. (9.74), each oscillator produces a wave amplitude

$$\delta \bar{h}_{xx} = -2\Omega^2 m l_0 R \cos\left[\Omega(r-t)-\phi\right]/r \tag{9.104}$$

at any point a distance r away. (We call it $\delta\bar{h}_{xx}$ to indicate that it is small compared to the incident wave.) It is a simple matter to get the total radiated field by adding up the contributions due to all the oscillators. In Fig. 9.5, consider a point P a distance z downstream from the plane

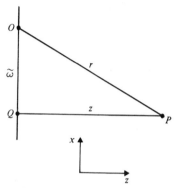

Fig. 9.5   Geometry for calculating the field at P due to an oscillator at O.

of oscillators. Set up polar coordinates $(\tilde{\omega}, \phi)$ in the plane, centered at Q beneath P. A typical oscillator O at a distance $\tilde{\omega}$ from Q contributes a field, Eq. (9.104), at P, with $r = (\tilde{\omega}^2 + z^2)^{1/2}$. Since the number of such oscillators between $\tilde{\omega}$ and $\tilde{\omega}+d\tilde{\omega}$ is $2\pi\sigma\tilde{\omega}\, d\tilde{\omega}$, the total oscillator-produced field at P is

$$\delta\bar{h}_{xx}^{\text{total}} = -2m\Omega^2 l_0 R 2\pi \int_0^\infty \sigma \cos\left[\Omega(r-t)-\phi\right]\frac{\tilde{\omega}\, d\tilde{\omega}}{r}.$$

But we may change the integration variable to r,

$$\tilde{\omega}\, d\tilde{\omega} = r\, dr,$$

obtaining

$$\delta\bar{h}_{xx}^{\text{total}} = -2m\Omega^2 l_0 R 2\pi \int_z^\infty \sigma \cos\left[\Omega(r-t)-\phi\right] dr. \tag{9.105}$$

If σ were constant, this would be trivial to integrate, but its value would be undefined at $r=\infty$. Physically, we should expect that the distant oscillators play no real role, so we adopt the device of assuming that σ is proportional to $\exp(-\varepsilon r)$ and allowing ε to tend to zero after the integration. The result is

$$\delta\bar{h}_{xx}^{\text{total}} = 4\pi\sigma m\Omega l_0 R \sin\left[\Omega(z-t)-\phi\right]. \tag{9.106}$$

So the plane of oscillators sends out a net plane wave. To compare this to the incident wave we must put Eq. (9.106) in the same TT gauge (recall that Eq. (9.74) is *not* in the TT gauge), with the result (Exer. 39, § 9.6)

$$\delta\bar{h}_{xx}^{TT} = -\delta\bar{h}_{yy}^{TT} = 2\pi\sigma m\Omega l_0 R \sin[\Omega(z-t)-\phi]. \qquad (9.107)$$

If we now add this to the incident wave, Eq. (9.98), we get the net result, to first order in R,

$$\bar{h}_{xx}^{net} = \bar{h}_{xx}^{TT} + \delta\bar{h}_{xx}^{TT}$$

$$= (A - 2\pi\sigma m\Omega l_0 R \sin\phi)\cos[\Omega(z-t)-\psi], \qquad (9.108)$$

where

$$\tan\psi = \frac{2\pi\sigma m\Omega l_0 R}{A}\cos\phi. \qquad (9.109)$$

Apart from a small phase shift ψ, the net effect is a reduction in the amplitude A by

$$\delta A = -2\pi\sigma m\Omega l_0 R \sin\phi. \qquad (9.110)$$

This reduction must be responsible for the decrease in flux F downstream. Dividing Eq. (9.102) by Eq. (9.110) and using Eqs. (9.45) and (9.46) to eliminate R and ϕ gives the remarkably simple result

$$\frac{\delta F}{\delta A} = \frac{1}{16\pi}\Omega^2 A. \qquad (9.111)$$

This is our key result. It says that a change δA in the amplitude A of a wave of frequency Ω changes its flux F (averaged over one period) by an amount depending only on Ω, A, and δA. The oscillators helped us to derive this result from conservation of energy, but they have dropped out completely! Eq. (9.111) is a property of the wave itself. We can 'integrate' Eq. (9.111) to get the total flux of a wave of frequency Ω and amplitude A:

$$F = \frac{1}{32\pi}\Omega^2 A^2. \qquad (9.112)$$

Since the average of the square of the wave, Eq. (9.98), is

$$\langle(\bar{h}_{xx}^{TT})^2\rangle = \tfrac{1}{2}A^2$$

(again, angle brackets denote an average over one period), and since there are only two nonvanishing components of $\bar{h}_{\mu\nu}^{TT}$, we can also write Eq. (9.112) as

$$\blacklozenge \qquad F = \frac{1}{32\pi}\Omega^2\langle\bar{h}_{\mu\nu}^{TT}\bar{h}^{TT\mu\nu}\rangle. \qquad (9.113)$$

This form is invariant under background Lorentz transformations, but not under gauge changes. Since one polarization can be transformed into another by a background Lorentz transformation (a rotation), Eq. (9.113) applies to all polarizations and hence to arbitrary plane waves of frequency Ω. In fact, since it gives an energy rate per unit area, it applies to *any* wavefront, either plane waves or the spherical expanding ones of § 9.3: one can always look at a small enough area that the curvature of the wavefront is not noticeable. The generalization to arbitrary waves (no single frequency) is in Exer. 40, § 9.6.

The reader who remembers the discussion of energy in § 7.3 may object that this whole derivation is suspect because of the difficulty of defining energy in GR. Indeed, we have not *proved* that energy is conserved, that the energy put into the oscillators must equal the decrease in flux; we have simply assumed this in order to derive the flux. Our proof may be turned around, however, to argue that the flux we have constructed is the only acceptable definition of energy for the waves, since our calculation shows it is conserved when added to other energies, to lowest order in $h_{\mu\nu}$. The qualification 'to lowest order' is important, since it is precisely because we are almost in flat spacetime that, at lowest order, we can construct conserved quantities. At higher order, away from linearized theory, local energy cannot be so easily defined, because the time dependence of the true metric becomes important. These questions are among the most fundamental in relativity, and are discussed in detail in any of the advanced texts. Our equations should be used *only* in linearized theory.

Energy lost by a radiating system. Consider a general isolated system, radiating according to Eqs. (9.73)–(9.78). By integrating Eq. (9.113) over a sphere surrounding the system, we can calculate its net energy loss rate. For example, at a distance r along the z axis, Eq. (9.113) is

$$F = \frac{\Omega^6}{32\pi r^2} \langle 2(I_{xx} - I_{yy})^2 + 8I^{xy} \rangle. \tag{9.114}$$

Use of the identity

$$I^i_i = I_{xx} + I_{yy} + I_{zz} = 0 \tag{9.115}$$

(which follows from Eq. (9.78)) gives, after some algebra,

$$F = \frac{\Omega^6}{16\pi r^2} \langle 2I_{ij}I^{ij} - 4I_{zj}I_z{}^j + I_{zz}^2 \rangle. \tag{9.116}$$

Now, the index z appears here only because it is the direction from the center of the coordinates, where the radiation comes from. It is the only part of F which depends on the location on the sphere of radius r about the source, since all the components of I_{ij} depend on time but not position. Therefore we can generalize Eq. (9.116) to arbitrary locations on the sphere by using the unit vector normal to the sphere,

$$n^j = x^j / r. \tag{9.117}$$

We get for F

$$F = \frac{\Omega^6}{16\pi r^2} \langle 2 I_{ij} I^{ij} - 4 n^j n^k I_{ji} I_k{}^i + n^i n^j n^k n^l I_{ij} I_{kl} \rangle. \tag{9.118}$$

The total luminosity of the source is the integral of this over the sphere of radius r. In Exer. 42, § 9.6 we prove the following integrals

$$\int n^j n^k \sin \theta \, d\theta \, d\phi = \frac{4\pi}{3} \delta^{jk}, \tag{9.119}$$

$$\int n^i n^j n^k n^l \sin \theta \, d\theta \, d\phi = \frac{4\pi}{15} (\delta^{ij} \delta^{kl} + \delta^{ik} \delta^{jl} + \delta^{il} \delta^{jk}). \tag{9.120}$$

It then follows that the luminosity L of a source of gravitational waves is

$$\int F r^2 \sin \theta \, d\theta \, d\phi = \tfrac{1}{4} \Omega^6 \langle 2 I_{ij} I^{ij} - \tfrac{4}{3} I_{ij} I^{ij}$$

$$+ \tfrac{1}{15} (I^i_i I^k_k + I^{ij} I_{ij} + I^{ij} I_{ij}) \rangle, \tag{9.121}$$

◆ $$L = \tfrac{1}{5} \Omega^6 \langle I_{ij} I^{ij} \rangle. \tag{9.122}$$

The generalization to cases where I_{ij} has a more general time dependence is

$$L = \tfrac{1}{5} \langle \dddot{I}_{ij} \dddot{I}^{ij} \rangle, \tag{9.123}$$

where dots denote time derivatives.

It must be stressed that Eqs. (9.112) and (9.123) are accurate only for weak gravitational fields and slow velocities. They can at best give only order-of-magnitude results for highly relativistic sources of gravitational waves. But in the spirit of our derivation and discussion of the order-of-magnitude estimate of h_{ij} in Eq. (9.61), we can still learn something about strong sources from Eq. (9.112). Since I_{jk} is of order MR^2, Eq. (9.112) tells us that $L \sim M^2 R^4 \Omega^6 \sim (M/R)^2 (R\Omega)^6 \sim \phi_0^2 v^6$. The luminosity is a very sensitive function of the velocity. The largest velocities one should expect are of the order of the velocity of free fall, $v^2 \sim \phi_0$, so we should expect

$$L \lesssim (\phi_0)^5. \tag{9.124}$$

Since $\phi_0 \lesssim 1$, the luminosity in geometrized units should never substantially exceed one. In ordinary units this is

$$L \lesssim 1 = c^5 / G \approx 3.6 \times 10^{52} \text{ W}. \tag{9.125}$$

We can understand why this particular luminosity is an upper limit by the following simple argument. The radiation field inside a source of size R and luminosity L has energy density $\geqslant L/R^2$ (because $|T^{0i}| \sim |v^i|T^{00} = cT^{00} = T^{00}$), which is the flux across its surface. The total energy in radiation is therefore $\geqslant LR$. The Newtonian potential of the radiation alone is therefore $\geqslant L$. We shall see in the next chapter that anything whose Newtonian potential substantially exceeds 1 must form a black hole: its gravitational field will be so strong that no radiation will escape at all. Therefore $L \sim 1$ is the largest luminosity any source can have. This argument applies equally well to all forms of radiation, electromagnetic as well as gravitational. Quasars, which are the most luminous class of object known, have a (geometrized) luminosity $\leqslant 10^{-7}$.

An example: the binary pulsar. In §9.3 we calculated I_{ij} for a binary system consisting of two stars of equal mass M in circular orbits a distance l_0 apart. If we use the real part of Eq. (9.88) in Eq. (9.123), we get

$$L = \tfrac{8}{5}M^2 l_0^4 \omega^6. \tag{9.126}$$

In terms of m and ω, this is

$$L = \frac{32}{5\sqrt[3]{4}}(M\omega)^{10/3} \approx 4.0\,(M\omega)^{10/3}. \tag{9.127}$$

This expression illustrates two things: first, that L is dimensionless in geometrized units and, second, that it is almost always easier to compute in geometrized units, and then convert back at the end. The conversion is

$$L \text{ (SI units)} = \frac{c^5}{G} L \text{ (geometrized)}$$

$$= 3.63 \times 10^{52} \text{ J s}^{-1} \times L \text{ (geometrized)}. \tag{9.128}$$

So for the binary pulsar system described in §9.3, if its orbit were circular, we would have $\omega = 2\pi/P = 7.5049 \times 10^{-13} \text{ m}^{-1}$ and

$$L = 1.71 \times 10^{-29} \tag{9.129}$$

in geometrized units. We can, of course, convert this to watts, but a more meaningful procedure is to compare this with the Newtonian energy of the system, which is (defining the orbital radius $r = \tfrac{1}{2}l_0$),

$$E = \tfrac{1}{2}M\omega^2 r^2 + \tfrac{1}{2}M\omega^2 r^2 - \frac{M^2}{2r}$$

$$= \frac{M}{r}(\omega^2 r^3 - \tfrac{1}{2}M) = -\frac{M^2}{4r}$$

$$= -4^{-2/3}M^{5/3}\omega^{2/3} \approx -0.40 M^{5/3}\omega^{2/3} \tag{9.130}$$

$$= -1.11 \times 10^{-3} \text{ m}. \tag{9.131}$$

The physical question is, how long does it take to change this? Put differently, the energy radiated in waves must change the orbit by decreasing its energy, which makes $|E|$ larger and hence ω larger and the period smaller. What change in the period do we expect in, say, one year?

From Eq. (9.130), by taking logarithms and differentiating, we get

$$\frac{1}{E}\frac{dE}{dt} = \frac{2}{3}\frac{1}{\omega}\frac{d\omega}{dt} = -\frac{2}{3}\frac{1}{P}\frac{dP}{dt}. \tag{9.132}$$

Since dE/dt is just $-L$, we can solve for dP/dt:

$$dP/dt = (3PL)/(2E) \approx -15\ \mathrm{PM}^{-1}\,(M\Omega)^{8/3} \tag{9.133}$$
$$= -2.0\times10^{-13},$$

which is dimensionless in any system of units. It can be reexpressed in seconds per year:

$$dP/dt = -6.0\times10^{-6}\ \mathrm{s\ yr}^{-1}. \tag{9.134}$$

This estimate needs to be revised to allow for the eccentricity of the orbit, which is considerable: $e = 0.617$. The correct formula is derived in Exer. 46, § 9.6. The result is that the true rate of energy loss is some 12 times our estimate, Eq. (9.129). This is such a large factor because the stars' maximum angular velocity (when they are closest) is larger than the mean value we have used for Ω, and since L depends on the angular velocity to a high power, a small change in the angular velocity accounts for this rather large factor of 12. So the relativistic prediction is:

$$dP/dt = -2.4\times10^{-12}, \tag{9.135}$$
$$= -7.2\times10^{-5}\ \mathrm{s\ yr}^{-1}. \tag{9.136}$$

The observed value as of 1982 is (Taylor & Weisberg 1982, Boriakoff *et al.* 1982)

$$dP/dt = -(2.30\pm0.22)\times10^{-12}. \tag{9.137}$$

9.5   Bibliography

Joseph Weber's early thinking about detectors is in Weber (1961). There have been several reviews of the state-of-the-art in detector design and operation: Press & Thorne (1972), Bertotti (1974, 1977), Tyson & Giffard (1978), Smarr (1979a), Douglass & Braginsky (1979), and Weber (1980). One of the most interesting theoretical developments stimulated by research into gravitational wave detection has been the design of so-called 'quantum nondemolition' detectors: methods of measuring aspects of the excitation of a vibrating bar to arbitrary precision without disturbing the quantity being measured, even when the bar is excited only at the one- or two-quantum (phonon) level. See Thorne *et al.* (1979) and Caves *et al.* (1980).

A full discussion of the wave equation is beyond our scope here, but is amply treated in many texts on electromagnetism, such as Jackson (1975). A simplified discussion of gravitational waves is in Schutz (1984).

Generation of radiation in the full nonlinear theory is very much more difficult to analyze than in the linearized theory. See Smarr (1979*a*) and Deruelle & Piran (1983) for a discussion of the difficulties that exist and attempts at their solution. These problems affect derivations of the energy emitted and of the reaction in the source. For weakly self-gravitating sources, there is general agreement on the answers but not on the validity of various derivations (see Thorne 1980*a,b* and Walker & Will 1980). For estimates of waves from strong sources, see Ostriker (1979).

The interaction of gravitational waves with matter has been well studied for detector theory, but less is known about astrophysical situations. See Grishchuck & Polnarev (1980).

9.6 Exercises

1 A function $f(s)$ has derivative $f'(s) = df/ds$. Prove that $\partial f(k_\mu x^\mu)/\partial x^\nu = k_\nu f'(k_\mu x^\mu)$. Use this to prove Eq. (9.4) and the one following it.

2 Show that the real and imaginary parts of Eq. (9.2) at a fixed spatial position $\{x^i\}$ oscillate sinusoidally in time with frequency $\omega = k^0$.

3 Let $\bar{h}^{\alpha\beta}(t, x^i)$ be any solution of Eq. (9.1), which has the property $\int dx^\alpha |\bar{h}^{\mu\nu}|^2 < \infty$, for the integral over any particular x^α holding other coordinates fixed. Define the Fourier transform of $\bar{h}^{\alpha\beta}(t, x^i)$ as

$$\bar{H}^{\alpha\beta}(\omega, k^i) = \int \bar{h}^{\alpha\beta}(t, x^i) \exp(i\omega t - ik_j x^j)\, dt\, d^3x.$$

Show, by transforming Eq. (9.1), that $\bar{H}^{\alpha\beta}(\omega, k^i)$ is zero except for those values of ω and k^i that satisfy Eq. (9.10). By applying the inverse transform, write $\bar{h}^{\alpha\beta}(t, x^i)$ as a superposition of plane waves.

4 Derive Eqs. (9.16) and (9.17).

5(a) Show that $A_{\alpha\beta}^{(\mathrm{NEW})}$, given by Eq. (9.17), satisfies the gauge condition $A^{\alpha\beta}k_\beta = 0$ if $A_{\alpha\beta}^{(\mathrm{OLD})}$ does.

   (b) Use Eq. (9.18) for $A_{\alpha\beta}^{(\mathrm{NEW})}$ to constrain B^μ.

   (c) Show that Eq. (9.19) for $A_{\alpha\beta}^{(\mathrm{NEW})}$ imposes only three constraints on B^μ, not the four that one might expect from the fact that the free index α can take any values from 0 to 3. Do this by showing that the particular linear combination $k^\alpha(A_{\alpha\beta}U^\beta)$ vanishes for any B^μ.

(d) Using (b) and (c), solve for B^μ as a function of k^μ, $A^{(OLD)}_{\alpha\beta}$, and U^μ. These determine B^μ: there is no further gauge freedom.

(e) Show that it is possible to choose ξ^β in Eq. (9.15) to make any super-position of plane waves satisfy Eqs. (9.18) and (9.19), so that these are generally applicable to gravitational waves of any sort.

(f) Show that one cannot achieve Eqs. (9.18) and (9.19) for a static solution, i.e. one for which $\omega = 0$.

6 Fill in all the algebra implicit in the paragraph leading to Eq. (9.21).

7 Give a more rigorous proof that eqs. (9.22) and (9.23) imply that a free particle initially at rest in the TT gauge remains at rest.

8 Does the free particle of the discussion following Eq. (9.23) *feel* any acceleration? For example, if the particle is a bowl of soup (whose diameter is much less than a wavelength), does the soup slosh about in the bowl as the wave passes?

9 Does the free particle of the discussion following Eq. (9.23) *see* any acceleration? To answer this, consider the two particles whose relative proper distance is calculated in Eq. (9.24). Let the one at the origin send a beam of light towards the other, and let it be reflected by the other and received back at the origin. Calculate the amount of proper time elapsed at the origin between the emission and reception of the light (you may assume that the particles' separation is much less than a wavelength). By monitoring changes in this time, the particle at the origin can 'see' the relative acceleration of the two particles.

10(a) We have seen that
$$h_{yz} = A \sin \omega(t-x), \quad \text{all other } h_{\mu\nu} = 0,$$
with A and ω constants, $|A| \ll 1$, is a solution to Eqs. (9.1) and (9.11). For this metric tensor, compute all the components of $R_{\alpha\beta\mu\nu}$ and show that some are not zero, so that the spacetime is not flat.

(b) Another metric is given by
$$h_{yz} = A \sin \omega(t-x), \qquad h_{tt} = 2B(x-t),$$
$$h_{tx} = -B(x-t), \quad \text{all other } h_{\mu\nu} = 0.$$
Show that this also satisfies the field equations and the gauge conditions.

(c) For the metric in (b) compute $R_{\alpha\beta\mu\nu}$. Show that it is the *same* as in (a).

(d) From (c) we conclude that the geometries are identical, and that the difference in the metrics is due to a small coordinate change. Find a ξ^μ such that
$$h_{\mu\nu}(\text{part } a) - h_{\mu\nu}(\text{part } b) = -\xi_{\mu,\nu} - \xi_{\nu,\mu}.$$

11(a) Derive Eq. (9.27).

(b) Solve Eqs. (9.28a) and (9.28b) for the motion of the test particles in the polarization rings shown in Fig. 9.1.

12 Do calculations analogous to those leading to Eqs. (9.28) and (9.29) to show that the separation of particles in the z direction (the direction of travel of the wave) is unaffected.

13 One kind of background Lorentz transformation is a simple 45° rotation of the x and y axes in the $x-y$ plane. Show that under such a rotation from (x, y) to (x', y'), we have $h^{TT}_{x'y'} = h^{TT}_{xx}$, $h^{TT}_{x'x'} = -h^{TT}_{xy}$. This is consistent with Fig. 9.1.

14(a) A wave is said to be circularly polarized in the $x-y$ plane if $h^{TT}_{yy} = -h^{TT}_{xx}$ and $h^{TT}_{xy} = \pm i h^{TT}_{xx}$ Show that for such a wave, the ellipse in Fig. 9.1 rotates without changing shape.

(b) A wave is said to be elliptically polarized with principal axes x and y if $h^{TT}_{xy} = \pm i a h^{TT}_{xx}$, where a is some real number, and $h^{TT}_{yy} = -h^{TT}_{xx}$. Show that if $h^{TT}_{xy} = \alpha h^{TT}_{xx}$, where α is a complex number (the general case for a plane wave), new axes x' and y' can be found for which the wave is elliptically polarized with principal axes x' and y'. Show that circular and linear polarization are special cases of elliptical.

15 Two plane waves with TT amplitudes, $A^{\mu\nu}$ and $B^{\mu\nu}$, are said to have orthogonal polarizations if $(A^{\mu\nu})^* B_{\mu\nu} = 0$, where $(A^{\mu\nu})^*$ is the complex conjugate of $A^{\mu\nu}$. Show that if $A^{\mu\nu}$ and $B^{\mu\nu}$ are orthogonal polarizations, a 45° rotation of $B^{\mu\nu}$ makes it proportional to $A^{\mu\nu}$.

16 Find the transformation from the coordinates (t, x, y, z) of Eqs. (9.30)–(9.33) to the local inertial frame of Eq. (9.34). Use this to verify Eq. (9.35).

17 Prove Eq. (9.36).

18 Use the sum of Eqs. (9.37) and (9.38) to show that the center of mass of the spring remains at rest as the wave passes.

19 Derive Eq. (9.41) from Eq. (9.40), and then prove Eq. (9.42).

20 Generalize Eq. (9.42) to the case of a plane wave with arbitrary elliptical polarization (Exer. 14) traveling in an arbitrary direction relative to the separation of the masses.

21 Consider the equation of geodesic deviation, Eq. (6.87), from the point of view of the geodesic at the center of mass of the detector of Eq. (9.42). Show that the vector ξ as we have defined it in Eq. (9.39) is twice the connecting vector from the center of mass to one of the masses, as defined in Eq. (6.83). Show that the tidal force as measured by the center of mass leads directly to Eq. (9.42).

22 Derive Eqs. (9.45) and (9.46), and derive the general solution of Eq. (9.42) for arbitrary initial data at $t = 0$, given Eq. (9.43).

23   Prove Eq. (9.50).

24   Derive Eq. (9.53) from the given definition of Q.

25(a)   Reconstruct $\bar{h}_{\mu\nu}$ as in Eq. (9.57), using Eq. (9.59), and show that surfaces of constant phase of the wave move outwards for the $A_{\mu\nu}$ term and inwards for $Z_{\mu\nu}$.

(b)   Fill in the missing algebra in Eqs. (9.60)–(9.62).

26   Eq. (9.58) in the vacuum region outside the source – i.e., where $S_{\mu\nu} = 0$ – can be solved by separation of variables. Assume a solution for $\bar{h}_{\mu\nu}$ has the form $\sum_{lm} A^{lm}_{\mu\nu} f_l(r) Y_{lm}(\theta, \phi)/\sqrt{r}$, where Y_{lm} is the spherical harmonic.

(a)   Show that $f_l(r)$ satisfies the equation

$$f_l'' + \frac{1}{r}f_l' + \left[\Omega^2 - \frac{(l+\frac{1}{2})^2}{r^2}\right]f_l = 0.$$

(b)   Show that the most general spherically symmetric solution is given by Eq. (9.59).

(c)   Substitute the variable $s = \Omega r$ to show that f_l satisfies the equation

$$s^2 \frac{d^2 f_l}{ds^2} + s\frac{df_l}{ds} + [s^2 - (l+\frac{1}{2})^2]f_l = 0. \tag{9.138}$$

This is known as Bessel's equation, whose solutions are called Bessel functions of order $l+\frac{1}{2}$. Their properties are explored in most text-books on mathematical physics.

(d)   Show, by substitution into Eq. (9.138), that the function $f_l/\sqrt{s}$ is a linear combination of what are called the spherical Bessel and spherical Neumann functions

$$j_l(s) = (-1)^l s^l \left(\frac{1}{s}\frac{d}{ds}\right)^l \left(\frac{\sin s}{s}\right), \tag{9.139}$$

$$n_l(s) = (-1)^{l+1} s^l \left(\frac{1}{s}\frac{d}{ds}\right)^l \left(\frac{\cos s}{s}\right). \tag{9.140}$$

(e)   Use Eqs. (9.139) and (9.140) to show that for $s \gg l$, the dominant behavior of j_l and n_l is

$$j_l(s) \sim \frac{1}{s}\sin\left(s - \frac{l\pi}{2}\right), \tag{9.141}$$

$$n_l(s) \sim -\frac{1}{s}\cos\left(s - \frac{l\pi}{2}\right). \tag{9.142}$$

(f)   Similarly, show that for $s \ll l$ the dominant behavior is

$$j_l(s) \sim s^l/(2l+1)!!, \tag{9.143}$$

$$n_l(s) \sim -(2l-1)!!/s^{l+1}, \tag{9.144}$$

where we use the standard double factorial notation

$$(m)!! = m(m-2)(m-4)\cdots 3\cdot 1 \tag{9.145}$$

for odd m.

(g) Show from (e) that the outgoing-wave vacuum solution of eq. (9.58) for any fixed l and m is

$$(\bar{h}_{\mu\nu})_{lm} = A_{\mu\nu}^{lm} h_l^{(1)}(\Omega r)\, e^{-i\Omega t} Y_{lm}(\theta, \phi), \tag{9.146}$$

where $h_l^{(1)}(\Omega r)$ is called the spherical Hankel function of the first kind,

$$h_l^{(1)}(\Omega r) = j_l(\Omega r) + i n_l(\Omega r). \tag{9.147}$$

(h) Repeat the calculation of Eqs. (9.60)–(9.65), only this time multiply Eq. (9.58) by $j_l(r\Omega) Y_{lm}^*(\theta, \phi)$ before performing the integrals. Show that the left-hand side of Eq. (9.58) becomes, when so integrated, exactly

$$\varepsilon^2 \left(j_l(\Omega\varepsilon) \frac{d}{dr} B_{\mu\nu}(\varepsilon) - B_{\mu\nu}(\varepsilon) \frac{d}{dr} j_l(\Omega\varepsilon) \right),$$

and that when $\Omega\varepsilon \ll l$ this becomes (with the help of Eqs. (9.146) and (9.143)–(9.144) above, since we assume $r = \varepsilon$ is outside the source) simply $i A_{\mu\nu}^{lm}/\Omega$. Similarly, show that the right-hand side of Eq. (9.58) integrates to $-16\pi\Omega^l \int T_{\mu\nu} r^l Y_{lm}^*(\theta, \phi)\, d^3x/(2l+1)!!$ in the same approximation.

(i) Show, then, that the solution is Eq. (9.146), with

$$A_{\mu\nu}^{lm} = 16\pi i \Omega^{l+1} J_{\mu\nu}^{lm}/(2l+1)!!, \tag{9.148}$$

where

$$J_{\mu\nu}^{lm} = \int T_{\mu\nu} r^l Y_{lm}^*(\theta, \phi)\, d^3x. \tag{9.149}$$

(j) Let $l = 0$ and deduce Eq. (9.64) and (9.65).

(k) Show that if $J_{\mu\nu}^{lm} \neq 0$ for some l, then the terms neglected in Eq. (9.148) because of the approximation $\Omega\varepsilon \ll 1$, are of the same order as the dominant terms in Eq. (9.148) for $l+1$. In particular, this means that if $J_{\mu\nu} \neq 0$ in Eq. (9.63) any attempt to get a more accurate answer than Eq. (9.65) must take into account not only the terms for $l > 0$ but also neglected terms in the derivation of Eq. (9.65), such as Eq. (9.60).

27   Re-write Eq. (9.73a) for a set of N discrete point particles, whose masses are $\{m_{(A)}, A = 1, \ldots, N\}$ and whose positions are $\{x_{(A)}^i\}$.

28   Calculate the quadrupole tensor I_{jk} and its traceless counterpart $\bar{I}_{jk}$ (Eq. (9.78)) for the following mass distributions.

(a) A spherical star whose density is $\rho(r, t)$. Take the origin of the coordinates in Eq. (9.73) to be the center of the star.

(b) The star in (a), but with the origin of the coordinates at an arbitrary point.

(c) An ellipsoid of uniform density ρ and semiaxes of length a, b, c oriented along the x, y, and z axes respectively. Take the origin to be at the center of the ellipsoid.

(d) The ellipsoid in (c), but rotating about the z axis with angular velocity ω.

(e) Four masses m located respectively at the points $(a, 0, 0), (0, a, 0)$, $(-a, 0, 0), (0, -a, 0)$.

(f) The masses as in (e), but all moving counter-clockwise about the z axis on a circle of radius a with angular velocity ω.

(g)   Two masses m connected by a massless spring, each oscillating on the x axis with angular frequency ω and amplitude A about mean equilibrium positions a distance l_0 apart, keeping their center of mass fixed.

(h)   Unequal masses m and M connected by a spring of spring constant k and equilibrium length l_0, oscillating (with their center of mass fixed) at the natural frequency of the system, with amplitude $2A$ (this is the total stretching of the spring). Their separation is along the x axis.

29   This exercise develops the TT gauge for spherical waves.

(a)   In order to transform Eq. (9.74) to the TT gauge, use a gauge transformation generated by a vector $\xi^\alpha = B^\alpha(x^\mu)\,e^{i\Omega(r-t)}/r$, where B^α is a slowly varying function of x^μ. Find the general transformation law to order $1/r$.

(b)   Demand that the new $\bar{h}_{\alpha\beta}$ satisfy three conditions to order $1/r$: $\bar{h}_{0\mu} = 0$, $\bar{h}^\alpha{}_\alpha = 0$, and $\bar{h}_{\mu j}n^j = 0$, where $n^j \equiv x^j/r$ is the unit vector in the radial direction. Show that it is possible to find functions B^α which accomplish such a transformation *and* which satisfy $\Box\,\xi^\alpha = 0$ to order $1/r$.

(c)   Show that Eqs. (9.75)–(9.78) hold in the TT gauge.

30(a)   Let n^j be a unit vector in three-dimensional Euclidean space. Show that $P^j{}_k = \delta^j{}_k - n^j n_k$ is the projection tensor orthogonal to n^j, i.e. show that for any vector V^j (i) $P^j{}_k V^k$ is orthogonal to n^j, and (ii) $P^j{}_k P^k{}_l V^l = P^j{}_k V^k$.

(b)   Show that the TT gauge $\bar{h}_{ij}^{\text{TT}}$ of Eqs. (9.75)–(9.77) is related to the original $\bar{h}_{kl}$ of Eq. (9.74) by

$$\bar{h}_{ij}^{\text{TT}} = P^k{}_i P^l{}_j \bar{h}_{kl} - \tfrac{1}{2} P_{ij}(P^{kl}\bar{h}_{kl}), \tag{9.150}$$

where n^j points in the z direction.

31   Show that I_{jk} is trace free, i.e. $I^l_l = 0$.

32   For the systems described in Exer. 28, calculate the transverse-traceless quadrupole radiation field, Eqs. (9.76)–(9.77) or (9.150), along the x, y, and z axes. In Eqs. (9.76)–(9.77) be sure to change the indices appropriately when doing the calculation on the x and y axes, as in the discussion leading to Eq. (9.82).

33   Use Eq. (9.150) or a rotation of the axes in Eqs. (9.76)–(9.77) to calculate the amplitude and orientation of the polarization ellipse of the radiation from the simple oscillator, Eq. (9.79), traveling at an angle θ to the x axis.

34   The ω and 2ω terms in Eq. (9.84) are qualitatively different, in that the 2ω term depends only on the amplitude of oscillator A, while the ω term depends on both A and the separation of the masses l_0. Why should l_0 be involved – the masses don't move over that distance? The answer is that stresses *are* transmitted over that distance by the spring, and stresses cause the radiation. To see this, do an analogous calculation for a similar system, in which stresses are *not* passed over large distances. Consider a system consisting of two pairs of masses. Each pair has one particle of mass m and another of mass $M \gg m$. The masses within each pair are connected by a short spring whose natural frequency is ω. The pairs' centers of mass are at rest relative to one another. The springs oscillate with equal amplitude in such a way that each mass m oscillates sinusoidally with amplitude A, and the centers of oscillation of the masses are separated by $l_0 \gg A$. The masses oscillate out of phase. Use the calculation of Exer. 28h to show that the radiation field of the system is Eq. (9.84) without the ω term. The difference between this system and that in Eq. (9.84) may be thought to be the origin of the stresses to maintain the motion of the masses m.

35 Do the same as Exer. 9.33 for the binary system, Eqs. (9.89)–(9.90), but instead of finding the orientation of the linear polarization, find the orientation of the ellipse of elliptical polarization.

36 Let two spherical stars of mass m and M be in elliptical orbit about one another in the x–y plane. Let the orbit be characterized by its total energy E and its angular momentum L.

(a) Use Newtonian gravity to calculate the equation of the orbits of both masses about their center of mass. Express the orbital period P, minimum separation a, and eccentricity e as functions of E and L.

(b) Calculate $\mathcal{I}_{kj}$ for this system.

(c) Calculate from Eq. (9.97) the TT radiation field along the x and z axes. Show that your result reduces to Eqs. (9.89)–(9.90) when $m = M$ and the orbits are circular.

37 Show from Eq. (9.92) that spherically symmetric motions produce no gravitational radiation.

38 Derive Eq. (9.106) from Eq. (9.105) in the manner suggested in the text.

39(a) Derive Eq. (9.107).

(b) Derive Eqs. (9.108) and (9.109) by superposing Eqs. (9.98) and (9.107) and assuming R is small.

(c) Derive Eq. (9.111) in the indicated manner.

40 Show that if we define an averaged stress–energy tensor for the waves

$$T_{\alpha\beta} = \ll \bar{h}^{TT}_{\mu\nu,\alpha} \bar{h}^{TT\mu\nu}{}_{,\beta} \gg /32\pi \qquad (9.151)$$

(where $\ll\ \gg$ denotes an average over both one period of oscillation in time and one wavelength of distance in all spatial directions), then the flux F of Eqs. (9.113) is the component T^{0z} for that wave. A more detailed argument shows that Eq. (9.151) can in fact be regarded as the stress–energy tensor of any wave packet, provided the averages are defined suitably. This is called the Isaacson stress–energy tensor. See Misner *et al.* (1973) for details.

41(a) Derive Eq. (9.116) from Eq. (9.114).

(b) Justify Eq. (9.118) from Eq. (9.116).

(c) Derive Eq. (9.118) from Eq. (9.113) using Exer. 30b.

42(a) Consider the integral in Eq. (9.119). We shall do it by the following method. (i) Argue on grounds of symmetry that $\int n^j n^k \sin \theta \, d\theta \, d\phi$ must be proportional to δ^{jk}. (ii) Evaluate the constant of proportionality by explicitly doing the case $j = k = z$.

  (b) Follow the same method for Eq. (9.120). In (i) argue that the integral can depend only on δ^{ij}, and show that the given tensor is the only one constructed only from δ^{ij} which has the symmetry of being unchanged when the values of any two of its indices are exchanged.

43 Derive Eqs. (9.121) and (9.122), remembering Eq. (9.115) and the fact that I_{ij} is symmetric.

44(a) Recall that the angular momentum of a particle is p_ϕ. It follows that the angular momentum flux of a continuous system across a surface $x^i = $ const. is $T_{i\phi}$. Use this and Exer. 40 to show that the total z component of angular momentum radiated by a source of gravitational waves (which is the integral over a sphere of large radius of $T_{r\phi}$ in Eq. (9.151)) is

$$F_J = -\tfrac{1}{5}(\ddot{I}_{xl}\ddot{I}_y{}^l - \ddot{I}_{yl}\ddot{I}_x{}^l).\tag{9.152}$$

  (b) Show that if $\bar{h}_{\mu\nu}^{TT}$ depends on t and ϕ only as $\cos(\Omega t - m\phi)$, then the ratio of the total energy radiated to the total angular momentum radiated is Ω/m.

45 Calculate Eq. (9.126).

46 For the arbitrary binary system of Exer. 36:
  (a) Show that the average energy loss rate over one orbit is

$$\langle dE/dt \rangle = -\frac{32}{5}\frac{\mu^2(m+M)^3}{a^5(1-e^2)^{7/2}}\left(1+\frac{73}{24}e^2+\frac{37}{96}e^4\right)\tag{9.153}$$

and from the result of Exer. 44a

$$\langle dL/dt \rangle = -\frac{32}{5}\frac{\mu^2(m+M)^{5/2}}{a^{7/2}(1-e^2)^2}\left(1+\frac{7}{8}e^2\right),\tag{9.154}$$

where $\mu = mM/(m+M)$ is the reduced mass;
  (b) Show that

$$\langle da/dt \rangle = -\frac{64}{5}\frac{\mu(m+M)^2}{a^3(1-e^2)^{7/2}}\left(1+\frac{73}{24}e^2+\frac{37}{96}e^4\right),\tag{9.155}$$

$$\langle de/dt \rangle = -\frac{304}{15}\frac{\mu(m+M)^2 e}{a^4(1-e^2)^{5/2}}\left(1+\frac{121}{304}e^2\right),\tag{9.156}$$

$$\langle dP/dt \rangle = -\frac{192\pi}{5}\frac{\mu(m+M)^{3/2}}{a^{5/2}(1-e^2)^{7/2}}\left(1+\frac{73}{24}e^2+\frac{37}{96}e^4\right);\tag{9.157}$$

  (c) Verify Eq. (9.135). (Do parts (b) and (c) even if you can't do (a).) These were originally derived by Peters (1964).

10

Spherical solutions for stars

10.1 Coordinates for spherically symmetric spacetimes

For our first study of strong gravitational fields in GR we will consider spherically symmetric systems. They are reasonably simple, yet physically very important, since very many objects of importance in astrophysics appear to be nearly spherical. We begin by choosing the coordinate system to reflect the assumed symmetry.

Flat space in spherical coordinates. By defining the usual coordinates (r, θ, ϕ), the line element of Minkowski space can be written

$$ds^2 = -dt^2 + dr^2 + r^2(d\theta^2 + \sin^2 \theta \, d\phi^2). \tag{10.1}$$

Each surface of constant r and t is a sphere or, more precisely, a two-sphere, a two-dimensional spherical surface. Distances along curves confined to such a sphere are given by the above equation with $dt = dr = 0$:

$$dl^2 = r^2(d\theta^2 + \sin^2 \theta \, d\phi^2) \equiv r^2 \, d\Omega^2, \tag{10.2}$$

which defines the symbol $d\Omega^2$. We note that such a sphere has circumference $2\pi r$ and area $4\pi r^2$, i.e. 2π times the square root of the coefficient of $d\Omega^2$ and 4π times the coefficient of $d\Omega^2$ respectively. Conversely, any two-surface whose line element is Eq. (10.2) with r^2 independent of θ and ϕ has the intrinsic geometry of a two-sphere.

Two-spheres in a curved spacetime. The statement that a spacetime is spherically symmetric can now be made more precise: it implies that every point of spacetime is on a two-surface which is a two-sphere, i.e. whose line element is

$$dl^2 = f(r', t)(d\theta^2 + \sin^2\theta\, d\phi^2),\tag{10.3}$$

where $f(r', t)$ is an unknown function of the two other coordinates of our manifold r' and t. The area of each sphere is $4\pi f(r', t)$. We *define* the radial coordinate r of our spherical geometry such that $f(r', t) \equiv r^2$. This represents a coordinate transformation from (r', t) to (r, t). Then any surface $r = $ const., $t = $ const. is a two-sphere of area $4\pi r^2$ and circumference $2\pi r$. This coordinate r is called the 'curvature coordinate' or 'area coordinate' because it defines the radius of curvature and area of the spheres. There is *no a priori* relation between r and the proper distance from the center of the sphere to its surface. This r is defined only by the properties of the spheres themselves. Since their 'centers' (at $r = 0$ in flat space) are not points on the spheres themselves, the statement that a spacetime is spherically symmetric does not require even that there *be* a point at the center. A simple counter-example of a two-space in which there are *circles* but no point at the center of them is in Fig. 10.1. The

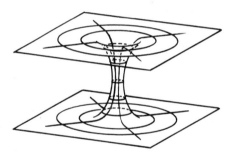

Fig. 10.1   Two plane sheets connected by a circular throat: there is circular (axial) symmetry, but the center of any circle is not in the two-space.

space consists of two sheets which are joined by a 'throat'. The whole thing is symmetric about an axis along the middle of the throat, but the points on this axis – which are the 'centers' of the circles – are *not* part of the two-dimensional surface illustrated. Yet if ϕ is an angle about the axis, the line element on each circle is just $r^2 d\phi^2$, where r is a constant labeling each circle. This r is the same sort of coordinate that we use in our spherically symmetric spacetime.

Meshing the two-spheres into a three-space for $t = const.$ Consider the spheres at r and $r + dr$. Each has a coordinate system (θ, ϕ), but up to now we have not required any relation between them. That is, one could conceive of having the pole for the sphere at r in one orientation, while that for $r + dr$ was in another. The sensible thing is to say that a line of $\theta = $ const., $\phi = $ const. is *orthogonal* to the two-spheres. Such a line has, by definition, a tangent $\vec{e}_r$. Since the vectors $\vec{e}_\theta$ and $\vec{e}_\phi$ lie in the spheres, we require $\vec{e}_r \cdot \vec{e}_\theta = \vec{e}_r \cdot \vec{e}_\phi = 0$. This means $g_{r\theta} = g_{r\phi} = 0$. (Recall Eqs. (3.3) and (3.21).) This is a definition of the coordinates, allowed by spherical symmetry. We thus have restricted the metric to the form

$$ds^2 = g_{00}\, dt^2 + 2g_{0r}\, dr\, dt + 2g_{0\theta}\, d\theta\, dt$$
$$+ 2g_{0\phi}\, d\phi\, dt + g_{rr}\, dr^2 + r^2\, d\Omega^2. \tag{10.4}$$

Spherically symmetric spacetime. Since not only the spaces $t = $ const. are spherically symmetric, but also the whole spacetime, we must have that a line $r = $ const., $\theta = $ const., $\phi = $ const. is also orthogonal to the two-spheres. Otherwise there would be a preferred direction in space. This means that $\vec{e}_t$ is orthogonal to $\vec{e}_\theta$ and $\vec{e}_\phi$, or $g_{t\theta} = g_{t\phi} = 0$. So now we have

$$ds^2 = -g_{00}\, dt^2 + 2g_{0r}\, dr\, dt + g_{rr}\, dr^2 + r^2\, d\Omega^2. \tag{10.5}$$

This is the general metric of a spherically symmetric spacetime, where g_{00}, g_{0r}, and g_{rr} are functions of r and t. We have used our coordinate freedom to reduce it to the simplest possible form.

10.2 Static spherically symmetric spacetimes

The metric. Clearly, the simplest physical situation we can describe is a quiescent star or black hole – a static system. We *define* a static spacetime to be one in which we can find a time coordinate t with two properties: (i) all metric components are independent of t, and (ii) the geometry is unchanged by time reversal, $t \to -t$. The second condition means that a film made of the situation looks the same when run backwards. This is not logically implied by (i), as the example of a rotating star makes clear: time reversal changes the sense of rotation, but the metric components will be constant in time. (A spacetime with property (i) but not necessarily (ii) is said to be *stationary*.)

Condition (ii) has the following implication. The coordinate transformation $(t, r, \theta, \phi) \to (-t, r, \theta, \phi)$ has $\Lambda^0{}_0 = -1$, $\Lambda^i{}_j = \delta^i{}_j$, and we find

$$\left. \begin{aligned} g_{\bar{0}\bar{0}} &= (\Lambda^0{}_{\bar{0}})^2 g_{00} = g_{00}, \\ g_{\bar{0}\bar{r}} &= \Lambda^0{}_{\bar{0}} \Lambda^r{}_{\bar{r}} g_{0r} = -g_{0r}, \\ g_{\bar{r}\bar{r}} &= (\Lambda^r{}_{\bar{r}})^2 g_{rr} = g_{rr}. \end{aligned} \right\} \tag{10.6}$$

Since the geometry must be unchanged (i.e. since $g_{\bar{\alpha}\bar{\beta}} = g_{\alpha\beta}$), we must have $g_{0r} \equiv 0$. Thus, the metric of a *static*, spherically symmetric spacetime is

$$ds^2 = -e^{2\Phi}\,dt^2 + e^{2\Lambda}\,dr^2 + r^2\,d\Omega^2, \tag{10.7}$$

where we have introduced $\Phi(r)$ and $\Lambda(r)$ in place of the two unknowns $g_{00}(r)$ and $g_{rr}(r)$. This replacement is acceptable provided $g_{00} < 0$ and $g_{rr} > 0$ everywhere. We shall see below that these conditions do hold inside stars, but they break down for black holes. When we study black holes in the next chapter we shall have to look carefully again at our coordinate system.

If we are interested in stars, which are bounded systems, we are entitled to demand that, far from the star, spacetime is flat. This means that we can impose the following boundary conditions (or asymptotic regularity conditions) on Einstein's equations:

$$\lim_{r \to \infty} \Phi(r) = \lim_{r \to \infty} \Lambda(r) = 0. \tag{10.8}$$

Physical interpretation of metric terms. Since we have constructed our coordinates to reflect the physical symmetries of the spacetime, the metric components have useful physical significance. The proper radial distance from any radius r_1 to another radius r_2 is

$$l_{12} = \int_{r_1}^{r_2} e^{\Lambda}\,dr, \tag{10.9}$$

since the curve is one on which $dt = d\theta = d\phi = 0$. More important is the significance of g_{00}. Since the metric is independent of t, we know from Ch. 7 that any particle following a geodesic has constant momentum component p_0, which we can define to be the constant $-E$:

$$p_0 \equiv -E. \tag{10.10}$$

But a local *inertial* observer at rest (momentarily) at any radius r of the spacetime measures a different energy. His four-velocity must have $U^i = dx^i/d\tau = 0$ (since he is momentarily at rest), and the condition $\vec{U} \cdot \vec{U} = 1$ implies $U^0 = e^{-\Phi}$. The energy he measures is

$$E^* = -\vec{U} \cdot \vec{p} = e^{-\Phi} E. \tag{10.11}$$

We therefore have found that a particle whose geodesic is characterized by the constant E has energy $e^{-\Phi}E$ relative to a locally inertial observer at rest in the spacetime. Since $e^{-\Phi} = 1$ far away, we see that E is the energy a distant observer would measure if the particle gets far away. We call it the energy at infinity. Since $e^{-\Phi} > 1$ everywhere else (this will be clear later), we see that the particle has larger energy relative to inertial observers that it passes elsewhere. This extra energy is just the kinetic energy it gains by falling in a gravitational field. The energy is studied in more detail in Exer. 3, § 10.9.

This is particularly significant for photons. Consider a photon emitted at radius r_1 and received very far away. If its frequency in the local inertial frame is ν_{em} (which would be determined by the process emitting it; e.g. a spectral line), then its local energy is $h\nu_{em}$ (h being Planck's constant) and its conserved constant E is $h\nu_{em} \exp[\Phi(r_1)]$. When it reaches the distant observer it is measured to have energy E, and hence frequency $E/h \equiv \nu_{rec} = \nu_{em} \exp[\Phi(r_1)]$. The *redshift* of the photon, defined by

$$z = \frac{\lambda_{rec} - \lambda_{em}}{\lambda_{em}} = \frac{\nu_{em}}{\nu_{rec}} - 1, \tag{10.12}$$

is therefore

◆ $\quad z = e^{-\Phi(r_1)} - 1. \tag{10.13}$

This important equation attaches physical significance to e^Φ. (Compare this calculation with the one in Ch. 2.)

The Einstein tensor. One can show that for the metric given by Eq. (10.7), the Einstein tensor has components

$$G_{00} = \frac{1}{r^2} e^{2\Phi} \frac{d}{dr} [r(1 - e^{-2\Lambda})], \tag{10.14}$$

$$G_{rr} = -\frac{1}{r^2} e^{2\Lambda} (1 - e^{-2\Lambda}) + \frac{2}{r} \Phi', \tag{10.15}$$

$$G_{\theta\theta} = r^2 e^{-2\Lambda} [\Phi'' + (\Phi')^2 + \Phi'/r - \Phi'\Lambda' - \Lambda'/r], \tag{10.16}$$

$$G_{\phi\phi} = \sin^2\theta G_{\theta\theta}. \tag{10.17}$$

where $\Phi' \equiv d\Phi/dr$, etc. All other components vanish.

10.3 Static perfect fluid Einstein equations

Stress–energy tensor. We are interested in static stars, in which the fluid has no motion. The only nonzero component of $\vec{U}$ is therefore U^0. What is more, the normalization condition

$$\vec{U} \cdot \vec{U} = -1 \tag{10.18}$$

implies, as we have seen before,

$$U^0 = e^{-\Phi}, \qquad U_0 = -e^\Phi. \tag{10.19}$$

Then **T** has components given by Eq. (4.31):

$$T_{00} = \rho \, e^{2\Phi}, \tag{10.20}$$

$$T_{rr} = p \, e^{2\Lambda}, \tag{10.21}$$

$$T_{\theta\theta} = r^2 p, \tag{10.22}$$

$$T_{\phi\phi} = \sin^2 \theta \, T_{\theta\theta}. \tag{10.23}$$

All other components vanish.

Equation of state. The stress–energy tensor involves both p and ρ, but these may be related by an equation of state. For a simple fluid in local thermodynamic equilibrium there always exists a relation of the form

$$p = p(\rho, S), \tag{10.24}$$

which gives the pressure in terms of the energy density and specific entropy. One often deals with situations in which the entropy can be considered to be a constant (in particular, negligibly small), so that one has a relation

$$p = p(\rho). \tag{10.25}$$

These relations will of course have different functional forms for different fluids. We will suppose that *some* such relation always exists.

Equations of motion. The conservation laws are (Eq. (7.6))

$$T^{\alpha\beta}{}_{;\beta} = 0. \tag{10.26}$$

These are four equations, one for each value of the free index α. Because of the symmetries, only one of these does not vanish identically: the one for which $\alpha = r$. It implies

$$\blacklozenge \qquad (\rho + p)\frac{d\Phi}{dr} = -\frac{dp}{dr}. \tag{10.27}$$

This is the equation which tells us what pressure gradient is needed to keep the fluid static in the gravitational field, whose effect depends on $d\Phi/dr$.

Einstein equations. The $(0, 0)$ component of Einstein's equations can be found from Eqs. (10.14) and (10.20). It is convenient at this point to replace $\Lambda(r)$ with a different unknown function $m(r)$ defined as

$$\blacklozenge \qquad m(r) \equiv \tfrac{1}{2}r(1 - e^{-2\Lambda}), \tag{10.28}$$

or

$$\blacklozenge \qquad g_{rr} = e^{2\Lambda} = \frac{1}{1 - \dfrac{2m(r)}{r}}. \tag{10.29}$$

Then the $(0, 0)$ equation implies

$$\blacklozenge \qquad \frac{dm(r)}{dr} = 4\pi r^2 \rho. \tag{10.30}$$

This has the same form as the Newtonian equation which calls $m(r)$ the mass inside the sphere of radius r. Therefore in relativity we call $m(r)$ the mass function, but it cannot be interpreted as the mass energy inside r since total energy is not localizable in GR. We shall explore the Newtonian analogy in § 10.5 below.

The (r, r) equation, from Eqs. (10.15) and (10.21), can be cast in the form

$$\blacklozenge \qquad \frac{d\Phi}{dr} = \frac{m(r) + 4\pi r^3 p}{r[r - 2m(r)]}. \qquad (10.31)$$

If one has an equation of state of the form Eq. (10.25) then Eqs. (10.25), (10.27), (10.30) and (10.31) are four equations for the four unknowns Φ, m, ρ, p. If the more general equation of state, Eq. (10.24), is needed, then S is a completely arbitrary function. There is *no* additional information contributed by the (θ, θ) and (ϕ, ϕ) Einstein equation, because (i) it is clear from Eqs. (10.16), (10.17), (10.22) and (10.23) that the two equations are essentially the same, and (ii) the Bianchi identities ensure that this equation is a consequence of Eqs. (10.26), (10.30) and (10.31).

10.4 The exterior geometry

Schwarzschild metric. In the region outside the star we have $\rho = p = 0$, and we get the two equations

$$\frac{dm}{dr} = 0, \qquad (10.32)$$

$$\frac{d\Phi}{dr} = \frac{m}{r(r - 2m)}. \qquad (10.33)$$

These have the solutions

$$m(r) = M = \text{const.}, \qquad (10.34)$$

$$e^{2\Phi} = 1 - \frac{2M}{r}, \qquad (10.35)$$

where the requirement that $\Phi \to 0$ as $r \to \infty$ has been applied. We therefore see that the exterior metric has the following form, called the *Schwarzschild metric*:

$$\blacklozenge \qquad ds^2 = -\left(1 - \frac{2M}{r}\right) dt^2 + \frac{dr^2}{1 - \dfrac{2M}{r}} + r^2 d\Omega^2. \qquad (10.36)$$

For large r this becomes

$$ds^2 \approx -\left(1 - \frac{2M}{r}\right) dt^2 + \left(1 + \frac{2M}{r}\right) dr^2 + r^2 d\Omega^2. \qquad (10.37)$$

One can find coordinates (x, y, z) such that this becomes

$$ds^2 \approx -\left(1 - \frac{2M}{R}\right) dt^2 + \left(1 + \frac{2M}{R}\right) (dx^2 + dy^2 + dz^2), \qquad (10.38)$$

where $R \equiv (x^2 + y^2 + z^2)^{1/2}$. We see that this is the far-field metric of a star of total mass M (see Eq. (8.60)). This justifies the definition, Eq. (10.28), and the choice of the symbol M.

Generality of the metric. A more general treatment, as in Misner *et al.* (1973), establishes *Birkhoff's theorem*, that the Schwarzschild solution, Eq. (10.36), is the only spherically symmetric, asymptotically flat solution to Einstein's vacuum field equations, even if we drop our initial assumptions that the metric is static, i.e. if we start with Eq. (10.5). This means that even a radially pulsating or collapsing star will have a static exterior metric of constant mass M. One conclusion one can draw from this is that there are no gravitational waves from pulsating spherical systems. (This has an analogy in electromagnetism: there is no 'monopole' electromagnetic radiation either.) We found this result from linearized theory in Exer. 37, § 9.6.

10.5   The interior structure of the star

Inside the star, we have $\rho \neq 0$, $p \neq 0$, and so we can divide Eq. (10.27) by $(\rho + p)$ and use it to eliminate Φ from Eq. (10.31). The result is called the Oppenheimer–Volkov (O–V) equation:

$$\blacklozenge \qquad \frac{dp}{dr} = -\frac{(\rho + p)(m + 4\pi r^3 p)}{r(r - 2m)}. \qquad (10.39)$$

Combined with Eq. (10.30) for dm/dr and an equation of state of the form of Eq. (10.25), this gives three equations for m, ρ, and p. We have reduced Φ to a subsidiary position; it can be found from Eq. (10.27) once the others have been solved.

General rules for integrating the equations. Since there are two first-order differential equations, Eqs. (10.30) and (10.39), they require two constants of integration, one being $m(r = 0)$ and the other $p(r = 0)$. We now argue that $m(r = 0) = 0$. A tiny sphere of radius $r = \varepsilon$ has circumference $2\pi\varepsilon$, and proper radius $|g_{rr}|^{1/2}\varepsilon$ (from the line element). Thus a small circle about $r = 0$ has ratio of circumference to radius of $2\pi|g_{rr}|^{-1/2}$. But if spacetime is locally flat at $r = 0$, as it must be at *any* point of the manifold, then a *small* circle about $r = 0$ must have ratio of circumference to radius of 2π. Therefore $g_{rr}(r = 0) = 1$, and so as r goes to zero, $m(r)$ must also go to zero, in fact *faster* than r. The other constant of integration, $p(r = 0) \equiv p_c$ or, equivalently, ρ_c, from the equation of state, simply defines the stellar model. *For a given equation of state $p = p(\rho)$, the set of all spherically symmetric static stellar models forms a one-parameter sequence, the parameter being the central density.* This result follows only from the standard uniqueness theorems for first-order ordinary differential equations.

Once $m(r)$, $p(r)$ and $\rho(r)$ are known, the *surface* of the star is defined as the place where $p = 0$. (Notice that, by Eq. (10.39), the pressure decreases monotonically outwards from the center.) The reason $p = 0$ marks the surface is that p must be continuous everywhere, for otherwise there would be an infinite pressure gradient and infinite forces on fluid elements. Since $p = 0$ in the vacuum outside the star, the surface must also have $p = 0$. Therefore one stops integrating the interior solution there and requires that the exterior metric should be the Schwarzschild metric. Let the radius of the surface be R. Then in order to have a smooth geometry the metric functions must be continuous at $r = R$. Inside the star we have

$$g_{rr} = \left(1 - \frac{2m(r)}{r}\right)^{-1}$$

and outside we have

$$g_{rr} = \left(1 - \frac{2M}{r}\right)^{-1}.$$

Continuity clearly defines the constant M to be

$$M \equiv m(R). \tag{10.40}$$

Thus the total mass of the star as determined by distant orbits is found to be the integral

♦ $$M = \int_0^R 4\pi r^2 \rho \, dr, \tag{10.41}$$

just as in Newtonian theory. This analogy is rather deceptive, however, since the integral is over the volume element $4\pi r^2 \, dr$, which is *not* the element of *proper* volume. Proper volume in the hypersurface $t = $ const. is given by

$$|-g|^{1/2} d^3 x = e^{\Phi + \Lambda} r^2 \sin\theta \, dr \, d\theta \, d\phi, \tag{10.42}$$

which, after doing the (θ, ϕ) integration, is just $4\pi r^2 e^{\Phi + \Lambda} \, dr$. Thus M is not in any sense just the sum of all the proper energies of the fluid elements. The difference between the proper and coordinate volume elements is where the 'gravitational potential energy' contribution to the total mass is placed in these coordinates. We need not look in more detail at this; it only illustrates the care one must take in applying Newtonian interpretations to relativistic equations.

Having obtained M, this determines g_{00} outside the star, and hence g_{00} at its surface:

$$g_{00}(r = R) = -\left(1 - \frac{2M}{R}\right). \tag{10.43}$$

This serves as the integration constant for the final differential equation, the one which determines Φ inside the star: Eq. (10.27). We thereby obtain the complete solution.

Notice that solving for the structure of the star is the first place where we have actually assumed that the point $r = 0$ is contained in the spacetime. We had earlier argued that it need not be, and the discussion *before* the interior solution made no such assumptions. We make the assumption here because we want to talk about 'ordinary' stars, which we feel must have the same global topology as Euclidean space, differing from it only by being curved here and there. However, the exterior Schwarzschild solution is independent of assumptions about $r = 0$, and when we discuss black holes we shall see how different $r = 0$ can be.

Notice also that for our ordinary stars we always have $2m(r) < r$. This is certainly true near $r = 0$, since we have seen that we need $m(r)/r \to 0$ at $r = 0$. If it ever happened that near some radius r_1 we had $r - 2m(r) = \varepsilon$, with ε small and decreasing with r, then by the O–V equation, Eq. (10.39), the pressure gradient would be of order $1/\varepsilon$ and negative. This would cause the pressure to drop so rapidly from any finite value that it would pass through zero *before* ε reached zero. But as soon as p vanishes we have reached the surface of the star. Outside that point, m is constant and r increases. So nowhere in the spacetime of an ordinary star can $m(r)$ reach $\frac{1}{2}r$.

The structure of Newtonian stars. Before looking for solutions, we shall briefly look at the Newtonian limit of these equations. In Newtonian situations we have $p \ll \rho$, so we also have $4\pi r^3 p \ll m$. Moreover, the metric must be nearly flat, so in Eq. (10.29) we require $m \ll r$. These inequalities simplify Eq. (10.39) to

$$\frac{dp}{dr} = -\frac{\rho m}{r^2}. \tag{10.44}$$

This is exactly the same as the equation of hydrostatic equilibrium for Newtonian stars (see Chandrasekhar 1939), a fact which should not surprise us in view of our earlier interpretation of m and of the trivial fact that the Newtonian limit of ρ is just the mass density. Comparing Eq. (10.44) with its progenitor, Eq. (10.39), shows that all the relativistic corrections tend to steepen the pressure gradient relative to the Newtonian one. In other words, for a fluid to remain static it must have stronger internal forces in GR than in Newtonian gravity. This can be interpreted loosely as indicating a stronger field. An extreme instance of this is gravitational collapse: a field so strong that the fluid's pressure cannot resist it. We shall discuss this more fully in § 10.7 below.

10.6   Exact interior solutions

In Newtonian theory, Eqs. (10.30) and (10.44) are very hard to solve analytically for a given equation of state. Their relativistic counterparts are worse.[1] We shall discuss two interesting exact solutions to the relativistic equations, one due to Schwarzschild and a much more recent one by Buchdahl (1981).

The Schwarzschild constant-density interior solution. To simplify the task of solving Eqs. (10.30) and (10.39), we make the assumption

$$\rho = \text{const.} \tag{10.45}$$

This replaces the question of state. There is no physical justification for it, of course. In fact, the speed of sound, which is proportional to $(\mathrm{d}p/\mathrm{d}\rho)^{1/2}$, is infinite! Nevertheless, the interiors of dense neutron stars are of *nearly* uniform density, so this solution has some interest for us in addition to its pedagogic value as an example of the method one uses to solve the system.

We can integrate Eq. (10.30) immediately:

$$m(r) = 4\pi\rho r^3/3, \qquad r \le R, \tag{10.46}$$

where R is the star's as yet undetermined radius. Outside R the density vanishes, so $m(r)$ is constant. By demanding continuity of g_{rr} we find that $m(r)$ must be continuous at R. This implies

$$m(r) = 4\pi\rho R^3/3 \equiv M, \qquad r \ge R, \tag{10.47}$$

where we denote this constant by M, the Schwarzschild mass.

---

1   If one does not restrict the equation of state, then Eqs. (10.44) and (10.39) are easier to solve. For example, one can arbitrarily assume a function $m(r)$, deduce $\rho(r)$ from it via Eq. (10.30), and hope to be able to solve Eqs. (10.44) or (10.39) for p. The result, two functions $p(r)$ and $\rho(r)$, implies an 'equation of state' $p = p(\rho)$ by eliminating r. This is unlikely to be physically realistic, so most exact solutions obtained in this way do not interest the astrophysicist.

We can now solve the O–V equation, Eq. (10.39):

$$\frac{dp}{dr} = -\tfrac{4}{3}\pi r \frac{(\rho + p)(\rho + 3p)}{1 - 8\pi r^2 \rho/3}.$$ (10.48)

This is easily integrated from an arbitrary central pressure p_c to give

$$\frac{\rho + 3p}{\rho + p} = \frac{\rho + 3p_c}{\rho + p_c}\left(1 - 2\frac{m}{r}\right)^{1/2}.$$ (10.49)

From this it follows that

$$R^2 = \frac{3}{8\pi\rho}[1 - (\rho + p_c)^2/(\rho + 3p_c)^2]$$ (10.50)

or

$$p_c = \rho[1 - (1 - 2M/R)^{1/2}]/[3(1 - 2M/R)^{1/2} - 1].$$ (10.51)

Replacing p_c in Eq. (10.49) by this gives

$$p = \rho \frac{(1 - 2Mr^2/R^3)^{1/2} - (1 - 2M/R)^{1/2}}{3(1 - 2M/R)^{1/2} - (1 - 2Mr^2/R^3)^{1/2}}.$$ (10.52)

Notice that Eq. (10.51) implies $p_c \to \infty$ as $M/R \to 4/9$. We will see later that this is a very general limit on M/R, even for more realistic stars.

We complete the uniform-density case by solving for Φ from Eq. (10.27). Here we know the value of Φ at R, since it is implied by continuity of g_{00}:

$$g_{00}(R) = -(1 - 2M/R).$$ (10.53)

Therefore we find

$$\exp(\Phi) = \tfrac{3}{2}(1 - 2M/R)^{1/2} - \tfrac{1}{2}(1 - 2Mr^2/R^3)^{1/2}, \qquad r \leqslant R. \quad (10.54)$$

Note that Φ and m are monotonically increasing functions of r, while p decreases monotonically.

Buchdahl's interior solution. Buchdahl (1981) found a solution for the equation of state

$$\rho = 12(p_* p)^{1/2} - 5p,$$ (10.55)

where p_* is an arbitrary constant. While this equation has no particular physical basis, it does have two nice properties: (i) it can be made causal everywhere in the star by demanding that the local sound speed $(dp/d\rho)^{1/2}$ be less than one; and (ii) for small p it reduces to

$$\rho = 12(p_* p)^{1/2},$$ (10.56)

which, in the Newtonian theory of stellar structure, is called an $n = 1$ polytrope. The $n = 1$ polytrope is one of the few exactly solvable Newtonian systems (see Exer. 14, § 10.9), so Buchdahl's solution may be regarded as its relativistic generalization. The causality requirement demands

$$p < p_*, \qquad \rho < 7p_*. \qquad (10.57)$$

Like most exact solutions[2] this one is difficult to deduce from the standard form of the equations. In this case, one requires a different radial coordinate r'. This is defined, in terms of the usual r, implicitly by Eq. (10.59) below, which involves a second arbitrary constant β, and the function[3]

$$u(r') \equiv \beta \, \frac{\sin Ar'}{Ar'}, \qquad A^2 \equiv \frac{288\pi p_*}{1-2\beta}. \qquad (10.58)$$

Then we write

$$r(r') = r' \, \frac{1-\beta+u(r')}{1-2\beta}. \qquad (10.59)$$

Rather than demonstrate how to obtain the solution (see Buchdahl 1981), we shall content ourselves simply to write it down. In terms of the metric functions defined in Eq. (10.7) we have, for $Ar' \le \pi$,

$$\exp(2\Phi) = (1-2\beta)(1-\beta-u)(1-\beta+u)^{-1}, \qquad (10.60)$$

$$\exp(2\Lambda) = (1-2\beta)(1-\beta+u)(1-\beta-u)^{-1}(1-\beta+\beta\cos Ar')^{-2}, \qquad (10.61)$$

$$p(r) = A^2(1-2\beta)u^2[8\pi(1-\beta+u)^2]^{-1}, \qquad (10.62)$$

$$\rho(r) = 2A^2(1-2\beta)u(1-\beta-\tfrac{3}{2}u)[8\pi(1-\beta+u)^2]^{-1}, \qquad (10.63)$$

where $u = u(r')$. The surface $p = 0$ is where $u = 0$, i.e. at $r' = \pi/A \equiv R'$. At this place we have

$$\exp(2\Phi) = \exp(-2\Lambda) = 1-2\beta, \qquad (10.64)$$

$$R \equiv r(R') = \pi(1-\beta)(1-2\beta)^{-1}A^{-1}. \qquad (10.65)$$

Therefore β is the value of M/R on the surface, which in the light of Eq. (10.13) is related to the surface redshift of the star by

$$z_s = (1-2\beta)^{-1/2} - 1. \qquad (10.66)$$

Clearly the nonrelativistic limit of this sequence of models is the limit $\beta \to 0$. The mass of the star is given by

$$M = \frac{\pi\beta(1-\beta)}{(1-2\beta)A} = \left[\frac{\pi}{288p_*(1-2\beta)} \right]^{1/2} \beta(1-\beta). \qquad (10.67)$$

2 An exact solution is one which can be written in terms of simple functions of the coordinates, such as polynomials and trigonometric functions. Finding such solutions is an art which requires the successful combination of useful coordinates, simple geometry, good intuition, and in most cases luck. See Kramer *et al.* (1981) for a recent review of the subject.

3 Buchdahl uses different notation for his parameters.

Since β alone determines how relativistic the star is, the constant p_* (or A) simply gives an overall dimensional scaling to the problem. It can be given any desired value by an appropriate choice of the unit for distance. It is β, therefore, whose variation produces nontrivial changes in the structure of the model. The lower limit on β is, as we remarked above, zero. The upper limit comes from the causality requirement, Eq. (10.57), and the observation that Eqs. (10.62) and (10.63) imply

$$p/\rho = \tfrac{1}{2}u(1 - \beta - \tfrac{3}{2}u)^{-1},$$

(10.68)

whose maximum value is at the center, $r = 0$:

$$p_c/\rho_c = \beta(2 - 5\beta)^{-1}.$$

(10.69)

Demanding that this be less than $\tfrac{1}{7}$ gives

$$0 < \beta < \tfrac{1}{6}.$$

(10.70)

This range spans a spectrum of physically reasonable models from the Newtonian ($\beta \approx 0$) to the very relativistic (surface redshift 0.22).

10.7 Realistic stars and gravitational collapse

Buchdahl's theorem. We have seen in the previous section that there are no uniform-density stars with radii smaller than $(9/4)M$, because to support them in a static configuration requires pressures larger than infinite! This is in fact true of *any* stellar model, and is known as Buchdahl's theorem (Buchdahl 1959). Suppose one constructs a star of radius $R = 9M/4$, and then gives it a (spherically symmetric) inward push. It has no choice but to collapse inwards: it cannot reach a static state again. But during its collapse, the metric outside it is just the Schwarzschild metric. What it leaves, then, is the vacuum Schwarzschild geometry outside. This is the metric of a black hole, and we will study it in detail later. First we look at some causes of gravitational collapse.

Stellar evolution. Any realistic appraisal of the chances of forming a black hole must begin with an understanding of the way stars evolve. We give a brief summary here. See, for example, Clayton (1968).

An ordinary star like our Sun derives its luminosity from nuclear reactions, mainly the conversion of hydrogen to helium. Because a star is always radiating energy, it needs the nuclear reactions to replace that energy in order to remain static. When the original supply of hydrogen is converted to helium, this energy source turns off, and the inner region (core) of the star begins to shrink as it gradually radiates energy away. This shrinking compresses and heats the core, until the temperatures are high enough to ignite another reaction which converts helium into carbon and oxygen, releasing more energy. In order to cope with this new energy the outer layers of the star actually expand, and the star becomes a sort of 'core-halo' structure, called a red giant. Eventually the helium is exhausted, and the star may then go through phases of turning carbon into silicon, and silicon into iron. Eventually, every star must run out of energy, since iron is the most stable of all nuclei – any reaction converting iron into something else absorbs energy rather than releasing it. The precise time when this happens and the subsequent evolution of the star depend mainly on three things: the star's mass, angular momentum, and magnetic field.

First consider slowly rotating stars, for which rotation is an insignificant factor in their structure. A star of the Sun's mass will find itself evolving smoothly to a state in which it is called a white dwarf. This is a star whose pressure comes not from thermal effects but from quantum mechanical ones, which we discuss later. The point about relatively low-mass stars like our Sun is that they don't have strong enough gravitational fields to overwhelm these quantum effects or to cause rapid contraction earlier on in their history. A higher-mass star will also evolve smoothly through the hydrogen-burning phase (which is called the 'main-sequence' phase of its life), but what happens after that is not completely understood. Massive stars are known to have strong stellar winds which can cause the loss of considerable fractions of their mass during the main-sequence stage. Stars in close binary systems may also pass considerable mass to their companions as they evolve off the main sequence. If a star loses enough mass, its subsequent evolution may be quiet, like that expected for our Sun. But it seems that not all stars follow this route. At some point in the nuclear cycle, one of two things may happen: either the star experiences a run-away nuclear explosion (stars up to mass $8–10\ M_\odot$) or the core of the star becomes hydrodynamically unstable and collapses to a compact object (a neutron star or black hole), releasing the gravitational binding energy of that object (affects stars above $10\ M_\odot$?). In either case, the result is believed to be what we observe as supernova explosions.

This picture can be substantially altered by rotation and magnetic fields, but we have very little understanding yet of how. Rotation may induce currents that change the main-sequence evolution by mixing inner and outer layers of the star. In the collapse phase, rotation becomes extremely important if angular momentum is conserved by the collapsing core. But substantial magnetic fields may allow transfer of angular momentum from the core to the rest of the star, permitting a more spherical collapse.

One thing is certain: supernova explosions often leave behind rapidly rotating neutron stars, which are observed as pulsars. In fact, one of the problems theoreticians must face is that the pulsar birth-rate may be as large as the supernova rate, so all these added complications must still somehow produce neutron stars. Black holes produced by supernovae would be much harder to observe unless they were part of a binary system which survived the explosion and in which the other star was not so highly evolved. One such system is the X-ray source Cygnus X-1, in which the compact object is thought to have a mass larger than 10 $M_\odot$ (Bahcall 1978). This exceeds, by a large margin, the theoretical upper limit on the mass of a neutron star, so there is general confidence that this is a black hole. But one example is too small a sample for statistical analysis, so it is very uncertain what observational constraints are set on the black-hole birth-rate.

Even the theoretical limit to the mass of a neutron star is very uncertain. For nonrotating stars (and éven for pulsars, which rotate but not fast enough to affect their structure much), the limit appears to be less than 2 $M_\odot$ (Hartle 1978), and there is some observational evidence for 1.4 $M_\odot$: the binary pulsar system contains two stars of that mass (see Ch. 9), one of which is certainly a neutron star and the other probably is one; and those binary X-ray systems in which the compact star's mass can be estimated either give about 1.4 $M_\odot$ or very much larger masses (as for Cygnus X-1).

Rotation can, in principle, considerably increase the upper limit on stellar masses, at least until rotation-induced relativistic instabilities set in (Durisen 1975, Friedman & Schutz 1978). This probably doesn't allow more than a factor of 3 in mass.

Quantum mechanical pressure. We shall now give an elementary discussion of the forces that support white dwarfs and neutron stars. Consider an electron in a box of volume V. Because of the Heisenberg uncertainty principle, its momentum is uncertain by an amount of the order of

$$\Delta p = hV^{-1/3}, \tag{10.71}$$

where h is Planck's constant. If its momentum has magnitude between p and $p + dp$, it is in a region of momentum space of volume $4\pi p^2 \, dp$. The number of 'cells' in this region of volume Δp is

$$N = 4\pi p^2 \, dp/(\Delta p)^3 = \frac{4\pi p^2 \, dp}{h^3} V. \tag{10.72}$$

Since it is impossible to define the momentum of the electron more precisely than Δp, this is the number of possible momentum states with momentum between p and $p + \mathrm{d}p$ in a box of volume V. Now, electrons are Fermi particles, which means that they have the remarkable property that no two of them can occupy exactly the same state. (This is the basic reason for the variety of the periodic table and the solidity and relative impermeability of matter.) Electrons have spin $\frac{1}{2}$, which means that for each momentum state there are two spin states ('spin-up' and 'spin-down'), so there are a total of

$$V \frac{8\pi p^2 \, \mathrm{d}p}{h^3} \tag{10.73}$$

states, which is then the *maximum* number of electrons that can have momenta between p and $p + \mathrm{d}p$ in a box of volume V.

Now suppose we cool off a gas of electrons as far as possible, which means reducing each electron's momentum as far as possible. If there is a total of N electrons, then they are as cold as possible when they fill all the momentum states from $p = 0$ to some upper limit p_f, determined by the equation

$$\frac{N}{V} = \int_0^{p_f} \frac{8\pi p^2 \, \mathrm{d}p}{h^3} = \frac{8\pi p_f^3}{3h^3}. \tag{10.74}$$

Since N/V is the number density, we get that a cold electron gas obeys the relation

$$n = \frac{8\pi p_f^3}{3h^3}, \qquad p_f = \left(\frac{3h^3}{8\pi}\right)^{1/3} n^{1/3}. \tag{10.75}$$

The number p_f is called the Fermi momentum. Notice that it depends only on the number of particles per unit volume, not on their masses.

Each electron has mass m and energy $E = (p^2 + m^2)^{1/2}$. Therefore the total energy density in such a gas is

$$\rho = \frac{E_{\text{TOTAL}}}{V} = \int_0^{p_f} \frac{8\pi p^2}{h^3} (m^2 + p^2)^{1/2} \, \mathrm{d}p. \tag{10.76}$$

The pressure can be found from Eq. (4.22) with ΔQ set to zero, since we are dealing with a closed system:

$$p = -\frac{\mathrm{d}}{\mathrm{d}V}(E_{\text{TOTAL}}) = -V \frac{8\pi p_f^2}{h^3} (m^2 + p_f^2)^{1/2} \frac{\mathrm{d}p_f}{\mathrm{d}V} - \rho.$$

For a constant number of particles N, we have

$$V \frac{\mathrm{d}p_f}{\mathrm{d}V} = -n \frac{\mathrm{d}p_f}{\mathrm{d}n} = \frac{1}{3} \left(\frac{3h^3}{8\pi}\right)^{1/3} n^{1/3} = \frac{1}{3} p_f,$$

and we get

$$p = \left(\frac{8\pi}{3h^3}\right) p_f^3 (m^2 + p_f^2)^{1/2} - \rho. \tag{10.77}$$

For a very relativistic gas where $p_f \gg m$ (which will be the case if the gas is compressed to small V) we have

$$\rho \approx \frac{2\pi p_f^4}{h^3} \tag{10.78}$$

$$p \approx \tfrac{1}{3}\rho. \tag{10.79}$$

This is the equation of state for a 'cold' electron gas. So the gas has a pressure comparable to its density even when it is as cold as possible. In Exer. 22, § 4.10 we saw that Eq. (10.79) is also the relation for a photon gas. The reason that the relativistic Fermi gas behaves like a photon gas is essentially that the energy of each electron far exceeds its rest mass; the rest mass is unimportant, so setting it to zero changes little.

White dwarfs. When an ordinary star is compressed, it reaches a stage where the electrons are free of the nuclei, and one has two gases, one of electrons and one of nuclei. Since they have the same temperatures, and hence the same energies per particle, the less-massive electrons have far less momentum per particle. Upon compression the Fermi momentum rises (Eq. (10.5)) until it becomes comparable with the momentum of the electrons. They are then effectively a cold electron gas, and supply the pressure for the star. The nuclei have momenta well in excess of p_f, so they are a classical gas, but they supply little pressure. On the other hand, the nuclei supply most of the gravity, since there are more neutrons and protons than electrons, and they are much more massive. So the mass density for Newtonian gravity (which is adequate here) is

$$\rho = \mu m_p n_e, \tag{10.80}$$

where μ is the ratio of number of nucleons to the number of electrons (of the order of 1 or 2), m_p is the proton mass, and n_e the number density of electrons. The relation between pressure and density for the whole gas when the electrons are relativistic is therefore

$$\left.\begin{aligned}
p &= k\rho^{4/3}, \\
k &= \frac{2\pi}{3h^3}\left(\frac{3h^3}{8\pi\mu m_p}\right)^{4/3}.
\end{aligned}\right\} \tag{10.81}$$

The Newtonian structure equations for the star are

$$\frac{dm}{dr} = 4\pi r^2 \rho,$$

$$\frac{dp}{dr} = -\rho \frac{m}{r^2}.$$

In order of magnitude, these are (for a star of mass M, radius R, typical density $\bar{\rho}$ and typical pressure $\bar{p}$)

$$\left.\begin{array}{l} M = R^3 \bar{\rho}, \\[2mm] \dfrac{\bar{p}}{R} = \bar{\rho} \dfrac{M}{R^2}. \end{array}\right\} \tag{10.82}$$

Setting $\bar{p} = k\bar{\rho}^{-4/3}$, from Eq. (10.81), gives

$$k\bar{\rho}^{1/3} = \frac{M}{R}. \tag{10.83}$$

Using Eq. (10.82) in this, we see that R cancels out and we get an equation only for M:

$$M = \left(\frac{3k^3}{4\pi}\right)^{1/2} = \frac{1}{32\mu^2 m_p^2}\left(\frac{6h^3}{\pi}\right)^{1/2}. \tag{10.84}$$

Using geometrized units, one finds (with $\mu = 2$)

$$M = 0.47 \times 10^5 \text{ cm} = 0.32 \text{ M}_\odot. \tag{10.85}$$

From our derivation we should expect this to be the order of magnitude of the maximum mass supportable by a relativistic electron gas, when most of the gravity comes from a cold nonrelativistic gas of nuclei. This is called the *Chandrasekhar limit*, and a more precise calculation puts it at $M \approx 1.3$ M$_\odot$. Any star more massive than this cannot be supported by electron pressure, and so cannot be a white dwarf. In fact, the upper mass limit is marginally smaller, occurring at central densities of about 10^{10} kg m^{-3}, and is caused by the instability described below.

Neutron stars. If the material is compressed further than that characteristic of a white dwarf (which has $\rho \lesssim 10^{10}$ kg m^{-3}), it happens that the kinetic energy of electrons gets so large that if they combine with a proton to form a neutron, energy can be released. So compression results in the loss of electrons from the gas which is providing pressure: pressure does not build up rapidly enough, and the star is unstable. There are no stable stars again until central densities reach the region of 10^{16} kg m^{-3}. By this density, all the electrons are united with the protons to form a gas of pure neutrons. These are also Fermi particles, and so obey exactly the same quantum equation of state as we derived for electrons, Eqs. (10.78) and (10.79). The differences between a neutron star and a white dwarf are two: firstly, a much higher density is required to push p_f up to the typical momentum of a neutron, which is more massive than an electron; and secondly, the total energy density is now provided by the neutrons themselves: there is no extra gas of ions providing most of the self-gravitation. So, here, the total equation of state at high compression is Eq. (10.79):

$$p = \tfrac{1}{3}\rho. \tag{10.86}$$

Unfortunately, there exist no simple arguments giving an upper mass in this case, since the fully relativistic equation must be used. One has to look at the upper limit given by instabilities, because as neutrons are compressed they can begin forming heavier baryons, by processes that are incompletely understood. The simple equation of state, Eq. (10.79), is inadequate for a study of instabilities, and the nuclear physics required for a proper study is incompletely understood. A review is Canuto (1974, 1975). It appears now that the upper mass limit for nonrotating stars is probably smaller than $2\,M_\odot$, and some observational evidence even suggests $1.4\,M_\odot$ (Taylor & Weisberg 1982).

10.8   Bibliography

Our construction of the spherical coordinate system is similar to that in most other texts, but it is not particularly systematic, nor is it clear how to generalize the method to other symmetries. Group theory affords a more systematic approach. See Kramer *et al.* (1981).

We have not tried to find simple ways to calculate the Riemann and Einstein tensors of any given metric, but labor-saving methods do exist. One, the Cartan approach, is described in Misner *et al.* (1973). Another, the use of modern algebraic computer programs, is described by d'Inverno (1980).

A full discussion of spherical stellar structure may be found in Thorne (1967). In deriving stellar solutions we demanded continuity of g_{00} and g_{rr} across the surface of the star. A full discussion of the correct 'junction conditions' across a surface of discontinuity is in Misner *et al.* (1973).

There are other exact compressible solutions for stars in the literature. See Kramer *et al.* (1981).

A more rigorous derivation of the equation of state of a Fermi gas may be found in quantum mechanics texts. See Chandrasekhar (1957) for a full derivation of his limit on white dwarf masses.

The instability which leads to the absence of stars intermediate in central density between white dwarfs and neutron stars is discussed in

Harrison *et al.* (1965), Shapiro & Teukolsky (1983), and Zel'dovich & Novikov (1971). It is generally agreed that pulsars are rapidly rotating neutron stars with strong magnetic fields. See Smith (1977).

Calculations of the collapse of stars to form black holes are reviewed by Miller & Sciama (1980).

10.9 Exercises

1   Starting with $ds^2 = \eta_{\alpha\beta}\, dx^\alpha dx^\beta$, show that the coordinate transformation $r = (x^2 + y^2 + z^2)^{1/2}$, $\theta = \text{arc cos}\,(z/r)$, $\phi = \text{arc tan}\,(y/x)$ leads to Eq. (10.1), $ds^2 = -dt^2 + dr^2 + r^2\,(d\theta^2 + \sin^2\theta\, d\phi^2)$.

2   In deriving Eq. (10.5) we argued that if $\vec{e}_t$ were not orthogonal to $\vec{e}_\theta$ and $\vec{e}_\phi$, the metric would pick out a preferred direction. To see this, show that under rotations that hold t and r fixed, the pair $(g_{\theta t}, g_{\phi t})$ transforms as a vector field. If these don't vanish, they thus define a vector field on every sphere. Such a vector field cannot be spherically symmetric unless it vanishes: construct an argument to this effect, perhaps by considering the discussion of parallel-transport on the sphere at the beginning of § 6.4.

3   The locally measured energy of a particle, given by Eq. (10.11), is the energy the same particle would have in SR if it passed the observer with the same speed. It therefore contains no information about gravity, about the curvature of spacetime. By referring to Eq. (7.34) show that the difference between E^* and E in the weak-field limit is, for particles with small velocities, just the gravitational potential energy.

4   Use the result of Exer. 35, § 6.9 to calculate the components of $G_{\mu\nu}$ in Eqs. (10.14–10.17).

5   Show that a static star must have $U^r = U^\theta = U^\phi = 0$ in our coordinates, by examining the result of the transformation $t \rightarrow -t$.

6(a)   Derive Eq. (10.19) from Eq. (10.18).
 (b)   Derive Eqs. (10.20)–(10.23) from Eq. (4.37).

7   Describe how to construct a static stellar model in the case that the equation of state has the form $p = p(\rho, S)$. Show that one must give an additional arbitrary function, such as $S(r)$ or $S(m(r))$.

8(a)   Prove that the expressions $T^{\alpha\beta}{}_{;\beta}$ for $\alpha = t$, θ, or ϕ must vanish by virtue of the assumptions of a static geometric and spherical symmetry. (Do *not* calculate the expressions from Eqs. (10.20)–(10.23). Devise a much shorter argument.)

(b)   Derive Eq. (10.27) from Eqs. (10.20)–(10.23).

(c)   Derive Eq. (10.30) from Eqs. (10.14), (10.20), (10.29).

(d)   Prove Eq. (10.31).

(e)   Derive Eq. (10.39).

9(a)   Define a new radial coordinate in terms of the Schwarzschild r by
$$r = \bar{r}(1 + M/2\bar{r})^2. \tag{10.87}$$
Notice that as $r \to \infty$, $\bar{r} \to r$, while at the horizon $r = 2M$, we have $\bar{r} = \frac{1}{2}M$. Show that the metric for spherical symmetry takes the form
$$ds^2 = -\left[\frac{1 - M/2\bar{r}}{1 + M/2\bar{r}}\right]^2 dt^2 + \left[1 + \frac{M}{2\bar{r}}\right]^4 [d\bar{r}^2 + \bar{r}^2 \, d\Omega^2]. \tag{10.88}$$

(b)   Define quasi-Cartesian coordinates by the usual equations $x = \bar{r} \cos \phi \sin \theta$, $y = \bar{r} \sin \phi \sin \theta$, and $z = \bar{r} \cos \theta$ so that (as in Exer. 1),
$$d\bar{r}^2 + \bar{r}^2 \, d\Omega^2 = dx^2 + dy^2 + dz^2.$$
Thus, the metric has been converted into coordinates (x, y, z) which are called *isotropic* coordinates. Now take the limit as $\bar{r} \to \infty$ and show
$$ds^2 = -\left[1 - \frac{2M}{\bar{r}} + 0\left(\frac{1}{\bar{r}^2}\right)\right] dt^2 + \left[1 + \frac{2M}{\bar{r}} + 0\left(\frac{1}{\bar{r}^2}\right)\right] (dx^2 + dy^2 + dz^2).$$
This proves Eq. (10.38).

10   Complete the calculation for the uniform-density star.

(a)   Integrate Eq. (10.48) to get Eq. (10.49) and fill in the steps leading to Eqs. (10.50)–(10.52) and (10.54).

(b)   Calculate e^{Φ} and the redshift to infinity from the center of the star if $M = 1 \, M_{\odot} = 1.47$ km and $R = 1 \, R_{\odot} = 7 \times 10^5$ km (a star like the Sun), and again if $M = 1 \, M_{\odot}$ and $R = 10$ km (typical of a neutron star).

(c)   Take $\rho = 10^{-11} \, \text{m}^{-2}$ and $M = 0.5 \, M_{\odot}$, and compute R, e^{Φ} at surface and center, and the redshift from the surface to the center. What is the density $10^{-11} \, \text{m}^{-2}$ in kg m^{-3}?

11   Derive the restrictions in Eq. (10.57).

12   Prove that Eqs. (10.60)–(10.63) do solve Einstein's equations, given by Eqs. (10.14)–(10.17) and (10.20)–(10.23) or (10.27), (10.30), and (10.39).

13   Derive Eqs. (10.66) and (10.67).

14   A Newtonian polytrope of index n satisfies Eqs. (10.30) and (10.44), with the equation of state $p = K\rho^{(1+1/n)}$ for some constant K. Polytropes are discussed in detail by Chandrasekhar (1957). Consider the case $n = 1$, to which Buchdahl's equation of state reduces as $\rho \to 0$.

(a)   Show that ρ satisfies the equation
$$\frac{1}{r^2} \frac{d}{dr}\left(r^2 \frac{d\rho}{dr}\right) + \frac{4\pi}{K} \rho = 0, \tag{10.89}$$

and show that its solution is

$$\rho = \alpha u(r), \qquad u(r) = \frac{\sin Ar}{Ar}, \qquad A^2 = \frac{2\pi}{K},$$

where α is an arbitrary constant.

(b) Find the relation of the Newtonian constants α and K to the Buchdahl constants β and p_* by examining the Newtonian limit $(\beta \to 0)$ of Buchdahl's solution.

(c) From the Newtonian equations find $p(r)$, the total mass M and the radius R, and show them to be identical to the Newtonian limits of Eqs. (10.62), (10.67), and (10.65).

15 Calculations of stellar structure more realistic than Buchdahl's solution must be done numerically. But Eq. (10.39) has a zero denominator at $r = 0$, so the numerical calculation must avoid this point. One approach is to find a power-series solution to Eqs. (10.30) and (10.39) valid near $r = 0$, of the form

$$m(r) = \sum_j m_j r^j,$$

$$p(r) = \sum_j p_j r^j, \qquad\qquad\qquad (10.90)$$

$$\rho(r) = \sum_j \rho_j r^j.$$

Assume that the equation of state $p = p(\rho)$ has the expansion near the central density ρ_c

$$p = p(\rho_c) + (p_c \Gamma_c / \rho_c)(\rho - \rho_c) + \cdots, \qquad\qquad (10.91)$$

where Γ_c is the adiabatic index $d(\ln p)/d(\ln \rho)$ evaluated at ρ_c. Find the first two nonvanishing terms in each power series in Eq. (10.90), and estimate the largest radius r at which these terms give an error no larger than 0.1% in any power series. Numerical integrations may be started at such a radius using the power series to provide the initial values.

16(a) The two simple equations of state derived in § 10.7, $p = k\rho^{4/3}$ (Eq. (10.81)) and $p = \rho/3$ (Eq. (10.86)), differ in a fundamental way: the first has an arbitrary dimensional constant k, the second doesn't. Use this fact to argue that a stellar model constructed using only the second equation of state can only have solutions in which $\rho = \mu / r^2$ and $m = \nu r$, for some constants μ and ν. The key to the argument is that $\rho(r)$ may be given any value by a simple change of the unit of length, but there are no other constants in the equations whose values are affected by such a change.

(b) Show from this that the only nontrivial solution of this type is for $\mu = 3/(56\pi)$, $\nu = 3/14$. This is physically unacceptable, since it is singular at $r = 0$ and it has no surface.

(c) Do there exist solutions which are nonsingular at $r = 0$ or which have finite surfaces?

17 (This problem requires access to a computer or a programmable calculator with a sizable memory.) Numerically construct a sequence of stellar models using the equation of state

$$p = \begin{cases} k\rho^{4/3}, & \rho \leqslant (27 \ k^3)^{-1}, \\ \frac{1}{3}\rho, & \rho \geqslant (27 \ k^3)^{-1}, \end{cases} \tag{10.92}$$

where k is given by Eq. (10.81). This is a crude approximation to a realistic 'stiff' neutron-star equation of state. Construct the sequence by using the following values for ρ_c: $\rho_c/\rho_* = 0.1, 0.8, 1.2, 2, 5, 10$, where $\rho_* = (27 \ k^3)^{-1}$. Use the power series developed in Exer. 15 to start the integration. Does the sequence seem to approach a limiting mass, a limiting value of M/R, or a limiting value of the central redshift?

18 Show that the remark made before Eq. (10.80), that the nuclei supply little pressure, is true for the regime under consideration, i.e. where $m_e < p_f^2/3kT < m_p$, where k is Boltzmann's constant (not the same k as in Eq. (10.81)). What temperature range is this for white dwarfs, where $n \approx 10^{37} \ m^{-3}$?

19 Our Sun has an equatorial rotation velocity of about $2 \ km \ s^{-1}$.
(a) Estimate its angular momentum, on the assumption that the rotation is rigid (uniform angular velocity) and the Sun is of uniform density. As the true angular velocity is likely to increase inwards, this is a lower limit on the Sun's angular momentum.
(b) If the Sun were to collapse to neutron-star size (say 10 km radius), conserving both mass and total angular momentum, what would its angular velocity of rigid rotation be? In nonrelativistic language, would the corresponding centrifugal force exceed the Newtonian gravitational force on the equator?
(c) A neutron star of 1 $M_\odot$ and radius 10 km rotates 30 times per second (typical of young pulsars). Again in Newtonian language, what is the ratio of centrifugal to gravitational force on the equator? In this sense the star is slowly rotating.
(d) Suppose a main-sequence star of 1 $M_\odot$ has a dipole magnetic field with typical strength 1 Gauss in the equatorial plane. Assuming flux conservation in this plane, what field strength should we expect if the star collapses to radius of 10 km? (The Crab pulsar's field is of the order of 10^{11} Gauss.)

11

Schwarzschild geometry and black holes

_____

11.1 Trajectories in the Schwarzschild spacetime

The 'Schwarzschild geometry' is the geometry of the vacuum spacetime outside a spherical star. It is determined by one parameter, the mass M, and has the line element

$$\mathrm{d}s^2 = -\left(1-\frac{2M}{r}\right)\mathrm{d}t^2 + \left(1-\frac{2M}{r}\right)^{-1}\mathrm{d}r^2 + r^2\,\mathrm{d}\Omega^2 \tag{11.1}$$

in the coordinate system developed in the previous chapter. Its importance is not just that it is the gravitational field of a star: we shall see that it is also the geometry of the spherical black hole. A careful study of its timelike and null geodesics – the paths of freely moving particles and photons – is the key to understanding the physical importance of this metric.

Conserved quantities. We have seen (Eq. (7.29) and associated discussion) that when a spacetime has a certain symmetry, then there is an associated conserved momentum component for trajectories. Because our space has so many symmetries – time independence and spherical symmetry – the values of the conserved quantities turn out to determine the trajectory completely. We shall treat 'particles' with mass and 'photons' without mass in parallel.

Time independence of the metric means that p_0 is constant on the trajectory. We define the related constants

particle: $\tilde{E} \equiv -p_0/m$;   photon: $E = -p_0$,   (11.2)

where m is the particle's rest mass. Independence of the metric of the angle ϕ about the axis implies that p_ϕ is constant. We again define

particle $\tilde{L} \equiv p_\phi/m$;   photon $L = p_\phi$.   (11.3)

Because of spherical symmetry, motion is always confined to a single plane, and we can choose that plane to be the equatorial plane. Then θ is constant ($\theta = \pi/2$) for the orbit, so $d\theta/d\lambda = 0$, where λ is any parameter on the orbit. But p^θ is proportional to this, so it also vanishes. The other components of momentum are:

$$\text{particle: } p^0 = g^{00}p_0 = m\left(1 - \frac{2M}{r}\right)^{-1}\tilde{E},$$

$$p^r = m\, dr/d\tau,$$

$$p^\phi = g^{\phi\phi}p_\phi = m\frac{1}{r^2}\tilde{L};\qquad (11.4)$$

$$\text{photon: } p^0 = \left(1 - \frac{2M}{r}\right)^{-1}E,$$

$$p^r = dr/d\lambda,$$

$$p^\phi = d\phi/d\lambda = L/r^2.\qquad (11.5)$$

The equation for a photon's p^r should be regarded as *defining* the affine parameter λ. The equation $\vec{p} \cdot \vec{p} = -m^2$ implies

particle:

$$-m^2\tilde{E}^2\left(1 - \frac{2M}{r}\right)^{-1} + m^2\left(1 - \frac{2M}{r}\right)^{-1}\left(\frac{dr}{d\tau}\right)^2$$

$$+\frac{m^2\tilde{L}^2}{r^2} = -m^2;\qquad (11.6)$$

photon:

$$-E^2\left(1 - \frac{2M}{r}\right)^{-1} + \left(1 - \frac{2M}{r}\right)^{-1}\left(\frac{dr}{d\lambda}\right)^2 + \frac{L^2}{r^2} = 0.\qquad (11.7)$$

These can be solved to give the basic equations for orbits,

$$\text{particle: } \left(\frac{dr}{d\tau}\right)^2 = \tilde{E}^2 - \left(1 - \frac{2M}{r}\right)\left(1 + \frac{\tilde{L}^2}{r^2}\right);\qquad (11.8)$$

$$\text{photon: } \left(\frac{dr}{d\lambda}\right)^2 = E^2 - \left(1 - \frac{2M}{r}\right)\frac{L^2}{r^2}.\qquad (11.9)$$

Types of orbits. Both equations have the same general form, and we define the effective potentials

particle: $\quad \tilde{V}^2(r) = \left(1 - \frac{2M}{r}\right)\left(1 + \frac{\tilde{L}^2}{r^2}\right);$ $\qquad$ (11.10)

photon: $\quad V^2(r) = \left(1 - \frac{2M}{r}\right)\frac{L^2}{r^2}.$ $\qquad$ (11.11)

Their typical forms are plotted in Figs. 11.1 and 11.2, in which various points have been labeled and possible trajectories drawn (dotted lines).

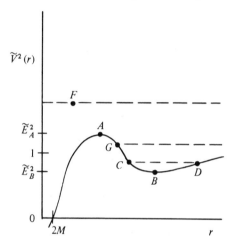

Fig. 11.1 Typical effective potential for a massive particle of fixed specific angular momentum in the Schwarzschild metric.

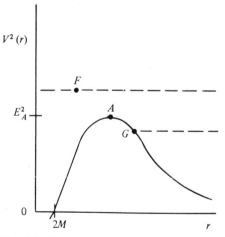

Fig. 11.2 The same as Fig. 11.1 for a massless particle.

Both Eq. (11.8) and Eq. (11.9) imply that, since the left side is positive or zero, the energy of a trajectory must not be less than the potential V. (Here and until Eq. (11.14) we will take E and V to refer to $\bar{E}$ and $\bar{V}$ as well, since the remarks for the two cases are identical.) So for an orbit of given E, the radial range is restricted to those radii for which V is smaller than E. For instance, consider the trajectory which has the value of E indicated by point G (in either diagram). If it comes in from $r = \infty$, then it cannot reach smaller r than where the dotted line hits the V^2 curve, at point G. Point G is called a *turning* point. At G, since $E^2 = V^2$ we must have $(dr/d\lambda)^2 = 0$, from Eq. (11.9). Similar conclusions apply to Eq. (11.8). To see what happens here we differentiate Eqs. (11.8) and (11.9). For particles, differentiating the equation

$$\left(\frac{dr}{d\tau}\right)^2 = \bar{E}^2 - \bar{V}^2(r)$$

with respect to τ gives

$$2\left(\frac{dr}{d\tau}\right)\left(\frac{d^2r}{d\tau^2}\right) = -\frac{d\bar{V}^2(r)}{dr}\frac{dr}{d\tau},$$

or

$$\text{particles:} \quad \frac{d^2r}{d\tau^2} = -\frac{1}{2}\frac{d}{dr}\bar{V}^2(r). \tag{11.12}$$

Similarly, the photon equation gives

$$\text{photons:} \quad \frac{d^2r}{d\lambda^2} = -\frac{1}{2}\frac{d}{dr}V^2(r). \tag{11.13}$$

These are the analogues in relativity of the equation

$$ma = -\nabla\phi,$$

where ϕ is the potential for some force. If we now look again at point G, we see that the radial acceleration of the trajectory is outwards, so that the particle (or photon) comes in to the minimum radius, but is accelerated outward as it turns around, and so it returns to $r = \infty$. This is a 'hyperbolic' orbit – the analogue of the orbits which are true hyperbolae in Newtonian gravity.

It is clear from Eq. (11.12) or (Eq. (11.13) that a circular orbit ($r = $ const.) is possible only at a minimum or maximum of V^2. These occur at points A and B in the diagrams (there is no point B for photons). A maximum is, however, unstable, since any small change in r results in an acceleration away from the maximum, by Eqs. (11.12) and (11.13). So for particles, there is one stable (B) and one unstable circular orbit (A) for this value of $\bar{L}$. For photons, there is only one unstable orbit for this L. We can be quantitative by evaluating

$$0 = \frac{d}{dr}\left[\left(1 - \frac{2M}{r}\right)\left(1 + \frac{\tilde{L}^2}{r^2}\right)\right]$$

and

$$0 = \frac{d}{dr}\left[\left(1 - \frac{2M}{r}\right)\frac{L^2}{r^2}\right].$$

These give, respectively

$$\text{particles:} \quad r = \frac{\tilde{L}^2}{2M}\left(1 \pm \sqrt{\left(1 - \frac{12M^2}{\tilde{L}^2}\right)}\right); \tag{11.14}$$

$$\text{photons:} \quad r = 3M. \tag{11.15}$$

For particles, there are two radii, as we expect, but only if $\tilde{L}^2 > 12M^2$. The two radii are identical for $\tilde{L}^2 = 12M^2$ and don't exist at all for $\tilde{L}^2 < 12M^2$. This indicates a qualitative change in the shape of the curve for $\tilde{V}^2(r)$ for small $\tilde{L}$. The two cases, $\tilde{L}^2 = 12M^2$ and $\tilde{L}^2 < 12M^2$, are illustrated in Fig. 11.3. Since there is a minimum $\tilde{L}^2$ for a circular particle

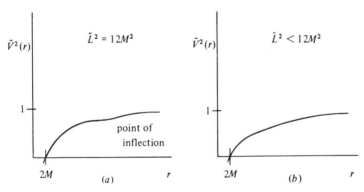

Fig. 11.3 As Fig. 11.1 for the indicated values of specific angular momentum.

orbit, there is also a minimum r, obtained by taking $\tilde{L}^2 = 12M^2$ in Eq. (11.14)

$$\text{particle:} \quad r_{\text{MIN}} = 6M. \tag{11.16}$$

For photons, the unstable circular orbit is always at the same radius, $r = 3M$, regardless of L.

The last kind of orbit we need consider is the one whose energy is given by the line which passes through the point F in Figs. 11.1 and 11.2. Since this nowhere intersects the potential curve, this orbit plunges right through $r = 2M$ and never returns. From Exer. 1, § 11.6, we see that for such an orbit the impact parameter (b) is small: it is aimed more directly at the hole than are orbits of smaller $\bar{E}$ and fixed $\bar{L}$.

Of course, if the geometry under consideration is a star, its radius R will exceed $2M$, and the potential diagrams, Figs. 11.1–11.3, will be valid only outside R. If a particle reaches R it will hit the star. Depending on R/M, then, only certain kinds of orbits will be possible.

Perihelion shift. A particle (or planet) in a (stable) *circular* orbit around a star will make one complete orbit and come back to the same point (i.e. same value of ϕ) in a fixed amount of coordinate time, which is called its period P. This period can be determined as follows. From Eq. (11.14) it follows that a stable circular orbit at radius r has angular momentum

$$\bar{L}^2 = \frac{Mr}{1 - \dfrac{3M}{r}}, \tag{11.17}$$

and since $\bar{E}^2 = \tilde{V}^2$ for a circular orbit, it also has energy

$$\bar{E} = \left(1 - \frac{2M}{r}\right)^2 \bigg/ \left(1 - \frac{3M}{r}\right). \tag{11.18}$$

Now, we have

$$\frac{d\phi}{d\tau} \equiv U^\phi = \frac{p^\phi}{m} = g^{\phi\phi}\frac{p_\phi}{m} = g^{\phi\phi}\bar{L} = \frac{1}{r^2}\bar{L} \tag{11.19}$$

and

$$\frac{dt}{d\tau} \equiv U^0 = \frac{p^0}{m} = g^{00}\frac{p_0}{m} = g^{00}(-\bar{E}) = \frac{\bar{E}}{1 - \dfrac{2M}{r}}. \tag{11.20}$$

We obtain the angular velocity by dividing these:

$$\frac{dt}{d\phi} = \frac{dt/d\tau}{d\phi/d\tau} = \left(\frac{r^3}{M}\right)^{1/2}. \tag{11.21}$$

The period, which is the time taken for ϕ to change by 2π, is

$$P = 2\pi\sqrt{\left(\frac{r^3}{M}\right)}. \tag{11.22}$$

This is the coordinate time, of course, not the particle's proper time. (But see Exer. 7, § 11.6: coordinate time *is* proper time far away.) It happens, coincidentally, that this is identical to the Newtonian expression.

Now, a slightly noncircular orbit will oscillate in and out about a central radius r. In Newtonian gravity the orbit is a perfect ellipse, which means, among other things, that it is *closed*: after a fixed amount of time it returns to the same point (same r and ϕ). In GR, this does not happen and a typical orbit is shown in Fig. 11.4. However, when the effects of relativity are small and the orbit is nearly circular, the relativistic orbit must be almost closed: it must look like an ellipse which slowly rotates

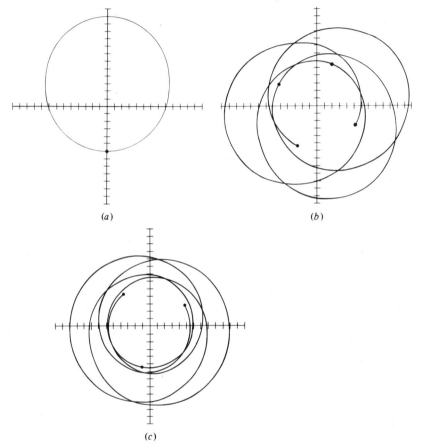

(a)

(b)

(c)

Fig. 11.4 (a) A Newtonian orbit is a closed ellipse. Grid marked in units of M. (b) An orbit in the Schwarzschild metric with pericentric and apcentric distances similar to those in (a). Pericenters (heavy dots) advance by about 97° per orbit. (c) A moderately more compact orbit than in (b) has a considerably larger pericenter shift, about 130°.

about the center. One way to describe this is to look at the *perihelion* of the orbit, the point of closest approach to the star. ('Peri' means closest and 'helion' refers to the Sun; for orbits about any old star the name 'periastron' is more appropriate. For orbits around Earth – 'geo' – one speaks of the 'perigee'. The opposite of 'peri' is 'ap': the furthest distance.

Thus, an orbit also has an aphelion, apastron, or apogee, depending on what it is orbiting around.) The perihelion will rotate around the star in some manner, and observers can hope to measure this. It has been measured for Mercury to be 43″/century, and we must try to calculate it. Note that all other planets are further from the Sun and therefore under the influence of significantly smaller relativistic corrections to Newtonian gravity. The measurement of Mercury's precession is a herculean task, first accomplished in the 1800s. Due to various other effects, such as the perturbations of Mercury's orbit due to the other planets, the observed precession is about 5600″/century. The 43″ is only the part not explainable by Newtonian gravity, and Einstein's demonstration that his theory predicts exactly that amount was the first evidence in favor of the theory.

To calculate the precession, let us begin by getting an equation for the particle's orbit. We have $dr/d\tau$ from Eq. (11.8). We get $d\phi/d\tau$ from Eq. (11.19) and divide to get

$$\left(\frac{dr}{d\phi}\right)^2 = \frac{\tilde{E}^2 - \left(1 - \frac{2M}{r}\right)\left(1 + \frac{\tilde{L}^2}{r^2}\right)}{\tilde{L}^2/r^4}. \tag{11.23}$$

It is convenient to define

$$u \equiv 1/r \tag{11.24}$$

and obtain

$$\left(\frac{du}{d\phi}\right)^2 = \frac{\tilde{E}^2}{\tilde{L}^2} - (1 - 2Mu)\left(\frac{1}{\tilde{L}^2} + u^2\right). \tag{11.25}$$

The *Newtonian* orbit is found by neglecting u^3 terms (see Exer. 11, § 11.6)

$$\text{Newt:} \quad \left(\frac{du}{d\phi}\right)^2 = \frac{\tilde{E}^2}{\tilde{L}^2} - \frac{1}{\tilde{L}^2}(1 - 2Mu) - u^2. \tag{11.26}$$

A circular orbit in Newtonian theory has $u = M/\tilde{L}^2$ (take the square root equal to 1 in Eq. (11.14)), so we define

$$y = u - \frac{M}{\tilde{L}^2}, \tag{11.27}$$

so that y represents the deviation from circularity. We then get

$$\left(\frac{dy}{d\phi}\right)^2 = \frac{\tilde{E}^2 - 1}{\tilde{L}^2} + \frac{M^2}{\tilde{L}^4} - y^2. \tag{11.28}$$

It is easy to see that this is satisfied by

$$\text{Newt:} \quad y = \left[\frac{\tilde{E}^2 + M^2/\tilde{L}^2 - 1}{\tilde{L}^2}\right]^{1/2} \cos(\phi + B), \tag{11.29}$$

where B is arbitrary. This is clearly periodic: as ϕ advances by 2π, y returns to its value and, therefore, so does r. The constant B just determines the initial orientation of the orbit. It is interesting, but unimportant for our purposes, that by solving for r we get

$$\text{Newt:} \quad \frac{1}{r} = \frac{M}{\tilde{L}^2} + \left[\tilde{E}^2 \frac{+ M^2/\tilde{L}^2 - 1}{\tilde{L}^2} \right]^{1/2} \cos(\phi + B), \tag{11.30}$$

which is the equation of an *ellipse*.

We now consider the relativistic case and make the same definition of y, but instead of throwing away the u^3 term in Eq. (11.25) we assume that the oribit is nearly circular, so that y is small, and we neglect only the terms in y^3. Then we get

Nearly circ:

$$\left(\frac{dy}{d\phi} \right)^2 = \frac{\tilde{E}^2 + M^2/\tilde{L}^2 - 1}{\tilde{L}^2} + \frac{2M^4}{\tilde{L}^6} + \frac{6M^3}{\tilde{L}^2} y + \left(\frac{6M^2}{\tilde{L}^2} - 1 \right) y^2.$$
$$\tag{11.31}$$

This can be made analogous to Eq. (11.28) by completing the square on the right-hand side. The result is the solution

$$y = y_0 + A \cos(k\phi + B), \tag{11.32}$$

where B is arbitrary and the other constants are

$$k = \left(1 - \frac{6M^2}{\tilde{L}^2} \right)^{1/2},$$

$$y_0 = 3M^3/k^2\tilde{L}^2,$$

$$A = \frac{1}{k} \left[\frac{\tilde{E}^2 + M^2/\tilde{L}^2 - 1}{\tilde{L}^2} + \frac{2M^4}{\tilde{L}^6} - y_0^2 \right]^{1/2}. \tag{11.33}$$

The appearance of the constant y_0 just means that the orbit oscillates not about $y = 0$ ($u = \tilde{M}/L^2$) but about $y = y_0$: Eq. (11.27) doesn't use the correct radius for a circular orbit in GR. The amplitude A is also somewhat different, but what is most interesting here is the fact that k is not 1. The orbit returns to the same r when $k\phi$ goes through 2π, from Eq. (11.32). Therefore the change in ϕ from one perihelion to the next is

$$\Delta\phi = \frac{2\pi}{k} = 2\pi \left(1 - \frac{6M^2}{\tilde{L}^2} \right)^{-1/2}, \tag{11.34}$$

which, for nearly Newtonian orbits, is

$$\Delta\phi \approx 2\pi \left(1 + \frac{3M^2}{\tilde{L}^2} \right). \tag{11.35}$$

The perihelion *shift*, then, from one orbit to the next, is

$$\Delta\phi = 6\pi M^2/\tilde{L}^2 \text{ radians per orbit.} \tag{11.36}$$

We can use Eq. (11.17) to obtain $\bar{L}$ in terms of r, since the corrections for noncircularity will make changes in Eq. (11.36) of the same order as terms we have already neglected. Moreover, if we consider orbits about a nonrelativistic star, we can approximate Eq. (11.17) by

$$\bar{L}^2 = \frac{Mr}{1 - \dfrac{3M}{r}} \approx Mr,$$

so that we get

$$\Delta\phi \approx 6\pi\frac{M}{r}. \tag{11.37}$$

For Mercury's orbit, $r = 5.55 \times 10^7$ km and $M = 1\,M_\odot = 1.47$ km, so that

$$(\Delta\phi)_{\text{Mercury}} = 4.99 \times 10^{-7} \text{ radians per orbit.} \tag{11.38}$$

Each orbit takes 0.24 yr, so the shift is

$$(\Delta\phi)_{\text{Mercury}} = .43''/\text{yr} = 43''/\text{century.} \tag{11.39}$$

The binary pulsar. Another system in which the pericenter shift is observable is the binary pulsar system discussed in Ch. 9. The stars have mean separation 1.2×10^9 m, so using Eq. (11.37) with $M = 1.4\,M_\odot = 2.07$ km gives a crude estimate of $\Delta\phi = 3.3 \times 10^{-5}$ radians per orbit $= 2°.1$ per year. This is much easier to measure than Mercury's shift! In fact, a more careful calculation, taking into account the high eccentricity of the orbit and the fact that the two stars are of comparable mass, predicts $4°.2$ per year.

For our purposes here we have calculated the periastron shift from the known masses of the star. But in fact the observed shift of $4.2261° \pm 0.0007$ per year is one of the data which enable us to calculate the masses. The other datum is another relativistic effect: a redshift of the signal which results from two effects. One is the special-relativistic 'transverse-Doppler' term: the $0(v^2)$ term in Eq. (2.39). The other is the changing gravitational redshift as the pulsar's eccentric orbit brings it in and out of its companion's gravitational potential. These two effects are observationally indistinguishable from one another, but their combined resultant redshift gives one more number which depends on the masses of the stars. Using it and the periastron shift and the Newtonian mass function for the orbit allows one to determine the stars' masses and the orbit's inclination (see Blandford & Teukolsky 1976 and Epstein 1977).

Gravitational deflection of light. In the previous section we treated particles only because photons do not have bound orbits in Newtonian gravity. In this section we treat the analogous effect for photons, their deflection from straight-line motion as they pass through a gravitational field. Historically, this was the first general-relativistic effect to have been predicted before it was observed and its confirmation in the eclipse of 1919 (see McCrae 1979) made Einstein an international celebrity. The fact that it was a British team (led by Eddington) who made the observations to confirm the theories of a German incidentally helped to alleviate post-war tension between the scientific communities of the two countries. In modern times, the light-deflection phenomenon remains important to astronomy, both because of observations of multiple images of the same quasar (gravitational lensing) and because corrections for deflection by the Sun and even by Jupiter will have to be applied to high-precision measurements of stellar positions that will be made by the Hipparcos satellite now being planned for launch in the mid-1980s.

We begin by calculating the trajectory of a photon in the Schwarzschild metric under the assumption that M/r is everywhere small along the trajectory. The equation of the orbit is the ratio of Eq. (11.5) to the sqaure root of Eq. (11.9):

$$\frac{d\phi}{dr} = \pm \frac{1}{r^2\left[\frac{1}{b^2} - \frac{1}{r^2}\left(1 - \frac{2M}{r}\right)\right]^{1/2}}, \tag{11.40}$$

where we have defined the *impact parameter*,

$$b \equiv L/E. \tag{11.41}$$

In Exer. 1, § 11.6, it is shown that b would be the minimum value of r in Newtonian theory, where there is no deflection. It therefore represents the 'offset' of the photon's initial trajectory from a parallel one moving purely radially. An incoming photon with $L > 0$ obeys the equation

$$\frac{d\phi}{du} = \frac{1}{\left(\frac{1}{b^2} - u^2 + 2Mu^3\right)^{1/2}}, \tag{11.42}$$

with the same definition as before,

$$u = 1/r. \tag{11.43}$$

If we neglect the u^3 term in Eq. (11.42), all effects of M disappear, and the solution is

$$r \sin(\phi - \phi_0) = b, \tag{11.44}$$

a straight line. This is, of course, the Newtonian result.

Suppose now we assume $Mu \ll 1$ but not entirely negligible. Then if we define

$$y \equiv u(1 - Mu), \qquad u = y(1 + My) + 0(M^2 u^2), \tag{11.45}$$

Eq. (11.42) becomes

$$\frac{d\phi}{dy} = \frac{(1 + 2My)}{\left(\dfrac{1}{b^2} - y^2\right)^{1/2}} + 0(M^2 u^2). \tag{11.46}$$

This can be integrated to give

$$\phi = \phi_0 + \frac{2M}{b} + \arcsin{(by)} - 2M\left(\frac{1}{b^2} - y^2\right)^{1/2}. \tag{11.47}$$

The initial trajectory has $y \to 0$, so $\phi \to \phi_0$: ϕ_0 is the incoming direction. The photon reaches its smallest r when $y = 1/b$, as one can see from setting $dr/d\lambda = 0$ in Eq. (11.19) and using our approximation $Mu \ll 1$. This occurs at the angle $\phi = \phi_0 + 2M/b + \pi/2$. It has thus passed through an angle $\pi/2 + 2M/b$ as it travels to its point of closest approach. By symmetry, it passes through a further angle of the same size as it moves outwards from its point of closest approach (see Fig. 11.5). It thus passes

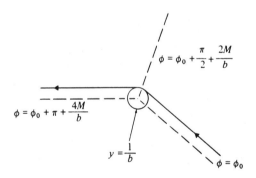

Fig. 11.5   Deflection of a photon.

through a total angle of $\pi + 4M/b$. If it were going on a straight line, this angle would be π, so the net deflection is

$$\Delta\phi = 4M/b. \tag{11.48}$$

To the accuracy of our approximations, we may use for b the radius of closest approach rather than the impact parameter L/E. For the sum, the maximum effect is for trajectories for which $b = R_\odot$, the radius of the Sun. Given $M = 1\ M_\odot = 1.47$ km and $R_\odot = 6.96 \times 10^5$ km, we find

$$(\Delta\phi)_{\odot,\text{max}} = 8.45 \times 10^{-6} \text{ rad} = 1''.74. \tag{11.49}$$

For Jupiter, with $M = 1.12 \times 10^{-3}$ km and $R = 6.98 \times 10^4$ km we have

$$(\Delta\phi)_{\Psi,\max} = 6.42 \times 10^{-8} \text{ rad} = 0''.013. \tag{11.50}$$

This is well above the limit of the accuracy anticipated for the Hipparcos satellite.

Of course, satellite observations of stellar positions are made from a position near Earth, and for stars that are not near the Sun in the sky the satellite will receive their light before the total deflection, given by Eq. (11.48), has taken place. This situation is illustrated in Fig. 11.6. An observer at *rest* at the position of the satellite observes an apparent position in the direction of the vector $\vec{a}$, tangent to the path of the light ray, and if he knows his distance r from the Sun he can calculate the true direction to the star, $\vec{s}$. Exer. 16, § 11.6, derives the general result.

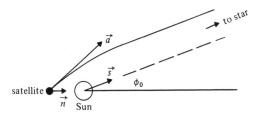

Fig. 11.6 An observation from Earth of a star not at the limb of the Sun does not need to correct for the full deflection of Fig. 11.5.

Fig. 11.7 Deflection can produce multiple images.

It may of course happen that photons from the same star will travel trajectories that pass on opposite sides of the deflecting star and intersect each other after deflection, as illustrated in Fig. 11.7. Rays 1 and 2 are essentially parallel if the star (*) is far from the deflecting object (S). An observer at position B would then see *two* images of the star, coming from apparently different directions. This phenomenon appears to have been observed in at least two cases where light from quasars has been deflected by a cluster of galaxies, so that on Earth we receive three or more images (see Young *et al.* 1981, Burke 1981). For a general discussion of the possibilities of observing lensing in a variety of astrophysical situations, see Cowling (1983).

11.2   Nature of the surface $r = 2M$

Coordinate singularities. It is clear that something funny goes wrong with the line element, Eq. (11.1) at $r = 2M$, but what is not clear is whether the problem is with the geometry or just with the coordinates. Coordinate singularities – places where the coordinates don't describe the geometry properly – are not unknown in ordinary calculus. Consider spherical coordinates at the poles. The north pole on a sphere has coordinates $\theta = 0$, $0 \leqslant \phi < 2\pi$. That is, although ϕ can have any value for $\theta = 0$, all values really correspond to a single point. We might draw a coordinate diagram of the sphere as follows (Fig. 11.8 – maps of the

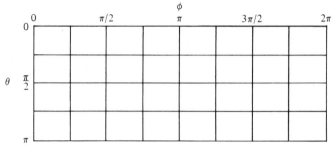

Fig. 11.8   One way of drawing as sphere on a flat piece of paper. Not only are $\phi = 0$ and $\phi = 2\pi$ really the same lines, but the lines $\theta = 0$ and $\theta = \pi$ are each really just one point. Spherical coordinates are therefore not faithful representations of the sphere everywhere.

globe are sometimes drawn this way), in which it would not be at all obvious that all points at $\theta = 0$ are really the same point. We *could*, however, convince ourselves of this by calculating the circumference of every circle of constant θ and verifying that these approached zero as $\theta \to 0$ and $\theta \to \pi$. That is, by asking questions that have an invariant geometrical meaning, one can tell if the coordinates are bad. For the sphere, the metric is positive-definite, so if two points have zero distance between them they are the same point (e.g. $\theta = 0$, $\phi = \pi$ and $\theta = 0$, $\phi = 2\pi$: see Exer. 18, § 11.6). In relativity, the situation is more subtle, since there are curves (null curves) where distinct points have zero invariant distance between them. In fact, the whole question of the nature of the surface $r = 2M$ is so subtle that it was not answered satisfactorily until 1960. (This was just in time, too, since black holes began to be of importance in astronomy within a decade as new technology made observations of quasars, pulsars, and X-ray sources possible.) We shall explore the problem by asking a few geometrical questions about the metric and then demonstrating a coordinate system which has no singularity at this surface.

Infalling particles. Let a particle fall to the surface $r = 2M$ from any finite radius R. How much proper time does that take? That is, how much time is elapsed on the particle's clock? The simplest particle to discuss is the one which falls in radially. Since $d\phi = 0$ we have $\tilde{L} = 0$ and, from Eq. (11.8),

$$\left(\frac{dr}{d\tau}\right)^2 = \tilde{E}^2 - 1 + \frac{2M}{r}, \tag{11.51}$$

or

$$d\tau = -\frac{dr}{\left(\tilde{E}^2 - 1 + \frac{2M}{r}\right)^{1/2}} \tag{11.52}$$

(the minus sign because the particle falls inward). It is clear that if $\tilde{E}^2 > 1$ (unbound particle) the integral of the right-hand side from R to $2M$ is finite. If $\tilde{E} = 1$ (particle falling from rest at ∞) the integral is simply

$$\Delta\tau = \frac{2M}{3}\left[\left(\frac{r}{2M}\right)^{3/2}\right]_{2M}^R, \tag{11.53}$$

which is again finite. And if $\tilde{E} < 1$, there is again no problem since the particle cannot be at larger r than where $1 - \tilde{E}^2 = 2M/r$ (see Eq. (11.51)). So the answer is that any particle can reach the horizon in a finite amount of proper time.

We now ask how much coordinate time elapses as the particle falls in. For this we use

$$U^0 = \frac{dt}{d\tau} = g^{00} U_0 = g^{00}\frac{p_0}{m} = -g^{00}\tilde{E} = \left(1 - \frac{2M}{r}\right)^{-1}\tilde{E}.$$

Therefore we have

$$dt = \frac{d\tau}{\tilde{E}\left(\frac{1-2M}{r}\right)} = -\frac{dr}{\tilde{E}\left(1 - \frac{2M}{r}\right)\left(\tilde{E}^2 - 1 + \frac{2M}{r}\right)^{1/2}}. \tag{11.54}$$

For simplicity, we consider the case $\tilde{E} = 1$ and examine this near $r = 2M$ by defining the new variable

$$\varepsilon \equiv r - 2M.$$

Then we get

$$dt = \frac{-(\varepsilon + 2M)^{3/2}\,d\varepsilon}{(2M)^{1/2}\varepsilon}. \tag{11.55}$$

It is clear that as $\varepsilon \to 0$ the integral of this goes like $\ln\varepsilon$, which diverges. One would also find this for $\tilde{E} \neq 1$, because the divergence comes from the $[1 - (2M/r)]^{-1}$ term, which doesn't contain $\tilde{E}$. Therefore a particle reaches the surface $r = 2M$ only after an infinite coordinate time has elapsed. Since the proper time is finite, the coordinate time must be behaving badly.

Inside r = 2M. To see just how badly it behaves, let us ask what happens to a particle after it reaches $r = 2M$. It must clearly pass to smaller r unless it is destroyed. This might happen if at $r = 2M$ there were a 'curvature singularity', where the gravitational forces grew strong enough to tear anything apart. But calculation of the components $R^\alpha{}_{\beta\mu\nu}$ of Riemannian tensor in the local inertial frame of the infalling particle shows them to be perfectly finite: Exer. 20, § 9.6. So we must conclude that the particle will just keep going. If we look at the geometry inside but near $r = 2M$, by introducing $\varepsilon \equiv 2M - r$, then the line element is

$$\mathrm{d}s^2 = \frac{\varepsilon}{2M - \varepsilon}\,\mathrm{d}t^2 - \frac{2M - \varepsilon}{\varepsilon}\,\mathrm{d}\varepsilon^2 + (2M - \varepsilon)^2\,\mathrm{d}\Omega^2. \tag{11.56}$$

Since $\varepsilon > 0$ inside $r = 2M$ we see that a line on which t, θ, ϕ are constant has $\mathrm{d}s^2 < 0$: it is timelike. Therefore ε (and hence r) is a *timelike* coordinate, while t has become spacelike: even more evidence for the funniness of t and r! Since the infalling particle must follow a timelike world line, it must constantly change r, and of course this means *decrease r*. So a particle inside $r = 2M$ will inevitably reach $r = 0$, and there a *true* curvature singulaity awaits it: sure destruction by infinite forces (Exer. 20, § 9.6). But what happens if the particle inside $r = 2M$ tries to send out a photon to someone outside $r = 2M$ in order to describe his impending doom? This photon, no matter how directed, must also go forward in 'time' as seen locally by the particle, and this means to decreasing r. So the photon will not get out either. Everything inside $r = 2M$ is trapped and, moreover, doomed to encounter the singularity at $r = 0$, since $r = 0$ is in the future of every timelike and null world line inside $r = 2M$. Once a particle crosses the surface $r = 2M$, it cannot be seen by an external observer, since to be seen means to send out a photon which reaches the external observer. This surface is therefore called a horizon, since a horizon on Earth has the same effect (for different reasons!). We shall henceforth refer to $r = 2M$ as the Schwarzschild horizon.

Coordinate systems. So far, our approach has been purely algebraic – we have no 'picture' of the geometry. To develop a picture we will first draw a coordinate diagram in Schwarzschild coordinates, and on it we will draw the light cones, or at least the paths of the radially ingoing and outgoing null lines emanating from certain events (Fig. 11.9). These light

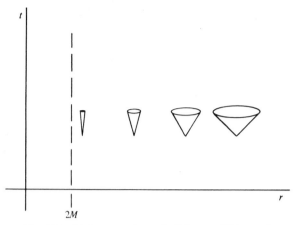

Fig. 11.9   Light cones drawn in Schwarzschild coordinates close up near the surface $r = 2M$.

cones may be calculated by solving $ds^2 = 0$ for θ and ϕ constant:

$$\frac{dt}{dr} = \pm \frac{1}{1 - \dfrac{2M}{r}}. \tag{11.57}$$

In a t–r diagram, these lines have slope ± 1 far from the star (standard SR light cone) but their slope approaches $\pm\infty$ as $r \to 2M$. This means that they become more vertical: the cone 'closes up'. Since particle world lines are confined within the local light cone (a particle must move slower than light) this closing up of the cones forces the world lines of particles to become more vertical: if they reach $r = 2M$, they reach it at $t = \infty$. This is the 'picture' behind the algebraic result that a particle takes infinite coordinate time to reach the horizon. Notice that *no* particle world line reaches the line $r = 2M$ for any finite value of t. This might suggest that the line ($r = 2M$, $-\infty < t < \infty$) is really not a line at all but a single point in spacetime. That is, our coordinates may go bad by expanding a single event into the whole line $r = 2M$, which would have the effect that if any particle reached the horizon after that event then it would have to cross $r = 2M$ 'after' $t = +\infty$. This singularity would then be very like the one in Fig. 11.8 for spherical coordinates at the pole: a whole line in the bad coordinates representing a point in the real space. Notice that the coordinate diagram in Fig. 11.9 makes no attempt to represent the *geometry* properly, only the coordinates. It clearly does a poor job on the geometry because the light cones close up. Since we have already decided that they *don't* really close up (particles reach the horizon at finite *proper* time and encounter a perfectly well-behaved geometry there), the remedy is to find coordinates which do not close up the light cones.

Kruskal–Szekeres coordinates. The search for these coordinates was a long and difficult one, and ended in 1960. The good coordinates are known as Kruskal–Szekeres coordinates, are called u and v, and are defined by

$$
\left.
\begin{aligned}
u &= \left(\frac{r}{2M} - 1\right)^{1/2} e^{r/4M} \cosh\frac{t}{4M}, \\
v &= \left(\frac{r}{2M} - 1\right)^{1/2} e^{r/4M} \sinh\frac{t}{4M},
\end{aligned}
\right\}
\tag{11.58}
$$

for $r > 2M$ and

$$
\left.
\begin{aligned}
u &= \left(1 - \frac{r}{2M}\right)^{1/2} e^{r/4M} \sinh\frac{t}{4M}, \\
v &= \left(1 - \frac{r}{2M}\right)^{1/2} e^{r/4M} \cosh\frac{t}{4M},
\end{aligned}
\right\}
\tag{11.59}
$$

for $r < 2M$. (This transformation is singular at $r = 2M$, but that is necessary in order to eliminate the coordinate singularity there.) The metric in these coordinates is found to be

$$
ds^2 = -\frac{32M^3}{r} e^{-r/2M}(dv^2 - du^2) + r^2\, d\Omega^2,
\tag{11.60}
$$

where, now, r is not to be regarded as a coordinate but as a function of u and v, given implicitly by the inverse of Eqs. (11.58) and (11.59):

$$
\left(\frac{r}{2M} - 1\right) e^{r/2M} = u^2 - v^2.
\tag{11.61}
$$

Notice several things about Eq. (11.60). There is nothing singular about any metric term at $r = 2M$. There *is*, however, a singularity at $r = 0$, where we expect it. A radial null line ($d\theta = d\phi = ds = 0$) is a line

$$
dv = \pm du.
\tag{11.62}
$$

This last result is very important. It means that in a (u, v) diagram, the light cones are all as open as in SR. This result makes these coordinates particularly useful for visualizing the geometry in a coordinate diagram. The (u, v) diagram is, then, given in Fig. 11.10. Compare this with the result of Exer. 21, § 5.9.

Much needs to be said about this. First, two light cones are drawn for illustration. *Any* 45° line is a radial null line. Second, only u and v are plotted: θ and ϕ are suppressed; therefore each point is really a two-

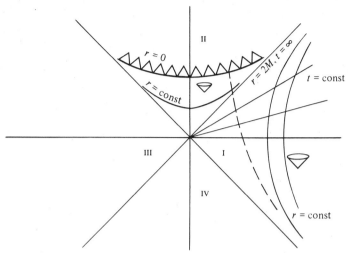

Fig. 11.10   Kruskal–Szekeres coordinates keep the light cones at 45° everywhere. The singularity at $r = 0$ (toothed line) bounds the future of all events inside (above) the line $r = 2M$, $t = +\infty$. Events outside this horizon have part of their future free of singularities.

sphere of events. Third, lines of constant r are hyperbolae, as is clear from Eq. (11.61). For $r > 2M$ these hyperbolae run roughly vertically, being asymptotic to the 45° line from the origin $u = v = 0$. For $r < 2M$ the hyperbolae run roughly horizontally, with the same asymptotes. This means that for $r < 2M$, a timelike line (confined within the light cone) cannot remain at constant r. This is the result we had before. The hyperbola $r = 0$ is the end of the spacetime, since a true singularity is there. Note that although $r = 0$ is a 'point' in ordinary space, it is a whole hyperbola here. However, not too much can be made of this, since it is a singularity of the geometry: we should not glibly speak of it as a part of spacetime with a well-defined dimensionality. Fourth, lines of constant t, being orthogonal to lines of constant r, are straight lines in this diagram, radiating outwards from the origin $u = v = 0$. (They are orthogonal to the hyperbolae $r = $ const. in the *spacetime* sense of orthogonality; recall our diagrams in § 1.7 of invariant hyperbolae in SR, which had the same property of being orthogonal to lines radiating out from the origin.) In the limit as $t \to \infty$, these lines approach the 45° line from the origin. Since all the lines $t = $ const. pass through the origin, the origin would be expanded into a whole line in a (t, r) coordinate diagram like Fig. 11.9, which is what we guessed after discussing that diagram. A world line crossing this $t = \infty$ line enters the region in which r is a time coordinate, and so cannot get out again. The true horizon, then, is this line $r = 2M$,

$t = +\infty$. Fifth, this horizon is itself a null line. This *must* be the case, since the horizon is the boundary between null rays that cannot get out and those that can. It is therefore the path of the 'marginal' null ray. Sixth, the 45° lines from the origin divide spacetime up into four regions, labeled I, II, III, IV. Region I is clearly the 'exterior', $r > 2M$, and region II is the interior of the horizon. But what about III and IV? To discuss them is beyond the scope of these lectures (see Misner *et al.* 1973, Box 33.2G and Ch. 34; and Hawking & Ellis 1973), but one remark must be made. Consider the dashed line in Fig. (11.10), which could be the path of an infalling particle. If this black hole were formed by the collapse of a star, then we know that outside the star the geometry is the Schwarzschild geometry, but inside it may be quite different. The dashed line may be taken to be the path of the *surface* of the collapsing star, in which case the region of the diagram to the right of it is *outside* the star and so correctly describes the spacetime geometry, but everything to the left would be inside the star (smaller r) and hence has possibly no relation to the true geometry of the spacetime. This includes all of regions III and IV, so they are to be ignored by the astrophysicist (though they can be interesting to the mathematician!). Note that parts of I and II are also to be ignored, but there is still a singularity and horizon outside the star. The seventh and last remark we will make is that the coordinates u and v are *not* particularly good for describing the geometry far from the star, where g_{uu} and g_{vv} fall off exponentially in r. The coordinates t and r are best there; indeed, they were constructed in order to be well behaved there. But if one is interested in the horizon, then one uses u and v.

11.3   More-general black holes

General theorems. The phenomenon of the formation of a horizon has to do with the collapse of a star to such small dimensions that the gravitational field traps everything within a certain region, which is called the interior of the horizon. We have explored the structure of the black hole in one particular case – the static, spherically symmetric situation – but the formation of a horizon is a much more general phenomenon. Just how general is still a matter for conjecture, since no one has determined a criterion which would decide exactly when a horizon must be present. Thorne has conjectured (see Misner *et al.* 1973, Box 32.3) that whenever matter of mass M is concentrated in a region whose circumference in any direction is always smaller than $2\pi(2M)$, then the mass will be inside a horizon. This conjecture is very vaguely worded, since 'mass' is not a locally determined quantity and since 'circumference' is ill defined in the absence of any symmetry, so it may be wiser to regard the conjecture as just a kind of physicist's rule of thumb for deciding in which physical situations a horizon is likely to appear.

The principal difficulty here is treating time-dependent situations. Since a horizon is the boundary between what can get out and what can't, its position in space can vary. For instance, a collapsing spherical star eventually produces a Schwarzschild black hole (after all of the star is inside the horizon), but there is an intermediate period of time in which the horizon is growing from zero radius to its full size. This is easy to see by considering Fig. 11.11, which illustrates (*very* schematically) the

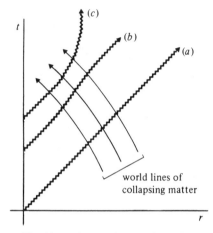

Fig. 11.11 Schematic spacetime diagram of spherical collapse. Light ray (*a*) hardly feels anything, (*b*) is delayed, and (*c*) is marginally trapped. The horizon is defined as the ray (*c*), so it grows continuously from zero radius as the collapse proceeds.

collapsing situation. (The time coordinate is a kind of Schwarzschild time, but it isn't to be taken too literally.) As matter falls in, the trajectories of outgoing photons (wavy lines) are more and more affected. Photon (*a*) gets out with little trouble, photon (*b*) has some delay, and photon (*c*) is the 'marginal' one, which just gets trapped and remains on the Schwarzschild horizon. Anything later than (*c*) is permanently trapped, anything earlier gets out. So photon (*c*) does in fact represent the *entire* horizon, by definition, since it is the boundary. Thus, one sees the horizon grow from zero radius to $2M$ by watching photon (*c*)'s progress outwards. For this spherically symmetric situation, if one knew the details of the collapse, one could easily determine the position of the horizon. But if there were no symmetry – particularly if the collapse produced a large amount of gravitational radiation – then the calculation would be far more difficult, and that is why it is only conjectural at this point. However, some results are known firmly:

(1) If the collapse is *nearly* spherical, then all nonspherical parts of the mass distribution – quadrupole moment, octopole moment – except for some angular momentum, are radiated away in gravitational waves, and a stationary black hole of the Kerr type (see below) is left behind. If there is no angular momentum, a Schwarzschild hole is left. This was proved by Price (1972*a*, *b*).

(2) It is felt that any horizon will eventually become stationary, after everything has settled down. The *stationary* horizons, by contrast with nonstationary ones, are completely known. The principal result is that a stationary black hole is characterized by four numbers: its total mass M, its total angular momentum J, its total electric charge Q, and its total magnetic monopole charge F. All four of these parameters are defined not by any integrals over the 'interior' of the horizon, but by the gravitational and electromagnetic fields far from the hole. We have defined the mass M of any metric in this fashion in Ch. 8, and in Exer. 19, § 8.6, we have seen how J can be similarly defined. The electric and magnetic charges are defined by Gauss' law integrals over surfaces surrounding the hole and far from it. Since magnetic monopoles (i.e. isolated north poles or south poles) are not known to exist in nature (though Maxwell's equations could be generalized to accommodate them), F is almost never considered. Also, it is felt that collapse is unlikely with significant charge Q, so we shall henceforth take only M and J to be nonzero. The unique stationary black hole (with $Q = F = 0$) is the Kerr black hole described below. This theorem results from work done by Hawking (1972), Carter (1973), and Robinson (1975). The Kerr hole for zero angular momentum is the Schwarzschild metric.

(3) One general result concerning nonstationary horizons is known, and is called Hawking's area theorem: in any physical process involving a horizon, the area of the horizon cannot decrease in time. We shall see below how to calculate the area of the Kerr horizon. This fundamental theorem has the result that, while two black holes can collide and coalesce, a single black hole can never bifurcate spontaneously into two smaller ones. (A restricted proof of this is in Exer. 26, § 11.6 using the Kerr area formula below; a full proof is outlined in Misner *et al.* 1973, Exer. 34.4,

and requires techniques beyond the scope of this book.) The theorem assumes that the local energy density of matter in spacetime (ρ) is positive. This can be violated in quantum mechanics, with fundamental consequences described in § 11.4 below.

Kerr black hole. The Kerr black hole is axially symmetric but not spherically symmetric (i.e. rotationally symmetric about one axis only, which is the angular-momentum axis), and is characterized by two parameters, M and J. Since J has dimension m^2, one conventionally defines

$$a \equiv J/M, \tag{11.63}$$

which then has the same dimensions as M. The line element is

$$ds^2 = -\frac{\Delta - a^2 \sin^2 \theta}{\rho^2} \, dt^2 - 2a \frac{2Mr \sin^2 \theta}{\rho^2} \, dt \, d\phi$$
$$+ \frac{(r^2 + a^2)^2 - a^2 \Delta \sin^2 \theta}{\rho^2} \sin^2 \theta \, d\phi^2 + \frac{\rho^2}{\Delta} \, dr^2 + \rho^2 \, d\theta^2, \tag{11.64}$$

where

$$\Delta \equiv r^2 - 2Mr + a^2,$$
$$\rho^2 \equiv r^2 + a^2 \cos^2 \theta. \tag{11.65}$$

The coordinates are called Boyer–Lindquist coordinates; ϕ is the angle around the axis of symmetry, t is the time coordinate in which everything is stationary, and r and θ are similar to the spherically symmetric r and θ but are not so readily associated to any geometrical definition. In particular, since there are no metric two-spheres, the coordinate r cannot be defined as an 'area' coordinate as we did before. The following points are important:

(1) Surfaces $t = \text{const.}$, $r = \text{const.}$ do not have the metric of the two-sphere, Eq. (10.2).

(2) The metric for $a = 0$ is identically the Schwarzschild metric.

(3) There is an off-diagonal term in the metric, in contrast to Schwarzschild:

$$g_{t\phi} = -a \frac{2Mr \sin^2 \theta}{\rho^2}, \tag{11.66}$$

which is $\frac{1}{2}$ the coefficient of $dt \, d\phi$ in Eq. (11.64) because the line element contains two terms,

$$g_{t\phi} \, dt \, d\phi + g_{\phi t} \, d\phi \, dt = 2g_{t\phi} \, d\phi \, dt,$$

by the symmetry of the metric. *Any* axially symmetric, stationary metric has preferred coordinates t and ϕ, namely those which have the property

$g_{\alpha\beta,t} = 0 = g_{\alpha\beta,\phi}$. But the coordinates r and θ are more-or-less arbitrary, except that they may be chosen to be (i) orthogonal to t and ϕ ($g_{rt} = g_{r\phi} = g_{\theta t} = g_{\theta\phi} = 0$) and (ii) orthogonal to each other ($g_{\theta r} = 0$). In general, one cannot choose t and ϕ orthogonal to each other ($g_{t\phi} \neq 0$). Thus Eq. (11.64) has the minimum number of nonzero $g_{\alpha\beta}$. (See Carter 1969.)

Dragging of inertial frames. The presence of $g_{t\phi} \neq 0$ in the metric introduces qualitatively new effects on particle trajectories. Because $g_{\alpha\beta}$ is independent of ϕ, a particle's trajectory still conserves p_ϕ. But now we have

$$p^\phi = g^{\phi\alpha}p_\alpha = g^{\phi\phi}p_\phi + g^{\phi t}p_t, \tag{11.67}$$

and similarly for the time components:

$$p^t = g^{t\alpha}p_\alpha = g^{tt}p_t + g^{t\phi}p_\phi. \tag{11.68}$$

Consider a zero angular-momentum particle, $p_\phi = 0$. Then, using the definitions (for nonzero rest mass)

$$p^t = m\,dt/d\tau, \qquad p^\phi = m\,d\phi/d\tau, \tag{11.69}$$

we find that the particle's trajectory has

$$\frac{d\phi}{dt} = \frac{p^\phi}{p^t} = \frac{g^{\phi t}}{g^{tt}} \equiv \omega(r,\theta). \tag{11.70}$$

This equation defines what we mean by ω, the angular velocity of a zero angular-momentum particle. We shall find ω explicitly for the Kerr metric when we obtain the contravariant components $g^{\phi t}$ and g^{tt} below. But it is clear that this effect will be present in any metric for which $g_{t\phi} \neq 0$, which in turn happens whenever the source is rotating (e.g. a rotating star as in Exer. 19, § 8.6). So we have the remarkable result that a particle dropped 'straight in' ($p_\phi = 0$) from infinity is 'dragged' just by the influence of gravity so that it acquires an angular velocity in the same sense as that of the source of the metric (we'll see below that, for the Kerr metric, ω has the same sign as a). This effect weakens with distance (roughly as $1/r^3$; see Eq. (11.83) below for the Kerr metric), and it makes the angular momentum of the source measurable in principle, although in most situations the effect is small, as we have seen in Exer. 19, § 8.6. (An experiment is being planned, however, to use precision gyroscopes in a satellite to measure the effect due to rotation of the earth. This is discussed in Misner *et al.* 1973, § 40.7). The effect is called – somewhat fancifully – the 'dragging of inertial frames'; the gyroscopic precession it produces is called the 'Lense–Thirring effect'.

Ergoregion. Consider photons emitted in the equatorial plane ($\theta = \pi/2$) at some given r. In particular, consider those initially going in the $\pm\phi$-direction, i.e. tangent to a circle of constant r. Then they generally have only dt and dϕ nonzero on the path at first and since d$s^2 = 0$, we have

$$0 = g_{tt}\,dt^2 + 2g_{t\phi}\,dt\,d\phi + g_{\phi\phi}\,d\phi^2$$

$$\Rightarrow \frac{d\phi}{dt} = -\frac{g_{t\phi}}{g_{\phi\phi}} \pm \sqrt{\left[\left(\frac{g_{t\phi}}{g_{\phi\phi}}\right)^2 - \frac{g_{tt}}{g_{\phi\phi}}\right]}. \tag{11.71}$$

Now, a remarkable thing happens if $g_{tt} = 0$: the two solutions are

$$\frac{d\phi}{dt} = 0 \quad \text{and} \quad \frac{d\phi}{dt} = -\frac{2g_{t\phi}}{g_{\phi\phi}}. \tag{11.72}$$

We will see below that for the Kerr metric the second solution gives dϕ/dt the *same* sign as the parameter a, and so represents the photon sent off in the same direction as the hole's rotating. The other solution means that the other photon – the one sent 'backwards' – initially doesn't move at all. The dragging of orbits has become so strong that this photon *cannot* move in the direction opposite the rotation. Clearly, any particle, which must move slower than a photon, will therefore have to rotate with the hole, even if it has an angular momentum arbitrarily large in the opposite sense to the hole's!

We shall see that the surface where $g_{tt} = 0$ lies outside the horizon; it is called the ergosphere. It is sometimes also called the 'static limit', since inside it no particle can remain at fixed r, θ, ϕ. From Eq. (11.64) we conclude that it occurs at

$$r_0 \equiv r_{\text{ergosphere}} = M + \sqrt{(M^2 - a^2\cos^2\theta)}. \tag{11.73}$$

Inside this radius, since $g_{tt} > 0$, all particles and photons must rotate with the hole.

Again, this effect can occur in other situations. Models for certain rotating stars are known where there are toroidal regions of space in which $g_{tt} > 0$ (Butterworth & Ipser 1976). These will have these super-strong frame-dragging effects. They are called ergoregions, and their boundaries are ergotoroids. They can exist in solutions which have no horizon at all.

The horizon. In the Schwarzschild solution the horizon was the place where $g_{tt} = 0$ and $g_{rr} = \infty$. In the Kerr solution the ergosphere occurs at $g_{tt} = 0$ and the horizon is at $g_{rr} = \infty$, i.e. where $\Delta = 0$:

$$r_+ \equiv r_{\text{horizon}} = M + \sqrt{(M^2 - a^2)}. \tag{11.74}$$

It is clear that the ergosphere lies outside the horizon except at the poles, where it is tangent to it. The full proof that this is the horizon is beyond our scope here: one needs to verify that no null lines can escape from inside r_+. We shall simply take it as given. (See the next section below for a partial justification.) Since the area of the horizon is important (Hawking's area theorem), we shall calculate it.

The horizon is a surface of constant r and t, by Eq. (11.74) and the fact that the metric is stationary. Any surface of constant r and t has an intrinsic metric whose line element comes from Eq. (11.64) with $dt = dr = 0$:

$$dl^2 = \frac{(r^2+a^2)^2 - a^2\Delta}{\rho^2} \sin^2\theta \, d\phi^2 + \rho^2 \, d\theta^2. \qquad (11.75)$$

The proper area of this surface is given by integrating the square root of the determinant of this metric over all θ and ϕ:

$$A(r) = \int_0^{2\pi} d\phi \int_0^{\pi} d\theta \, \sqrt{[(r^2+a^2)^2 - a^2\Delta]} \sin\theta. \qquad (11.76)$$

Since nothing in the square root depends on θ or ϕ, and since the area of a unit two-sphere is

$$4\pi = \int_0^{2\pi} d\phi \int_0^{\pi} d\theta \sin\theta,$$

we immediately conclude that

$$A(r) = 4\pi\sqrt{[(r^2+a^2)^2 - a^2\Delta]}. \qquad (11.77)$$

Since the horizon is defined by $\Delta = 0$, we get

$$A(\text{horizon}) = 4\pi(r_+^2 + a^2). \qquad (11.78)$$

Equatorial photon motion in the Kerr metric. A detailed study of the motion of photons in the equatorial plane gives insight into the ways in which 'rotating' metrics differ from nonrotating ones. First, we must obtain the inverse of the metric, Eq. (11.64), which we write in the general stationary, axially symmetric form:

$$ds^2 = g_{tt} \, dt^2 + 2g_{t\phi} \, dt \, d\phi + g_{\phi\phi} \, d\phi^2 + g_{rr} \, dr^2 + g_{\theta\theta} \, d\theta^2.$$

The only off-diagonal element involves t and ϕ; therefore

$$g^{rr} = \frac{1}{g_{rr}} = \Delta\rho^{-2}, \qquad g^{\theta\theta} = \frac{1}{g_{\theta\theta}} = \rho^{-2}. \qquad (11.79)$$

We need to invert the matrix

$$\begin{pmatrix} g_{tt} & g_{t\phi} \\ g_{t\phi} & g_{\phi\phi} \end{pmatrix}.$$

Calling its determinant D, the inverse is

$$\frac{1}{D}\begin{pmatrix} g_{\phi\phi} & -g_{t\phi} \\ -g_{t\phi} & g_{tt} \end{pmatrix}, \qquad D = g_{tt}g_{\phi\phi} - (g_{t\phi})^2. \tag{11.80}$$

Notice one important deduction from this. The angular velocity of the dragging of inertial frames is Eq. (11.70):

$$\omega = \frac{g^{\phi t}}{g^{tt}} = \frac{-g_{t\phi}/D}{g_{\phi\phi}/D} = -\frac{g_{t\phi}}{g_{\phi\phi}}. \tag{11.81}$$

This makes Eqs. (11.71) and (11.72) more meaningful. For the metric, Eq. (11.64), some algebra gives

$$D = -\Delta \sin^2\theta, \qquad g^{tt} = -\frac{(r^2+a^2)^2 - a^2\Delta\sin^2\theta}{\rho^2\Delta},$$

$$g^{t\phi} = -a\frac{2Mr}{\rho^2\Delta}, \qquad g^{\phi\phi} = \frac{\Delta - a^2\sin^2\theta}{\rho^2\Delta\sin^2\theta}. \tag{11.82}$$

Then the frame dragging is

$$\omega = \frac{2Mra}{(r^2+a^2)^2 - a^2\Delta\sin^2\theta}. \tag{11.83}$$

The denominator is positive everywhere (by Eq. 11.65), so this has the same sign as a, and it falls off for large r as r^{-3}, as we noted earlier.

A photon whose trajectory is in the equatorial plane has $d\theta = 0$; but, unlike the Schwarzschild case, this is only a special kind of trajectory: photons not in the equatorial plane may have qualitatively different orbits. Nevertheless, a photon for which $p^\theta = 0$ initially in the equatorial plane always has $p^\theta = 0$, since the metric is reflection symmetric through the plane $\theta = \pi/2$. By stationarity and axial symmetry the quantities $E = -p_t$ and $L = p_\phi$ are constants of the motion. Then the equation $\vec{p} \cdot \vec{p} = 0$ determines the motion. Denoting p^r by $dr/d\lambda$ as before, we get, after some algebra,

$$\left(\frac{dr}{d\lambda}\right)^2 = g^{rr}[(-g^{tt})E^2 + 2g^{t\phi}EL - g^{\phi\phi}L^2]$$

$$= g^{rr}(-g^{tt})\left[E^2 - 2\omega EL + \frac{g^{\phi\phi}}{g^{tt}}L^2\right]. \tag{11.84}$$

Using Eqs. (11.65), (11.79), and (11.83) for $\theta = \pi/2$, we get

$$\left(\frac{dr}{d\lambda}\right)^2 = \frac{(r^2+a^2)^2 - a^2\Delta}{r^4}\left[E^2 - \frac{4Mra}{(r^2+a^2)^2 - a^2\Delta}EL\right.$$

$$\left. - \frac{r^2 - 2Mr}{(r^2+a^2)^2 - a^2\Delta}L^2\right]. \tag{11.85}$$

This is to be compared with Eq. (11.9), to which it reduces when $a = 0$. Apart from the complexity of the coefficients, Eq. (11.85) differs from Eq. (11.9) in a qualitative way in the presence of a term in EL. So we cannot simply define an effective potential V^2 and write $(dr/d\lambda)^2 = E^2 - V^2$. What we can do is nearly as good. We can factor Eq. (11.84):

$$\left(\frac{dr}{d\lambda}\right)^2 = \frac{(r^2 + a^2)^2 - a^2\Delta}{r^4}(E - V_+)(E - V_-). \tag{11.86}$$

Then $V_\pm$, by Eqs. (11.84) and (11.85), are

$$V_\pm(r) = [\omega \pm (\omega^2 - g^{\phi\phi}/g^{tt})^{1/2}]L \tag{11.87}$$

$$= \frac{2Mra \pm r^2\Delta^{1/2}}{(r^2 + a^2)^2 - a^2\Delta}L. \tag{11.88}$$

This is to be compared to the *square root* of Eq. (11.11), to which it reduces when $a = 0$. Now, the square root of Eq. (11.11) becomes imaginary inside the horizon; similarly, Eq. (11.88) is complex when $\Delta < 0$. In each case the meaning is that in such a region there are *no* solutions to $dr/d\lambda = 0$, *no* turning points regardless of the energy of the photon. Once a photon crosses the line $\Delta = 0$ it *cannot* turn around and get back outside that line. Clearly, $\Delta = 0$ marks the *horizon* in the equatorial plane. What we haven't shown, but what is also true, is that $\Delta = 0$ marks the horizon for trajectories not in the equatorial plane.

We can discuss the qualitative features of photon trajectories by plotting $V_\pm(r)$. We choose first the case $aL > 0$ (angular momentum in the same sense as the hole), and of course we confine attention to $r \geq r_+$ (outside the horizon). Notice that for large r the curves (in Fig. 11.12)

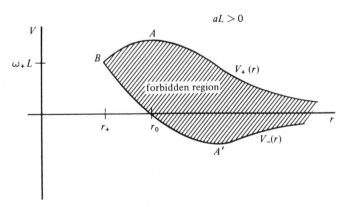

Fig. 11.12   Factored potential diagram for equatorial photon orbits of positive angular momentum in the Kerr metric. As $a \to 0$, the upper and lower curves approach the two square roots of Fig. 11.2 outside the horizon.

are asymptotic to zero, falling off as $1/r$. This is the regime in which the rotation of the hole makes almost no difference. For small r we see features not present without rotation: V_- goes through zero (easily shown to be at $r_0 = 2M$, the edge of the ergosphere) and meets V_+ at the horizon, both curves having the value $aL/2Mr_+ \equiv \omega_+ L$, where ω_+ is the value of ω on the horizon. From Eq. (11.86) it is clear that a photon can move only in regions where $E > V_+$ or $E < V_-$. We are used to photons with positive E: they may come in from infinity and either reach a minimum r or plunge in, depending on whether or not they encounter the hump in V_+. There is nothing qualitatively new here. But what of those for which $E \le V_-$? Some of these have $E > 0$. Are they to be allowed?

To discuss negative-energy photons we must digress a moment and talk about moving along a geodesic backwards in time. We have associated our particles' paths with the mathematical notion of a geodesic. Now a geodesic is a curve, and the path of a curve can be traversed in either of two directions; for timelike curves one is forwards and the other backwards in time. The tangents to these two motions are simply opposite in sign, so one will have four-momentum $\vec{p}$ and the other $-\vec{p}$. The energies measured by observer $\vec{U}$ will be $-\vec{U} \cdot \vec{p}$ and $+\vec{U} \cdot \vec{p}$. So one particle will have positive energy and another negative energy. In flat spacetime we conventionally take all particles to travel forwards in time; since all known particles have positive or zero rest mass, this causes them all to have positive energy relative to any Lorentz observer who also moves forwards in time. Conversely, if $\vec{p}$ has positive energy relative to some Lorentz observer, it has positive energy relative to *all* that go forwards in time. In the Kerr metric, however, it will not do simply to demand positive E. This is because E is the energy relative to an observer at infinity; the particle near the horizon is far from infinity, so the direction of 'forward time' isn't so clear. What we must do is set up some observer $\vec{U}$ near the horizon who will have a clock, and demand that $-\vec{p} \cdot \vec{U}$ be positive for particles that pass near him. A convenient observer (but any will do) is one who has zero angular momentum and resides at fixed r, circling the hole at the angular velocity ω. This zero angular-momentum observer (ZAMO) is not on a geodesic, so he must have a rocket engine to remain on this trajectory. (In this respect he is no different from us, who must use our legs to keep us at constant r in Earth's gravitational field.) It is easy to see that he has four-velocity $U^0 = A$, $U^\phi = \omega A$, $U^r = U^\theta = 0$, where A is found from the condition $\vec{U} \cdot \vec{U} = -1$: $A = g_{\phi\phi}/(-D)$. This is nonsingular for $r > r_+$. Then he measures the energy of a particle to be

$$E_{ZAMO} = -\vec{p} \cdot \vec{U} = -(p_0 U^0 + p_\phi U^\phi)$$
$$= A(E - \omega L). \tag{11.89}$$

This is the energy we must demand be positive-definite. Since A is positive, we require

$$E > \omega L. \tag{11.90}$$

From Eq. (11.88) it is clear that any photon with $E > V_+$ also satisfies Eq. (11.90) and so is allowed, while any with $E < V_-$ violates Eq. (11.90) and is moving backwards in time. So in Fig. 11.12 we consider only trajectories for which E lies above V_+; for these there is nothing qualitatively different from Schwarzschild.

For negative angular-momentum particles, however, new features do appear. If $aL < 0$ it is clear from Eq. (11.88) that the shape of the $V_\pm$ curves is just turned over, so they look like Fig. 11.13. Again, of course,

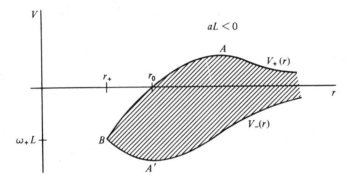

Fig. 11.13   As Fig. 11.12 for negative angular momentum photons.

condition Eq. (11.90) means that forward-going photons must lie above $V_+(r)$, but now some of these can have $E < 0$! This happens only for $r < r_0$, i.e. inside the ergoregion. Now we see the origin of the name ergoregion: it is from the Greek 'ergo-', meaning energy, a region in which energy has peculiar properties. It leads to the following interesting circumstance. At some point between r_0 and r_+ it is possible to create two photons, one having energy $+E$ and the other $-E$, so that their *total* energy is zero. Then the positive-energy photon could be directed in such a way as to leave the hole and reach infinity, while the negative-energy photon is necessarily trapped, and inevitably crosses the horizon. The net effect is that the positive-energy photon will leave the hole, carrying its energy to infinity, where it can be converted into useful work: energy, and therefore mass, has been extracted from the hole, at zero cost! By examining Figs. 11.12 and 11.13, one can convince oneself that this only works if the negative-energy particle has a negative L whose absolute value exceeds the value of L for the positive-energy particle, so that the process involves a decrease in the angular momentum of the hole. The energy extracted can therefore be thought of as coming from the rotational energy of the hole. This process is called the Penrose process, after its discoverer. It is not peculiar to the Kerr black hole; it happens whenever there is an ergoregion. If a rotating star has one without a horizon, the effective potentials look like Fig. 11.14, drawn for $aL < 0$ (Comins and Schutz 1978). (For $aL > 0$ the curves just turn over, of course.)

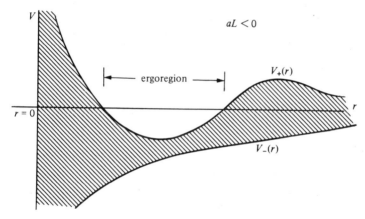

Fig. 11.14 As Fig. 11.13 for equatorial orbits in the spacetime of a star rotating rapidly enough to have an ergoregion.

The curve for V_+ dips below zero inside the ergoregion, which is toroidal. Outside the ergoregion it is positive, climbing to infinity as $r \to 0$. The curve for V_- never changes sign, and also goes to infinity as $r \to 0$. The Penrose process operates inside the ergoregion, where the negative-energy photons are trapped. This leads, in fact, to an instability in such stars (Friedman 1978).

11.4 Quantum mechanical emission of radiation by black holes: The Hawking process

In 1974 Hawking startled the physics community by proving that black holes aren't black: they radiate energy continuously! This doesn't come from any mistake in what we have already done; it arises in the application of quantum mechanics to electromagnetic fields near a black hole. We have until now spoken of photons as particles following a geodesic trajectory in spacetime; but according to the uncertainty principle these 'particles' cannot be localized to arbitrary precision. Near the horizon this markedly changes the behavior of 'real' photons from what we have already described for idealized null particles.

Hawking's calculation (Hawking 1975) uses the techniques of quantum field theory, but we can derive its main prediction very simply from elementary considerations. What follows, therefore, is a 'plausibility argument', not a rigorous discussion of the effect. One form of the uncertainty principle is $\Delta E \, \Delta t = \hbar$, where ΔE is the minimum uncertainty in a particle's energy which resides in a quantum mechanical state for a time Δt. According to quantum field theory, ordinary space is filled with 'vacuum fluctuations' in electromagnetic fields, which consist of pairs of photons being produced at one event and recombining at another. Such pairs violate conservation of energy, but if they last less than $\Delta t = \hbar / \Delta E$, where ΔE is the amount of violation, they violate no physical law. Thus, in the large, energy conservation holds rigorously, while, on a small scale, it is always being violated. Now, as we have emphasized before, spacetime near the horizon of a black hole is perfectly ordinary and, in particular, locally flat. Therefore these fluctuations will also be happening there. Consider a fluctuation which produces two photons, one of energy E and the other with energy $-E$. In flat spacetime the negative-energy photon would not be able to propagate freely, so it would necessarily recombine with the positive-energy one within a time $\hbar / E$. But if produced just outside the horizon, it has a chance of crossing the horizon before the time $\hbar / E$ elapses; once inside the horizon it *can* propagate freely, as we shall now show. Consider the Schwarzschild metric for simplicity, and recall from our discussion of orbits in the Kerr metric that negative energy is normally excluded because it corresponds to a particle that propagates backwards in time. Inside the event horizon, an observer going forwards in time is one going toward decreasing r. For simplicity let us choose one on a trajectory for which $p_0 = 0 = U^0 = 0$. Then U^r is the only nonzero component of $\vec{U}$, and by the normalization condition $\vec{U} \cdot \vec{U} = -1$ we find U^r:

$$U^r = -\left(\frac{2M}{r} - 1\right)^{1/2}, \qquad r < 2M, \tag{11.91}$$

negative because the observer is ingoing. Any photon orbit is allowed for which $-\vec{p} \cdot \vec{U} > 0$. Consider a zero angular-momentum photon, moving radially inside the horizon. By Eq. (11.9) with $L = 0$, it clearly has $E = \pm p^r$. Then its energy relative to the observer is

$$-\vec{p} \cdot \vec{U} = -p^r U^r g_{rr} = -\left(\frac{2M}{r} - 1\right)^{-1/2} p^r. \tag{11.92}$$

This is positive if and only if the photon is also ingoing: $p^r < 0$. But it sets *no* restriction at all on E. Photons may travel on null geodesics inside the horizon, which have either sign of E, as long as $p^r < 0$. (Recall that t is a spatial coordinate inside the horizon, so this result should not be surprising: E is a spatial momentum component there.)

Since a fluctuation near the horizon *can* put the negative-energy photon into a realizeable trajectory, the positive-energy photon is allowed to escape to infinity. Let us see what we can say about its energy. We first look at the fluctuations in a freely falling inertial frame, which is the one for which spacetime is locally flat and in which the fluctuations should look normal. A frame that is momentarily at rest at coordinate $2M + \varepsilon$ will immediately begin falling inwards, following the trajectory of a particle with $\tilde{L} = 0$ and $\tilde{E} = [1 - 2M/(2M + \varepsilon)]^{1/2} \approx (\varepsilon/2M)^{1/2}$, from Eq. (11.8). It reaches the horizon after a proper-time lapse $\Delta\tau$ obtained by integrating Eq. (11.52):

$$\Delta\tau = -\int_{2M+\varepsilon}^{2M} \frac{dr}{\left(\dfrac{2M}{r} - \dfrac{2M}{2M+\varepsilon}\right)^{1/2}}. \tag{11.93}$$

To first order in ε this is

$$\Delta\tau = 2(2M\varepsilon)^{1/2}. \tag{11.94}$$

We can find the energy $\mathscr{E}$ of the photon in this frame by setting this equal to the fluctuation time $\hbar/\mathscr{E}$. The result is

$$\mathscr{E} = \tfrac{1}{2}\hbar(2M\varepsilon)^{-1/2}. \tag{11.95}$$

This is the energy of the outgoing photon, the one which reaches infinity, as calculated on the local inertial frame. To find its energy when it gets to infinity we recall that

$$\mathscr{E} = -\vec{p} \cdot \vec{U},$$

with $-U_0 = \tilde{E} \approx (\varepsilon/2M)^{1/2}$. Therefore

$$\mathscr{E} = -g^{00} p_0 U_0 = U_0 g^{00} E, \tag{11.96}$$

where E is the conserved energy on the photon's trajectory, and is the energy it is measured to have when it arrives at infinity. Evaluating g^{00} at $2M + \varepsilon$ gives, finally,

$$E = \mathscr{E}(\varepsilon/2M)^{1/2} = \hbar/4M = h/8\pi M. \tag{11.97}$$

Remarkably, it doesn't matter where the photon originated: it always comes out with this characteristic energy!

The rigorous calculation which Hawking performed showed that the photons which come out have the spectrum characteristic of a black body with a temperature

$$T = \hbar/8\pi kM, \tag{11.98}$$

where k is Boltzmann's constant. Associated with this radiation is a typical photon energy

$$E = kT = \hbar/8\pi M, \tag{11.99}$$

fairly close to our crude result, Eq. (11.97). Our argument does not show that the photons should have a black-body spectrum; but the fact that the spectrum originates in random fluctuations, plus the fact that the black hole is, classically, a perfect absorber, makes this result plausible as well.

Notice that the temperature of the hole is proportional to M^{-1}. The rate of radiation from a black body is proportional to AT^4, where A is the area of the body, in this case of the horizon, which is proportional to M^2 (see Eq. (11.78)). So the luminosity of the hole is proportional to M^{-2}. This energy must come from the mass of the hole (every negative-energy photon falling into it decreases M), so we have

$$\left. \begin{array}{l} \dfrac{dM}{dt} \sim M^{-2}, \\[2mm] M^2 \, dM \sim dt, \end{array} \right\} \tag{11.100}$$

or the lifetime of the hole is

$$\tau \sim M^3. \tag{11.101}$$

The bigger the hole the longer it lives, and the cooler its temperature. The numbers work out that a hole of mass 10^{12} kg has a lifetime of 10^{10} yr, about the age of the universe. Thus

$$\left(\frac{\tau}{10^{10} \, \text{yr}}\right) = \left(\frac{M}{10^{12} \, \text{kg}}\right)^3. \tag{11.102}$$

Since a solar mass is about 10^{30} kg, black holes formed from stellar collapse are essentially unaffected by this radiation, which has a temperature of about 10^{-7} K. On the other hand, it is possible for holes of 10^{13} kg to form in the very early universe. To see the observable effect of their 'evaporation', let us calculate the energy radiated in the last second by setting $\tau = 1 \, \text{s} = (3 \times 10^7)^{-1}$ yr in Eq. (11.102). We get

$$M \approx 10^6 \, \text{kg} \sim 10^{23} \, \text{J}. \tag{11.103}$$

So for a brief second it would be comparable in luminosity to a small star, but in spectrum it would be very different. Its temperature would be 10^{11} K, emitting primarily in γ-rays! No such events have been identified.

It must be pointed out that all derivations of Hawking's result are valid only if the typical photon has $E \ll M$, since they involve treating the spacetime of the black hole as a fixed background in which one solves the equations of quantum mechanics, unaffected to first order by the propagation of these photons. This approximation fails for $M \approx h/M$, or for black holes of mass

$$M_p = h^{1/2} = 1.6 \times 10^{-35} \text{ m} = 2.2 \times 10^{-8} \text{ kg.} \tag{11.104}$$

This is called the Planck mass, since it is a mass derived only from Planck's constant (and c and G). To treat quantum effects involving such holes, one needs a consistent theory of quantum gravity, which is one of the most active areas of research in relativity today. All we can say here is that the search has not yet proved fully successful, but Hawking's calculation appears to have been one of the most fruitful steps.

Before leaving the Hawking effect, we shall show how it has provided a remarkable unification of gravity and thermodynamics. Consider Hawking's area theorem, which we may write as

$$\frac{dA}{dt} \ge 0. \tag{11.105}$$

For a Schwarzschild black hole,

$$A = 16\pi M^2,$$
$$dA = 32\pi M \, dM,$$

or

$$dM = \frac{1}{32\pi M} dA = \frac{h}{8\pi M} d\left(\frac{A}{4h}\right). \tag{11.106}$$

Since dM is the change in the hole's total energy, and since $h/8\pi M$ is its temperature, we may write Eq. (11.106) in the form

$$dE = T \, dS,$$

with

$$S = A/4h. \tag{11.107}$$

Since, by Eq. (11.105), this quantity S can never decrease, we have in Eqs. (11.106) and (11.105) the first and second laws of thermodynamics as they apply to black holes! That is, a black hole behaves in every respect as a thermodynamic black body with temperature $h/8\pi M$ and entropy $A/4h$. This analogy had been noticed as soon as the area theorem was discovered (see Misner *et al.* 1973, Box 33.4), but at that time it was thought to be an incomplete analogy because black holes did not have a temperature. The Hawking effect has fitted the missing piece into the puzzle.

11.5 Bibliography

The perihelion shift and deflection of light are the two classical tests of GR. Other theories predict different results: see Will (1972, 1979, 1982). A short, entertaining account of the observation of the deflection of light and its impact on Einstein's fame is in McCrea (1979).

The coordinate singularity at the horizon of the Schwarzschild solution has stimulated work toward a general method of determining whether a true singularty exists. At present no such method exists, and there is not even a fully satisfactory definition of a 'true' singularity. See the discussions in Misner *et al* (1973), Hawking & Ellis (1973), or Kramer *et al.* (1981). Recent work is reviewed by Geroch & Horowitz (1979) and Tipler *et al.* (1980).

The Kerr metric has less symmetry than the Schwarzschild metric, so it might be expected that particle orbits would have fewer conserved quantities and therefore be harder to calculate. This is, quite remarkably, false: even orbits out of the equator have three conserved quantities: energy, angular momentum, and a difficult-to-interpret quantity associated with the θ motion. The same remarkable property carries over to the wave equations that govern electromagnetic fields and gravitational waves in the Kerr metric: these equations separate completely in certain coordinate systems. See Teukolsky (1972) for the first general proof of this, or Benenti & Francaviglia (1980) and Chandrasekhar (1983) for full discussions.

We have barely touched upon the role that black holes play in modern astrophysics. Their combination of a strong gravitational field with a 'sink' at the center, into which matter can flow unimpeded, makes them ideal catalyzers of activity in galactic nuclei and X-ray binaries. Various reviews covering most applications are: Eardley & Press (1975), Novikov & Thorne (1973), Carr (1977), Rees (1977), Bahcall (1978), Blandford & Thorne (1979), Shapiro & Teukolsky (1983).

Black-hole thermodynamics is treated thoroughly in Carter (1979), while the related theory of quantum fields in curved spacetimes is reviewed by Gibbons (1979) and Davies (1980). To go beyond this and quantize the geometry of spacetime is the goal of all attempts to devise a self-consistent quantum gravity. Reviews of recent promising work in what is called supergravity appear in several articles in Held (1980a, b) and in the article by Weinberg (1979). Supergravity starts from weak gravity (linearized theory) and hopes to extrapolate to strong fields. More-general perspectives on quantum gravity research may be found in DeWitt (1979) and Hawking (1979).

One of the few strong-field, time-dependent situations to have been studied numerically is the collision of two black holes: Smarr (1979b) and references therein.

11.6 Exercises

1 Consider a particle or photon in an orbit in the Schwarzschild metric with a certain E and L, at a radius $r \gg M$. Show that if spacetime were really *flat*, the particle would travel a straight line which would pass a distance $b \equiv L/[E^2 - m^2]^{1/2}$ from the center of coordinates $r = 0$. This ratio b is called the *impact parameter*. Show also that photon orbits that follow from Eq. (11.9) depend *only* on b.

2 Prove Eqs. (11.14) and (11.15).

3 Plot $\tilde{V}^2$ against r/M for the three cases $\tilde{L}^2 = 25\,M^2$, $\tilde{L}^2 = 12\,M^2$, $\tilde{L}^2 = 9\,M^2$ and verify the qualitative correctness of Figs. 11.1 and 11.3.

4 What kind of orbits are possible outside a star of radius (a) $2.5\,M$, (b) $4\,M$, (c) $10\,M$?

5 It is possible that the centers of active galaxies and quasars may contain black holes of mass $10^6 - 10^9\,M_\odot$.
(a) Find the radius $R_{0.01}$ at which $-g_{00}$ differs from the 'Newtonian' value $1 - 2\,M/R$ by only 1%. (One may think of this as a kind of limit on the region in which relativistic effects are important.)
(b) A 'normal' star may have a radius of 10^{10} m. Approximately how many such stars could occupy the volume of space between the horizon $R = 2\,M$ and $R_{0.01}$?

6 Compute the wavelength of light that gets to a distant observer from the following sources.
(a) Light emitted with wavelength 6563 Å ($H\alpha$ line) by a source at rest where $\Phi = -10^{-6}$. (Typical star.)
(b) Same as (a) for $\Phi = -6 \times 10^{-5}$ (value for the white dwarf 40 Eridani B).
(c) Same as (a) for a source at rest at radius $r = 2.2\,M$ outside a black hole of mass $M = 1\,M_\odot = 1.47 \times 10^5$ cm.
(d) Same as (c) for $r = 2.02\,M$.

7   A clock is in a circular orbit at $r = 10\,M$ in a Schwarzschild metric.
(a)   How much time elapses on the clock during one orbit? (Integrate the proper time $d\tau = |ds^2|^{1/2}$ over an orbit.)
(b)   It sends out a signal to a distant observer once each orbit. What time interval does the distant observer measure between receiving any two signals?
(c)   A second clock is located at rest at $r = 10\,M$ next to the orbit of the first clock. (Rockets keep it there.) How much time elapses on it between successive passes of the orbiting clock?
(d)   Calculate (b) again in seconds for an orbit at $r = 6\,M$ where $M = 14\,\mathrm{M}_\odot$. This is the minimum fluctuation time one expects in the X-ray spectrum of Cyg X-1: why?
(e)   If the orbiting 'clock' is the twin Artemis, in the orbit in (d), how much does she age during the time her twin Diana lives 40 years far from the black hole and at rest with respect to it?

8(a)   Derive Eqs. (11.17) and (11.21).
(b)   Derive Eqs. (11.23) and (11.25).

9   (This problem requires access to a computer or a programmable calculator.)
(a)   Integrate numerically Eq. (11.23) or Eq. (11.25) for the orbit of a particle (i.e. for r/M as a function of ϕ) when $E^2 = 0.91$ and $(\tilde{L}/M)^2 = 13.0$. Compare the perihelion shift from one orbit to the next with Eq. (11.34).
(b)   Integrate again when $\tilde{E}^2 = 0.95$ and $(\tilde{L}/M)^2 = 13.0$. How much proper time does this particle require to reach the horizon from $r = 10\,M$ if its initial radial velocity is negative?

10(a)   For a given value of $\tilde{L}$, what is the minimum value of $\tilde{E}$ that permits a particle with $m \neq 0$ to reach the Schwarzschild horizon?
(b)   Express this result in terms of the *impact parameter b* (see Exer. 1).
(c)   Conversely, for a given value of b, what is the maximum value of $\tilde{L}$ that permits a particle wth $m \neq 0$ to reach the Schwarzschild horizon? Relate your result to Fig. 11.3.

11   The right-hand side of Eq. (11.25) is a polynomial in u. Trace the u^3 term back through the derivation and show that it would not be present if we had started with the Newtonian version of Eq. (11.6). Interpret this term as a redshift effect on the orbital kinetic energy. Show that it is responsible for the maximum in the curve in Fig. 11.1.

12(a)   Prove that Eq. (11.29) solves Eq. (11.28).
(b)   Derive Eq. (11.30) from Eq. (11.29) and show that it describes an ellipse by transforming to Cartesian coordinates.

13(a)   Derive Eq. (11.31) in the approximation that y is small. What must it be small compared to?

(b)   Derive Eqs. (11.32) and (11.33) from (11.31).
(c)   Verify the remark after Eq. (11.33) that $y = 0$ is not the correct circular orbit for the given $\bar{E}$ and $\bar{L}$ by using Eqs. (11.17) and (11.18) to find the correct value of y and comparing it to y_0 in Eq. (11.33).
(d)   Show from Eq. (11.10) that a particle which has an inner turning point in the 'Newtonian' regime, i.e. for $r \gg M$, has a value $\bar{L} \gg M$. Use this to justify the step from Eq. (11.34) to Eq. (11.35).

14   Compute the perihelion shift per orbit and per year for the following planets, given their distance from the Sun and their orbital period: Venus $(1.1 \times 10^{11}$ m, 1.9×10^{7} s); Earth $(1.5 \times 10^{11}$ m, 3.2×10^{7} s); Mars $(2.3 \times 10^{11}$ m, 5.9×10^{7} s).

15(a)   Derive Eq. (11.44) from (11.42), and show that it describes a straight line passing a distance b from the origin.
(b)   Derive Eq. (11.46) from (11.42).
(c)   Integrate Eq. (11.46) to get (11.47).

16   We calculate the observed deflection of a null geodesic anywhere on its path as follows. See Ward (1970).
(a)   Show that Eq. (11.47) may be solved to give

$$bu = \sin(\phi - \phi_0) + \frac{M}{b}[1 - \cos(\phi - \phi_0)]^2 + 0\left(\frac{M^2}{b^2}\right). \tag{11.108}$$

(b)   In Schwarzschild coordinates, the vector

$$\vec{v} \rightarrow -(0, 1, 0, d\phi/dr) \tag{11.109}$$

is tangent to the photon's path as seen by an observer at rest in the metric at the position r. Show that this observer measures the angle α in Fig. 11.15 to be

$$\cos \alpha = (\vec{v} \cdot \vec{e}_r)/(\vec{v} \cdot \vec{v})^{1/2}(\vec{e}_r \cdot \vec{e}_r)^{1/2}, \tag{11.110}$$

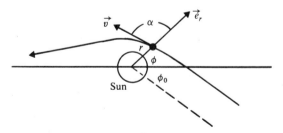

Fig. 11.15   The deflection of light by the Sun.

where $\vec{e}_r$ has components $(0, 1, 0, 0)$. Argue that $\phi - \pi + \alpha$ is the *apparent* angular position of the star, and show from Eq. (11.108) that if $M = 0$ (no deflection), $\phi - \pi + \alpha = \phi_0$.

(c)   When $M \neq 0$, calculate the deflection

$$\delta\phi \equiv (\phi - \pi + \alpha) - \phi_0 \tag{11.111}$$

to first order in M/b. Don't forget to use the Schwarzschild metric to compute the dot products in Eq. (11.110). Obtain

$$\delta\phi = \frac{2M}{b}[1 - \cos(\phi - \phi_0)], \tag{11.112}$$

which is, in terms of the position r of the observer,

$$\delta\phi = \frac{2M}{r} \frac{1 - \cos(\phi - \phi_0)}{\sin(\phi - \phi_0)}. \tag{11.113}$$

(d)   For $M = 1\ M_\odot = 1.47$ km, $r = 1$ AU $= 1.5 \times 10^6$ km, how far from the Sun on the sky can this deflection be detected if one can measure angles to an accuracy of 2×10^{-3} arcsec?

17   We can use Eq. (11.108) above on a different problem, namely to calculate the expected arrival times at a distant observer of pulses regularly emitted by a satellite in a circular orbit in the Schwarzschild metric. This is a simplified version of the timing problem of the binary pulsar system, which consists of two neutron stars of roughly equal mass in orbit about one another, one of which is a pulsar. See Blandford & Teukolsky (1976) and Epstein (1977).

(a)   Show that along the trajectory, Eq. (11.108), coordinate time elapses at the rate

$$dt/d\phi = b\left[(bu)^2\left(1 - \frac{2M}{b}bu\right)\right]^{-1}. \tag{11.114}$$

(b)   Integrate this to find the coordinate travel time for a photon emitted at the position u_E, ϕ_E and received at the position u_R, ϕ_R, where $u_R \ll u_E$.

(c)   Since Eq. (11.108) is satisfied at both (u_R, ϕ_R) and (u_E, ϕ_E), show that

$$\phi_R - \phi_0 = \frac{u_R}{u_E} \sin(\phi_E - \phi_R)\left\{1 + \frac{u_R}{u_E}\cos(\phi_E - \phi_R)\right.$$

$$\left. + Mu_E(1 - \cos[\phi_E - \phi_R])^2/\sin^2(\phi_E - \phi_R)\right\}, \tag{11.115}$$

to first order in M and u_R/u_E and that, similarly,

$$b = (1/u_R)\{\phi_R - \phi_0 + Mu_E[1 - \cos(\phi_E - \phi_R)]^2/\sin(\phi_E - \phi_R)\}. \tag{11.116}$$

(d)   Use these in your result in (b) to calculate the difference δt in travel time between pulses emitted at (u_E, ϕ_E) and at $(u_E, \phi_E + \delta\phi_E)$ to first order in $\delta\phi_E$. (The receiver is at fixed (u_R, ϕ_R).)

(e)   For an emitter in a circular orbit $u_E = $ const., $\phi_E = \Omega t_E$, plot the relativistic corrections to the arrival time interval between successive pulses as a function of observer 'time', Ωt_R. Comment on the use of this graph, in view of the original assumption $M/b \ll 1$.

18   Use the expression for distances on a sphere, Eq. (10.2), to show that all the points on the line $\theta = 0$ in Fig. 11.8 are the same physical point.

19 Derive Eqs. (11.52) and (11.53).

20(a) Using the Schwarzschild metric, compute all the nonvanishing Christoffel symbols:

$$\Gamma^t_{rt} = -\Gamma^r_{rr} = \frac{M}{r^2}\left(1 - \frac{2M}{r}\right)^{-1}; \qquad \Gamma^r_{tt} = \frac{M}{r^2}\left(1 - \frac{2M}{r}\right),$$

$$\Gamma^r_{\theta\theta} = \Gamma^r_{\phi\phi}/\sin^2\theta = -r\left(1 - \frac{2M}{r}\right),$$

$$\Gamma^\theta_{\theta r} = \Gamma^\phi_{\phi r} = \Gamma^\theta_{\phi\phi}/\sin^2\theta = \frac{1}{r},$$

$$\Gamma^\phi_{\theta\phi} = \cot\theta.$$

(11.117)

Show that all others vanish or are obtained from these by symmetry. (In your argument that some vanish, you should use the symmetries $t \to -t$, $\phi \to -\phi$, under either of which the metric is invariant.)

(b) Use (a) or the result of Exer. (6.35) to show that the only nonvanishing components of the Riemann tensor are

$$R^t_{rtr} = -2\frac{M}{r^3}\left(1 - \frac{2M}{r}\right)^{-1},$$

$$R^t_{\theta t\theta} = R^t_{\phi t\phi}/\sin^2\theta = M/r^5,$$

$$R^\theta_{\phi\theta\phi} = 2M\sin^2\theta/r^5,$$

$$R^r_{\theta r\theta} = R^r_{\phi r\phi}/\sin^2\theta = -M/r^5,$$

(11.118)

plus those obtained by symmetries of the Riemann tensor.

(c) Convert these components to an *orthonormal* basis aligned with the Schwarzschild coordinates. Show that all components fall off as r^{-3} for large r.

(d) Compute $R^{\alpha\beta\mu\nu}R_{\alpha\beta\mu\nu}$, which is independent of the basis, and show that it is singular as $r \to 0$.

21 A particle of $m \neq 0$ falls radially toward the horizon of a Schwarzschild black hole of mass M. The geodesic it follows has $\tilde{E} = 0.95$.

(a) Find the proper time required to reach $r = 2M$ from $r = 3M$.

(b) Find the proper time required to reach $r = 0$ from $r = 2M$.

(c) Find, on the Schwarzschild coordinate basis, its four-velocity components at $r = 2.001\,M$.

(d) As it passes $2.001\,M$, it sends a photon out radially to a distant stationary observer. Compute the redshift of the photon when it reaches the observer. Don't forget to allow for the Doppler part of the redshift caused by the particle's velocity.

22 A measure of the tidal force on a body is given by the equation of geodesic deviation, Eq. (6.87). If a human will be crushed when the acceleration gradient across its body is $400\ \text{m s}^{-2}$ per meter, calculate the minimum mass Schwarzschild black hole that would permit a human to survive long enough to reach the horizon on the trajectory in Exer. 21.

23　Prove Eq. (11.60).

24　Show that spacetime is locally flat at the center of the Kruskel–Szekeres coordinate system, $u = v = 0$ in Fig. 11.10.

25　Given a spherical star of radius $R \gg M$ and mean density ρ, estimate the tidal force across it which would be required to break it up. Use this as in Exer. 22 to define the tidal radius R_T of a black hole of mass M_H: the radius at which a star of density ρ near the hole will be torn apart. For what mass M_H is $R_T = 100\, M_H$ if $\rho = 10^3\, \text{kg m}^{-3}$, typical of our Sun? This illustrates that even some applications of black holes in astrophysical contexts require few 'relativistic' effects.

26　Given the area of a Kerr hole, Eq. (11.78), with r_+ defined in Eq. (11.74), show that any two holes with masses m_1 and m_2 and angular momenta $m_1 a_1$ and $m_2 a_2$ respectively have a total area less than that of a single hole of mass $m_1 + m_2$ and angular momentum $m_1 a_1 + m_2 a_2$.

27　Show that the 'static limit', Eq. (11.73), is a limit on the region of spacetime in which curves with r, θ, and ϕ constant are timelike.

28(a)　Prove Eq. (11.80).
　(b)　Derive Eq. (11.82).

29　In the Kerr metric, show (or argue on symmetry grounds) that a geodesic which passes through a point in the equatorial 'plane' ($\theta = \pi/2$) and whose tangent there is tangent to the plane ($p^\theta = 0$) remains always in the plane.

30　Derive Eqs. (11.84) and (11.85).

31　Show that a ZAMO has four-velocity components $U^0 = A$, $U^\phi = \omega A$, $U^r = U^\theta = 0$, $A = g_{\phi\phi}(-D)$, where D is defined in Eq. (11.80).

32　Show, as argued in the text, that the Penrose process decreases the angular momentum of the hole.

33　Derive Eq. (11.94) from Eq. (11.93).

34(a)　Use the area theorem to calculate the *maximum* energy released when two Schwarzschild black holes of mass M collide to form a Schwarzschild hole.
　(b)　Do the same for holes of mass m_1 and m_2, and express the result as a percentage of m_1 when $m_1 \to 0$ for fixed m_2.

35　The Sun rotates with a period of approximately 25 days.
(a)　Idealize it as a solid sphere rotating uniformly. Its moment of inertia is $\frac{2}{5} M_\odot R_\odot^2$, where $M_\odot = 2 \times 10^{30}\, \text{kg}$ and $R_\odot = 7 \times 10^8\, \text{m}$. In SI units compute $J_\odot$.

(b)   Convert this to geometrized units.

(c)   If the entire Sun suddenly collapsed into a black hole, it would form a Kerr hole of mass $M_\odot$ and angular momentum $J_\odot$. What would be the Kerr parameter, $a_\odot = J_\odot / M_\odot$, in cm? What is the ratio $a_\odot / M_\odot$? Physicists expect that a Kerr hole will *never* be formed with $a > M$, because centrifugal forces will halt the collapse or create a rotational instability. The result of this exercise is that even a quite ordinary star like the sun needs to get rid of angular momentum before forming a black hole.

(d)   Does an electron have too much angular momentum to form a Kerr hole with $a < M$? (Neglect its charge.)

36(a)   For a Kerr black hole, prove that for fixed M, the largest area is obtained for $a = 0$ (Schwarzschild).

(b)   Conversely, prove that for fixed area, the smallest mass is obtained for $a = 0$.

37(a)   An observer sits at constant r, ϕ in the equatorial plane of the Kerr metric ($\theta = \pi/2$) outside the ergoregion. He uses mirrors to cause a photon to circle the hole along a circular path of constant r in the equatorial plane. Its world line is thus a null line with $dr = d\theta = 0$, but it is not, of course, a geodesic. How much coordinate time t elapses between the emission of a photon in the direction of increasing ϕ and its receipt after it has circled the hole once? Answer the same for a photon sent off in the direction of decreasing ϕ, and show that this is a different amount of time. Does the photon return redshifted from its original frequency?

(b)   A different observer rotates about the hole on an orbit of $r =$ const. and angular velocity given by Eq. (11.70). Using the same arrangement of mirrors, he measures the coordinate time that elapses between his emission and his receipt of a photon sent in either direction. Show that in this case the two terms are *equal*. (This is a ZAMO, as defined in the text.)

38   Consider equatorial motion of particles with $m \neq 0$ in the Kerr metric. Find the analogues of Eqs. (11.84)–(11.88) using $\tilde{E}$ and $\tilde{L}$ as defined in Eqs. (11.2) and (11.3). Plot $\tilde{V}_\pm$ for $a = 0.5\ M$ and $\tilde{L}/M = 20$, 12, and 6. Discuss the qualitative features of the trajectories. For arbitrary a determine the relations among $\tilde{E}$, $\tilde{L}$, and r for circular orbits with either sense of rotation. What is the minimum radius of a stable circular orbit? What happens to circular orbits in the ergosphere?

39(a)   Derive Eq. (11.102) from Eq. (11.98) and the black-body law, luminosity $= \sigma A T^4$, where A is the area and σ is the Stefan–Boltzmann radiation constant, $\sigma = 0.567 \times 10^{-7}\ \mathrm{Wm^{-2}(K)^{-4}}$.

(b)   How small must a black hole be to be able to emit substantial numbers of electron–positron pairs?

12

Cosmology

12.1 What is cosmology?

The universe in the large. Newtonian theory is an adequate description of gravity as long as, roughly speaking, the mass M of a system is small compared to the size, $R : M/R \ll 1$. Conversely, GR may be expected to be important where either R becomes small faster than M or M becomes large faster than R. The first case is that of compact or collapsed objects: neutron stars and black holes have masses typical of ordinary stars (though larger black holes may, of course, exist), but much smaller radii. The second case is cosmology: if space is filled with matter of roughly the same density everywhere, then, as we consider volumes of larger and larger radius R, the mass increases as R^3 and M/R eventually must get so large that GR becomes important.

What length scale is this? Suppose we begin increasing R from the center of our Sun. The Sun is nowhere relativistic, and once R is larger than $R_{\odot}$, M hardly increases at all until the next star is reached. The system of stars of which the Sun is a minor member is a galaxy, and contains some 10^{11} stars in a radius of about 15 kpc. (One parsec, abbreviated pc, is about 3×10^{16} m.)

For this system, $M/R \sim 10^{-6}$, similar to that for the Sun itself. So galactic dynamics has no need for relativity. (This applies to the galaxy as a whole: small regions may be dominated by black holes or other relativistic objects.)

When we go to larger scales than the size of a galaxy we enter the domain of *cosmology*.

In the cosmological picture, galaxies are very small-scale structures, mere atoms in the larger universe. Even clusters of galaxies, which can have thousands of members, are mere density fluctuations. Our telescopes are capable of seeing to distances of the order of 10^{11} pc; the diameter of a cluster of galaxies, typically some 10^6 pc, is much smaller. One finds that, if one averages over distances of, say, 10^9 pc, the universe seems to be pretty much the same everywhere (Tammann *et al.* 1980). On the large scale, the universe seems to have roughly the same density everywhere. This density is very poorly measured. It is at least 10^{-28} kg m^{-3} but may be considerably larger. (It is uncertain because we can measure directly only the matter that sends electromagnetic radiation to us. There are indirect indications that the amount of 'dark' matter may substantially exceed 10^{-28} kg m^{-3}. See Peebles (1971 and 1980) for a discussion.) Taking this density, then, $M = 4\pi\rho R^3/3$ is equal to R for $R \sim 10^{27}$ m $\sim 10^4$ Mpc. We can certainly study objects, such as quasars, at distances of this order, so to understand the universe that our telescopes reveal to us, we need GR.

Homogeneity and isotropy of the universe. The simplest approach to applying GR is to use the remarkable large-scale uniformity we observe. We see, on scales of 10^3 Mpc, not only a uniform average density but uniformity in other properties: types of galaxies, their clustering densities, their chemical composition and stellar composition. We therefore conclude that, on the large scale, the universe is *homogeneous*. What is more, on this scale the universe seems to be *isotropic* about every point. What this means is that it is not possible by making local observations to distinguish one direction on the sky from another. (A universe *could* be homogeneous but anisotropic, if, for instance, it had a large-scale magnetic field which pointed in one direction everywhere and whose magnitude was the same everywhere. On the other hand, an inhomogeneous universe could not be isotropic about every point, since most – if not all – pieces in the universe would see a sky that is 'lumpy' in one direction and not in another.) A third feature of the observable universe is its expansion: all galaxies, on average, seem to be receding from us at a speed which is proportional to their distance from us. This is easily visualized in the 'balloon' model (see Fig. 12.1). Paint dots on a balloon and then inflate it. As it grows, the distance on the surface of the balloon between any two points grows at a rate proportional to that distance. Therefore *any* point will see all other points receding at a rate propor-

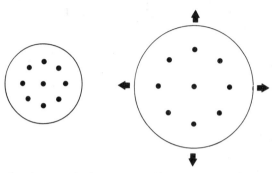

Fig. 12.1   As the figure is magnified, all relative distances increase at a rate proportional to their magnitudes.

tional to their distance. This recessional velocity, called the 'Hubble flow' after its discoverer, gives another opportunity for anisotropy. The universe would be homogeneous and anisotropic if every point saw a recessional velocity larger in, say, the x direction than in the y direction. This does not appear to be the case in our universe, however. The constant of proportionality H in the equation relating recessional velocity and distance,

$$v = Hd \tag{12.1}$$

is called Hubble's constant and has the value (75 ± 25) km/s/Mpc in the astronomer's peculiar but useful units. (In normal units it is 2.5×10^{-18} s^{-1}; in geometrized units 8.3×10^{-27} m^{-1}.) It has this value in all directions on the sky, to quite high accuracy. This gives a good argument for isotropy of the universe about us; if we are 'typical' (the Copernican viewpoint), then the universe should be isotropic about *every* point and, therefore, must be homogeneous.

One may object that the above discussion ignores the relativity of simultaneity. If the universe is changing in time – expanding – then it may be possible to find *some* definition of time such that hypersurfaces of constant time are homogeneous and isotropic, but this would not be true for other choices of a time coordinate. Moreover, Eq. (12.1) cannot be exact since, for $d > 1.2 \times 10^{26}$ m $= 4000$ Mpc, the velocity exceeds the velocity of light! These objections are right on both counts. Our discussion was a *local* one (applicable for recessional velocities $\ll 1$) and took the point of view of a particular observer, ourselves. Fortunately, the cosmological expansion is slow, so that over distances of 1000 Mpc, encompassing most visible ordinary galaxies, the velocities are essentially nonrelativistic. Moreover, the average random velocities of galaxies relative to their near neighbors is less than 500 km s^{-1}, which is very nonrelativistic.

So one can properly say that in our neighborhood of the universe there exists a preferred choice of time, whose hypersurfaces are homogeneous and isotropic, and with respect to which Eq. (12.1) is valid.

Models of the universe: the cosmological principle. If we are to make a large-scale model of the universe, we must make some assumption about regions that we have no way of seeing now because they are too distant for our telescope. One must in fact distinguish two different inaccessible regions of the universe. The first is the region which is so distant that no information (traveling on a null geodesic) could reach us from it no matter how early this information began traveling. Such regions usually exist if the universe has a finite age, as ours does. (See Fig. 12.2.) These

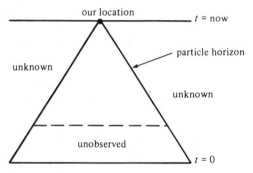

Fig. 12.2 Schematic spacetime diagram showing the past history of the Universe, back to $t = 0$. The 'unknown' regions have not had time to send us information; the 'unobserved' regions are obscured by intervening matter.

'unknown' regions are unimportant in one respect: they have no effect on the interior of our past light cone, so how we incorporate them into our model universe has no effect on the way it describes our history. On the other hand, our past light cone is a kind of horizon, called 'the particle horizon': every moment, more and more of the 'unknown' enters the interior and becomes known. So the unknown regions can have a real influence on our future. In this sense cosmology is a retrospective study: it reliably helps us understand only our past. It must be acknowledged, however, that if information began coming in tomorrow that yesterday's 'unknown' region was in fact very inhomogeneous, then we would be posed difficult physical and philosophical questions regarding the apparently special nature of our history until now. It is to avoid these difficulties that a good number of scientists believe in the homogeneity and isotropy of the unknown regions. This is called the cosmological principle, or the assumption of mediocrity: the ordinary-ness of our own location in the universe. It is, mathematically, a powerful assumption. We shall use it, bearing in mind its limited predictive power.

Another unknown region is that part of the interior of our past light cone which our instruments cannot get information about. This includes galaxies so distant that they are too dim to be seen; processes that give off radiation – like gravitational waves – which we do not have the sophistication to detect; and events that are masked from view, such as those which emitted electromagnetic radiation before recombination (see below). None of this is, in principle, unknowable, and the study of cosmology can in fact help us to understand what went on in regions not directly observable, by drawing conclusions from observable effects these events have had (i.e. indirect observations). But one must be careful of one fundamental limitation: the very early universe is not yet directly accessible to our instruments, so we have no direct knowledge that it had the high degree of isotropy and homogeneity that the present universe has. Any theory framed in a homogeneous, isotropic model must be treated cautiously. We do not have the time here to study anisotropic or inhomogeneous models of the early universe, but this is a very active field of research today.

12.2   General relativistic cosmological models

Robertson–Walker metrics. We shall adopt the following assumptions about the universe: (i) spacetime can be sliced into hypersurfaces of constant time which are perfectly homogeneous and isotropic; and (ii) the mean rest frame of the galaxies agrees with this definition of simultaneity. Let us therefore adopt comoving coordinates: each galaxy is idealized as having no random velocity, as it has a fixed set of coordinates $\{x^i, i = 1, 2, 3\}$. The time coordinate is t, the proper time for each galaxy. The expansion of the universe – the change of proper distance between galaxies – is represented by time-dependent metric coefficients. Thus, if at one moment, t_0, the hypersurface of constant time has the line element

$$\mathrm{d}l^2(t_0) = h_{ij}(t_0)\, \mathrm{d}x^i\, \mathrm{d}x^j \tag{12.2}$$

(these hs have nothing to do with linearized theory), then the expansion of the hypersurface can be represented by

$$\begin{aligned} \mathrm{d}l^2(t_1) &= f(t_1, t_0) h_{ij}(t_0)\, \mathrm{d}x^i\, \mathrm{d}x^j \\ &= h_{ij}(t_1)\, \mathrm{d}x^i\, \mathrm{d}x^j. \end{aligned} \tag{12.3}$$

Here we have assumed that all the h_{ij}s increase at the same rate; otherwise the expansion would be anisotropic (see Exer. 3, § 12.6). In general, then, Eq. (12.2) can be written

$$dl^2(t) = R^2(t)h_{ij} \, dx^i \, dx^j, \tag{12.4}$$

where R is an overall scale factor which equals 1 at t_0, and where h_{ij} is a constant metric equal to that of the hypersurface at t_0. We shall explore h_{ij} in detail in a moment.

First we extend the constant-time line element to a line element for the full spacetime. In general, it would be

$$ds^2 = -dt^2 + g_{0i} \, dt \, dx^i + R^2(t)h_{ij} \, dx^i \, dx^j, \tag{12.5}$$

where $g_{00} = -1$, because t is proper time along a line $dx^i = 0$. However, if the definitions of simultaneity given by $t = \text{const.}$, and by the local Lorentz frame of a galaxy, are to agree (assumption (ii) above), then $\vec{e}_0$ must be orthogonal to $\vec{e}_i$ in our comoving coordinates. This means that $g_{0i} = \vec{e}_0 \cdot \vec{e}_i$ must vanish, and we get

$$ds^2 = -dt^2 + R^2(t)h_{ij} \, dx^i \, dx^j. \tag{12.6}$$

What form can h_{ij} take? It must, first of all, be isotropic about every point, in particular spherically symmetric about the origin of the coordinates. When we discussed spherical stars we showed that such a metric always has the line element

$$dl^2 = e^{2\Lambda(r)} \, dr^2 + r^2 \, d\Omega^2. \tag{12.7}$$

This follows only from isotropy at one point. Now, isotropy about every point implies homogeneity. In particular, if the three-space is curved, the Ricci scalar curvature $R^i{}_i$ must have the same value at every point. We can calculate $R^i{}_i$ using Eqs. (10.15)–(10.17) of our discussion of spherically symmetric spacetimes in Ch. 10, by realizing that G_{ij} for the line element, Eq. (12.7), above is obtainable from G_{ij} for the line element, Eq. (10.7), of a spherical star by setting Φ to zero. One gets

$$\begin{aligned} G_{rr} &= -\frac{1}{r^2} e^{2\Lambda}(1 - e^{-2\Lambda}), \\ G_{\theta\theta} &= -r \, e^{-2\Lambda}\Lambda', \\ G_{\phi\phi} &= \sin^2 \theta G_{\theta\theta}. \end{aligned} \tag{12.8}$$

The Ricci scalar curvature is found from

$$\begin{aligned} G_{ij} &= R_{ij} - \tfrac{1}{2}g_{ij}R, \\ G^i{}_i &= R^i{}_i - \tfrac{1}{2}(3) R = -\tfrac{1}{2}R, \\ R &= -2G_{ij}g^{ij}. \end{aligned} \tag{12.9}$$

Then we get

$$R = -2\left[-\frac{1}{r^2}e^{2\Lambda}(1-e^{-2\Lambda})e^{-2\Lambda} - 2r\,e^{-2\Lambda}\Lambda' r^{-2}\right]$$

$$= \frac{2}{r^2} - \frac{2}{r^2}e^{-2\Lambda}(1-2r\Lambda')$$

$$= \frac{2}{r^2}[1-(r\,e^{-2\Lambda})'].\tag{12.10}$$

Demanding homogeneity means setting R to some constant k:

$$k = \frac{2}{r^2}[r(1-e^{-2\Lambda})]'.\tag{12.11}$$

This is easily integrated to give

$$g_{rr} = e^{2\Lambda} = \frac{1}{1-\frac{1}{6}kr^2 - A/r},\tag{12.12}$$

where A is a constant of integration. As in the case of spherical stars, we must demand local flatness at $r = 0$: $g_{rr}(r = 0) = 1$. This implies $A = 0$. Re-defining the constant k gives

$$g_{rr} = \frac{1}{1-kr^2}$$

$$dl^2 = \frac{dr^2}{1-kr^2} + r^2\,d\Omega^2.\tag{12.13}$$

We have not yet proved that this space is isotropic about every point; all we have shown is that Eq. (11.13) is the unique space which satisfies the necessary condition that its scalar curvature be homogeneous. Thus, if a space that is isotropic and homogeneous exists at all, it must have the metric, Eq. (12.13), for some k. In fact, the converse *is* true: the metric of Eq. (12.13) is homogeneous and isotropic for all k. General proofs can be found in, for example, Weinberg (1972) or Schutz (1980*b*). We will demonstrate it explicitly for positive, negative, and zero k separately in the next section. Therefore, the cosmological spacetime has the metric

$$ds^2 = -dt^2 + R^2(t)\left[\frac{dr^2}{1-kr^2} + r^2\,d\Omega^2\right].\tag{12.14}$$

This is called the Robertson–Walker metric. Notice that we can, without loss of generality, scale the coordinate r in such a way as to make k take one of the three values $+1$, 0, -1, For, consider $k = -3$. Then re-define $\bar{r} = \sqrt{3}\,r$ and $\bar{R} = 1/\sqrt{3}R$, and the line element becomes

$$ds^2 = -dt^2 + \bar{R}^2(t)\left[\frac{d\bar{r}^2}{1-\bar{r}^2} + \bar{r}^2\,d\Omega^2\right].\tag{12.15}$$

What one cannot do is change the sign of k. Therefore there are only three spatial hypersurfaces we need consider. We do this next.

Three types of universe. Consider first $k = 0$. Then, at any moment t_0, the line element with $dt = 0$ is

$$dl^2 = d\bar{r}^2 + \bar{r}^2 \, d\Omega^2, \tag{12.16}$$

with $\bar{r} = R(t_0)r$. This is obviously the metric of flat Euclidean space. This is the *flat* Robertson–Walker universe. That it is homogeneous and isotropic is obvious.

Consider, next, $k = +1$. Let us define a new coordinate $\chi(r)$ such that

$$d\chi^2 = \frac{dr^2}{1 - r^2}. \tag{12.17}$$

This implies that

$$r = \sin \chi \tag{12.18}$$

and that the line element for the space $t = t_0$ is

$$dl^2 = R^2(t_0)[d\chi^2 + \sin^2 \chi \, (d\theta^2 + \sin^2 \theta \, d\phi^2)]. \tag{12.19}$$

We showed in Exer. 33, § 6.9, that this is the metric of a three-sphere of radius $R(t_0)$. This model is called the *closed*, or *spherical* Robertson–Walker metric and the balloon analogy of cosmological expansion is particularly appropriate (Fig. 12.1) for it. Remember that the fourth spatial dimension – the radial direction to the center of the three-sphere – has *no* physical meaning to us: no known physical law permits any measurements in that dimension.

The final possibility is $k = -1$. A similar coordinate transformation (Exer. 7, § 12.6) gives the line element

$$dl^2 = R^2(t_0)(d\chi^2 + \sinh^2 \chi \, d\Omega^2). \tag{12.20}$$

This is called the *hyperbolic*, or *open*, Robertson–Walker model. Notice one peculiar property. As the proper radial coordinate χ increases away from the origin, the circumferences of spheres increase as $\sinh \chi$. Since $\sinh \chi > \chi$ for all $\chi > 0$, it follows that these circumferences increase *more* rapidly with proper radius than in flat space. For this reason this hypersurface is *not* realizable as a three-dimensional hypersurface in a four- or higher-dimensional Euclidean space. That is, there is no picture which we can easily draw like that for the three-sphere. The space is call 'open' because, unlike $k = +1$, circumferences of spheres increase monotonically with χ: there is no natural end to the space.

In fact, as we show in Exer. 7, § 12.6, this geometry *is* the geometry of a hypersurface in Minkowski space: specifically, a hypersurface of constant timelike interval from the origin. Since this hypersurface has the same interval from the origin in any Lorentz frame (intervals are Lorentz invariant), this hypersurface is indeed homogeneous and isotropic.

Dynamics of Robertson–Walker universes: the big bang. Until now we have simply cataloged possible geometries; now we ask how Einstein's equations predict how they should behave. We idealize the universe as filled with a perfect fluid $\rho = \rho(t)$, $p = p(t)$, etc. First, consider $T^{\mu\nu}{}_{;\nu} = 0$. Since there is spatial homogeneity, only the time component of this equation is nontrivial. It is easy to show that it gives

$$\frac{d}{dt}(\rho R^3) = -p\frac{d}{dt}(R^3),\tag{12.21}$$

where $R(t)$ is the cosmological expansion factor. This is easily interpreted: R^3 is proportional to the volume of any fluid element, so the left-hand side is the rate of change of its total energy, while the right-hand side is the work it does as it expands ($-p\,dV$). There are two cases of interest in cosmology, the matter-dominated and radiation-dominated eras. In the matter-dominated era, which is the present epoch, the main energy density is that of ordinary matter in galaxies, whose random velocities are small and which therefore behave like dust: $p = 0$. So we have

$$\text{Matter-dom:} \quad \frac{d}{dt}(\rho R^3) = 0.\tag{12.22}$$

In a radiation-dominated era (as we shall see, in the early universe) the principal energy density is in radiation or relativistic particles, which have an equation of state $p = \frac{1}{3}\rho$ (Exer. § 4.10). Then we get

$$\text{Radiation-dom:} \quad \frac{d}{dt}(\rho R^3) = -\frac{1}{3}\rho\frac{d}{dt}(R^3),\tag{12.23}$$

or

$$\text{Radiation-dom:} \quad \frac{d}{dt}(\rho R^4) = 0.\tag{12.24}$$

The Einstein equations are also easy to write down. The Einstein tensor for Eq. (12.14) has a time component,

$$G_{tt} = 3(\dot{R}/R)^2 + 3k/R^2\tag{12.25}$$

and also r–r, θ–θ, and ϕ–ϕ components. These last three are all proportional, by isotropy, so contribute only one equation. And even this equation gives no new information, since it is obtained from the G_{tt} equation via the Bianchi identities. (The same happened for the spherical star.) Therefore, besides Eqs. (12.22) or (12.24), we have only

$$G_{tt} = 8\pi T_{tt},\tag{12.26}$$

or

$$(\dot{R}/R)^2 = -k/R^2 + \tfrac{8}{3}\pi\rho.\tag{12.27}$$

If one wants to include the cosmological constant in Eq. (12.26) it is easy to do so; however, we will reserve this for an exercise (Exer. 9, § 12.6).

Let us see what the equations predict for each era. For matter-dominated universes we have $\rho = A/R^3$, where A is a constant, by Eq. (12.22). This transforms Eq. (12.27) into

$$(\dot{R}/R)^2 = (\tfrac{8}{3}\pi A)\frac{1}{R^3} - \frac{k}{R^2},$$

or

$$(\dot{R})^2 = \tfrac{8}{3}\pi A\frac{1}{R} - k.\tag{12.28}$$

We can use effective-potential techniques here. Define

$$(\dot{R})^2 = (-k) - V_M(R), \qquad V_M = -\tfrac{8}{3}\pi A/R;\tag{12.29}$$

$V_M(R)$ is plotted in Fig. 12.3. Then the universe can only exist in regions where $-k$ exceeds $V_M(R)$, so that $(\dot{R})^2 > 0$. Since we know that, at present,

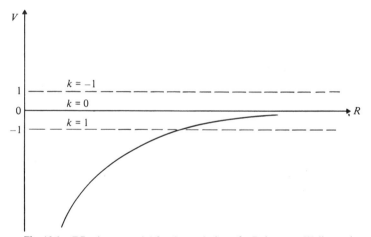

Fig. 12.3   Effective potential for the evolution of a Robertson–Walker universe. The different values of k play the role of different values of energy in analogous diagrams like Fig. 11.1.

$\dot{R} > 0$, then there are three possible futures: if $k = -1$, the universe expands to infinite radius with finite terminal velocity; if $k = 0$, the universe expands to infinite radius with ever decreasing speed; and if $k = +1$ it reaches a maximum radius $R = \tfrac{8}{3}\pi A$, at which it has a turning point, and then it re-collapses. Perhaps more interesting is the past history of all of these: *all* of them originate at $R = 0$; none has a turning point for small R. This is the 'Big Bang'.

If the universe is today described by Fig. 12.3, then, much earlier, it must have been smaller, denser and hotter. Since the density of matter is proportional to R^{-3}, while that of radiation to R^{-4} (Eqs. (12.22) and (12.24)), there will be some radius at which the two will have been equal. Earlier than that, we can, to a good approximation, use the radiation-dominated equation of state. Then we can write $\rho = BR^{-4}$, and Einstein's equations, Eq. (12.27), give

$$\dot{R}^2 = \tfrac{8}{3}\pi B / R^2 - k. \tag{12.30}$$

Qualitatively, this differs in no way from Eq. (12.28), so we are indeed permitted to conclude that the universe originated at $R = 0$, provided, of course, it is adequately described as homogeneous and isotropic back then. We'll come back to this point. First we must ask whether $R = 0$ was a finite or infinite proper time in the past. From Eq. (12.30) we can see that if R is sufficiently small, then k is negligible, i.e. all universes have the same initial dynamics, and we are dealing with

$$\dot{R}^2 = \tfrac{8}{3}\pi BR^{-2},$$

or

$$\frac{dR}{dt} = (\tfrac{8}{3}\pi B)^{1/2} R^{-1}. \tag{12.31}$$

This has the solution

$$R^2 = (\tfrac{32}{3}\pi B)^{1/2} t + \text{const.} \tag{12.32}$$

So, indeed, $R = 0$ was achieved at a finite time in the past, and we conventionally adjust our zero of time so that $R = 0$ at $t = 0$.

The conceptual and philosophical implications of this result are enormous; we have simply no framework within which we can meaningfully discuss earlier times in this model. How certain, then, is our conclusion that the universe began with a Big Bang? First, one must ask if isotropy and homogeneity were crucial; the answer is no. The 'singularity theorems' of Penrose and Hawking (see Hawking & Ellis 1973) have shown that our universe certainly had a singularity in its past, regardless of how asymmetric it may have been. But the theorems predict only the *existence* of the singularity: the nature of the singularity is unknown, except that it has the property that at least one particle in the present universe must have originated in it. Nevertheless, the evidence is strong indeed that we *all* originated in it. Another consideration however is that we don't know the laws of physics at the incredibly high densities ($\rho \to \infty$)

which existed in the early universe. The singularity theorems of necessity assume (1) something about the nature of $T^{\mu\nu}$, and (2) that Einstein's equations (without cosmological constant) are valid at all R. Almost certainly a proper theory of quantum gravity and an increasing understanding of high-energy physics will revise our view of the early universe to some degree. We will come back to this point at the end of § 12.4. For now, we will just remark that the very earliest epochs are not understood with any confidence.

12.3 Cosmological observations

Observations in an expanding universe. Before we discuss the physical history of the universe insofar as our understanding of physics allows, we must learn how observational data are gathered and what they tell us about the universe. The objects observed are on the observer's past light cone, not on his preferred $t = \text{const.}$ hypersurface, and this fact complicates the analysis of the observations in two ways. First, the universe changes along the past light cone: because of its expansion, it is homogeneous only on a $t = \text{const.}$ hypersurface. Second, astronomical objects evolve in time in ways which are very incompletely understood, so it is dangerous to assume that a distant object, say a giant elliptical galaxy, shares all the properties of similar nearby objects.

Deceleration parameter. To illustrate the importance of these two problems, and the general subtlety of cosmological analysis, we shall look at a fundamental problem: deducing the mass density of the universe from its deceleration. A fuller discussion is in Weinberg (1972). The general idea is that the Hubble law, Eq. (12.1), tells us how fast the universe is expanding; if we extend that law to very distant objects we should see the expansion changing with time, so we should be able to see the expansion slowing down. This in turn is related to the mass density through the time derivative of Eq. (12.28):

$$\dot{R}\ddot{R} = -\tfrac{4}{3}\pi A R^{-2}\dot{R} = -\tfrac{4}{3}\pi\rho\dot{R}R.$$

It is convenient to define the dimensionless *deceleration parameter*

$$q \equiv -R\ddot{R}/\dot{R}^2 \tag{12.33}$$

in terms of which we have

$$\rho = \frac{3}{4\pi}\, q(\dot{R}/R)^2. \tag{12.34}$$

Recall that the scale factor R contains an arbitrary multiplicative constant, so that only ratios in which that constant cancels out are measurable; q and ρ are evidently measurable. Notice also that Eq. (12.27) picks out a certain density as special: since $\dot{R}/R$ is measurable (it is just Hubble's constant, as we will see below), the kind of universe we live in (sign of k) is determined by the magnitude of ρ relative to the *critical density* ρ_c for which $k = 0$:

$$\rho_c = \frac{3}{8\pi}(\dot{R}/R)^2. \tag{12.35}$$

The ratio of the universe's true density to the critical is called Ω

$$\Omega = \rho/\rho_c = 2q. \tag{12.36}$$

So if we can measure q, the deceleration, we can deduce the total density of the universe ρ and, incidentally, the global structure of the universe (if the cosmological principle applies): if $\Omega < 1$ the universe is open, if $\Omega > 1$ it is closed! (If Ω is so close to 1 that local density fluctuations caused by clustering of galaxies make it larger than 1 in some places and smaller in others, then the simple Friedman model will break down eventually.)

Let us anticipate a later result by denoting

$$\dot{R}/R \equiv H(t). \tag{12.37}$$

If we Taylor-expand $R(t)$ about the present time t_0, then we may write

$$R(t) = R_0[1 + H_0(t - t_0) - \tfrac{1}{2}q_0 H_0^2(t - t_0)^2 + \cdots], \tag{12.38}$$

where subscripted zeroes denote quantities evaluated at t_0. What does Hubble's law, Eq. (12.1), look like to this accuracy? The recessional velocity v is deduced from the redshift of spectral lines, so it is more convenient to work directly with the redshift. Then Exer. 12, § 12.6 shows that

$$1 + z = R_0/R(t), \tag{12.39}$$

where t is the time of emission of the light observed now. From Eq. (12.38) we then have

$$z(t) = H_0(t_0 - t) + (1 + q_0/2)H_0^2(t_0 - t)^2 + \cdots. \tag{12.40}$$

This is not directly useful yet, since we have no independent information about the time t at which a galaxy emitted its light. Perhaps Eq. (12.40) is more useful when inverted:

$$t_0 - t = z(t)/H_0 - (1 + q_0/2)z(t)^2/H_0 + \cdots. \tag{12.41}$$

Luminosity distance. By analogy with Eq. (12.1), we would like to replace t in Eq. (12.40) with distance. But what distance? Not coordinate distance, which would be unmeasurable. What about proper distance? The proper distance between the events of emission and reception of the light is zero, since light travels on null lines. The proper distance between the emitting galaxy and us at the present time is also unmeasurable: in principle, the galaxy may not even exist now, perhaps because of a collision with another galaxy. To get out of this difficulty, let us ask how distance crept into Eq. (12.1) in the first place. Distances of nearby galaxies are always inferred from luminosity measurements. An object whose absolute luminosity L is known (say from a theory of its nature) is observed in a galaxy; its flux F is measured and its distance d deduced from the relation

$$L = 4\pi d^2 F. \tag{12.42}$$

The role of d in Eq. (12.1) is, then, as a replacement for the observable $(L/F)^{1/2}$. So what we should be aiming for is to replace t in Eq. (12.40) by L/F, which would then relate the observables L/F and z to the parameter q_0 that we wish to deduce.

We must therefore answer the question: What is the flux of an object of luminosity L, at time t, which is observed at time t_0? Suppose for simplicity that the object gives off only photons of frequency ν_e at time t_e. In a small interval of time δt_e it emits

$$N = L\delta t_e/h\nu_e \tag{12.43}$$

such photons. We must calculate the area of the sphere that these photons occupy at a later time. From the line element, Eq. (12.14), with the spatial origin at the emitting event, a radial null line obeys the equation

$$dr/dt = (1 - kr^2)^{1/2}/R(t). \tag{12.44}$$

In the spirit of our earlier Taylor expansion we write this as

$$R_0\, dr = [1 - H_0(t - t_0)]\, dt.$$

(We keep only the accuracy that we shall later require.) This can be integrated to give

$$R_0 r_0 = (t_0 - t_e)[1 + \tfrac{1}{2}H_0(t_0 - t_e) + \cdots], \tag{12.45}$$

where r_0 is the radial coordinate of the photon at t_0. The N photons are therefore spread out over an area given by the spherical part of the line element, Eq. (12.14):

$$A = 4\pi(R_0 r_0)^2. \tag{12.46}$$

The photons have been redshifted by the amount given in Eq. (12.39) to frequency ν_0:

$$h\nu_0 = h\nu_e/(1 + z). \tag{12.47}$$

Moreover, they arrive spread out over a time δt_0 which is also redshifted:

$$\delta t_0 = \delta t_e(1 + z). \tag{12.48}$$

The energy flux at t_0 is thus

$$F = Nh\nu_0/(A\delta t_0) = L/A(1 + z)^2. \tag{12.49}$$

Rather than use the ratio L/F directly, it is conventional to retain the language of distance and define the *luminosity distance* d_L of an object by Eq. (12.42):

$$d_L = (L/4\pi F)^{1/2}. \tag{12.50}$$

This is the distance at which the object would be if it were stationary in Euclidean space with luminosity L and flux F. Then we have

$$d_L = (1 + z)R_0 r_0. \tag{12.51}$$

We can now eliminate t from our formulae. If we substitute Eq. (12.41) for $t_0 - t_e$ into Eq. (12.45) for $R_0 r_0$, and put the result into Eq. (12.51) with $(1 + z)$ replaced using Eq. (12.40), we obtain

$$H_0 d_L = z + \tfrac{1}{2}(1 - q_0)z^2 + \cdots. \tag{12.52}$$

Remembering that, to lowest order, $z = v$ and $d_L = d$, we see that this is indeed the law which Hubble found empirically, Eq. (12.1). Incidentally, it justifies our earlier identification of H with $\dot{R}/R$ in Eq. (12.37).

It is clear from Eq. (12.52) that we can measure q_0 if we can find an object sufficiently far away for which d_L is known, i.e. whose intrinsic luminosity is known. This is not easy, since few objects are understood well enough theoretically. Many candidates have been explored (see Peebles 1980), but, so far, the best bet seems to be supernovae in very distant galaxies (Kirschner 1976, Branch 1977, Wagoner 1980a). The required observations do not appear to be practicable until the Space Telescope is launched. When we do eventually have a reliable value for q_0, the problem will then be to account for the required density of 'dark' matter: the excess of $\rho = 3q_0 H_0^2/4\pi$ over the density of luminous matter that we can observe.

Eq. (11.52) illustrates the two difficulties mentioned earlier, that of accounting for the effect on observations of the expanding universe and that of knowing enough about objects at early times to apply the mathematical formulae. Its derivation illustrates another point which we have encountered before: in the attempt to translate the nonrelativistic formula $v = Hd$ into relativistic language, we were forced to re-think the meaning of all the terms in the equations and to go back to the quantities which one can directly measure. If the study of GR teaches us only one thing, it should be that physics rests ultimately on *measurements*: concepts like distance, time, velocity, energy and mass are derived from measurements, but they are often not the quantities directly measured, and one's assumptions about their global properties must be guided by a careful understanding of how they are related to measurements.

The extragalactic distance scale. The observational task of deducing d_L is often called building up the cosmological distance scale. It works as follows. Suppose that a certain class of stars is known to have intrinsic luminosity L_* and is observable at d_L between r and $10r$. Suppose another kind of star is observed in physical association with the more distant stars of the first class. Then its intrinsic luminosity L'_* may be deduced from the ratio of the two fluxes. If this is considerably larger than L_*, then the second class of stars may be observable at $d_L \gg 10r$. If, in addition, there is good reason to believe that the more distant stars of the second class have the same luminosity L'_*, then their distance can be deduced from their flux, even where stars of the first class are unobservable. We say that distances to stars of the second kind have been calibrated in terms of distance to stars of the first kind.

The most uncertain step is, of course, the assumption that the intrinsic luminosities L_* and L'_* are uniform within each class of objects. Another source of uncertainty is the distance to (or intrinsic luminosity of) the nearest objects in the distance scale. In practice, several classes of objects are used at each distance (see Table 12.1) but, at present, estimates of H_0 vary from 50 km s^{-1} Mpc^{-1} (Tammann *et al.* 1980) to 95 km s^{-1} Mpc^{-1} (de Vaucouleurs 1982). In view of this disagreement, we adopt the value $H = 75$ km s^{-1} Mpc^{-1} in this book, but regard it as uncertain by 30% or so. There are no reliable values of q_0.

Satellite-based measurements in the late 1980s may improve the situation considerably. The European satellite Hipparcos may provide considerably better data on parallaxes for calibrating the nearest of the objects in the first group in Table 12.1. The Space Telescope may be able to use the supernova method mentioned earlier to give good data for H_0 and moderate accuracy for q_0 (Wagoner 1980*a*). It should be noted that this method is an angular-diameter method rather than a luminosity method. The effective distance d_A that it defines is the ratio of the true diameter D to the apparent angular size θ of the supernova. The relation between d_A and z is deduced in Exer. 16, § 12.6.

Table 12.1. *Distance indicators*

Class of object	Where calibrated	Maximum distance used
(1) Cephied variables RR Lyrae variables Novae AB supergiants Eclipsing binaries	Within our Galaxy, i.e., by parallax and stellar-model computations	Local group of galaxies, $d_L < 1$ Mpc
(2) Bright supergiant stars Supergiant variables Globular clusters HII regions	Within our Galaxy and the local group, using (1) above	Nearby galaxies, $d_L < 8$ Mpc
(3) Total magnitudes of certain spiral galaxies Visual diameters of certain spiral galaxies Brightest galaxies in clusters Supernova maxima Bright superassociations 21 cm line width	Nearby galaxies	$d_L < 100$ Mpc, sufficient for calculating H_0

12.4   Physical cosmology

Physical regimes. Observations of our 'neighborhood' can, in principle, tell us what sort of universe we live in, but in order to understand the early history of the universe, one must solve the dynamical equations of GR using realistic physics for the stress–energy tensor. Because a full discussion of the physics is inappropriate in this introductory textbook, our discussion will be largely descriptive. Going back in time from the present, there are a number of major milestones, each of which presents its own physical problems: galaxy and star formation, recombination of the ionized plasma into neutral hydrogen, production of deuterium and helium, thermal energies exceeding those achievable in present-day particle accelerators, hypothetical 'grand unification' energy scales, and the quantum gravity regime. We shall treat each regime in turn.

Galaxy formation. If the universe were perfectly homogeneous, galaxies could not form. But they did indeed form, probably between redshifts of 100 and 10. (It is customary to use redshift as a delimiter of epochs. Radius or time can be deduced from Eq. (12.39) or Eq. (12.41). Redshift is more natural than time, since most physical problems depend on the density of the universe, which is a function only of z and ρ_0; by contrast, an epoch's time is uncertain because H_0 is uncertain.) The initial density perturbations which produced galaxies had to be much greater than those which might be attributed to random placement of hydrogen atoms within an overall smooth density, and we have very little idea about how these perturbations arose. Recent work (Peebles 1980) suggests that if we accept an initial random density perturbation, then the observed properties of galaxies – their clustering and angular momenta – may not be too hard to explain. But much observational work is needed before we can infer the initial perturbation spectrum with confidence. It is not known whether stars formed before or after galaxies began condensing, and whether spiral and elliptical galaxies have similar or radically different histories.

Decoupling and the microwave background. One of the most important cosmological observations was the accidental discovery of the cosmic microwave background radiation by Penzias and Wilson in 1965. (See Penzias 1979 and Wilson 1979 for their fascinating accounts of the discovery in their Nobel-prize acceptance speeches.) It is a radiation bath of nearly uniform temperature 2.7 K in which we and, apparently, the whole universe, are immersed. This radiation had actually been predicted by Gamow (1948) and Alpher & Herman (1949), but the prediction had been ignored. It is usually interpreted as the remnant of the radiation present in a high-temperature phase in the early universe.

The present temperature of the microwave background is 2.7 K, and increases inversely as $R(t)$ (as for any adiabatic compression of a photon gas: see Exer. 8, § 12.6), hence as $1 + z$. For $z \sim 3 \times 10^3$ the temperature will be such that the energy kT is enough to ionize hydrogen. In fact, because there are about 10^9 black-body photons for each hydrogen atom, ionization will be complete at rather lower temperatures, such that the high-energy tail of the black-body distribution contains enough photons to ionize the gas. This happens at $z \sim 1600$, depending on Ω and H_0 (Peebles 1971: see Exer. 11, § 12.6). For z less than this, photons move through neutral hydrogen or, later, empty space, and have a very small chance of being scattered. This means that the observed microwave photons come to us directly from $z \sim 1600$, and their isotropy implies the isotropy of the universe at that epoch.

In fact, the microwave background is not perfectly isotropic, but, rather, has a *dipole* anisotropy: temperature a maximum in one direction and a minimum in the opposite direction. This is easily explained as an effect of Earth's motion relative to the cosmological rest frame, a motion toward the direction of greatest temperature. Measurements indicate a velocity of about 500 km s^{-1} roughly toward the Virgo cluster (Smoot & Lubin 1979). Since this is very large compared to velocities within our Galaxy, the whole Galaxy must be falling toward the massive Virgo cluster at this rate. The dipole anisotropy does not indicate an anisotropy in the geometry of the cosmological model: this would require a quadrupole anisotropy, which has not been observed.

The fact that the present microwave radiation is decoupled from the matter means that the two fluids have evolved independently since decoupling. The radiation gas has retained a Planck spectrum with decreasing temperature. The hydrogen gas initially remained in rough thermal equilibrium, but with its own temperature independent of that of the radiation. But, soon, density perturbations caused different regions to separate, and hydrogen is therefore no longer in a global thermal equilibrium state.

Element production. If we go earlier still, we reach temperatures at which nuclear reactions can occur. At still earlier times, thermal energies will be enough to break up any nuclei, so that for $T > 10^{10} \text{ K}$ ($z > 3 \times 10^9$), one expects only protons, neutrons, electrons, neutrinos, and radiation. As the universe expands, first neutrons and later deuterium and helium 'freeze out' (Peebles 1971). If the universe is assumed to be homogeneous at this early epoch (for which we have little evidence), then one can calculate the final abundances of the various elements. These numbers are, in principle, sensitive to the rate at which the universe was then expanding (and hence to the present density), since abundances freeze out when the decreasing density and temperature cause reaction times to become long compared to the expansion time. Calculations (Wagoner 1980b) show that the abundance of ^4He is relatively insensitive to Ω, being of the order of 20% by mass relative to ^1H, in good agreement with the present observed abundance. The abundance of deuterium, ^2H, however, is very sensitive to Ω. Since the subsequent nuclear history of the universe probably does not make more ^2H, the present abundance of $\sim 10^{-5}$, or so, implies $\Omega \lesssim 0.06$. This is one of the strongest arguments for a low-density, open universe, but it is subject to uncertainties, particularly regarding the validity of the assumption that the early universe was as homogeneous and isotropic as today's universe.

Limit of present-day experimental physics. In the era of helium production, the major uncertainty is the inhomogeneity, not the physics. But at earlier epochs we reach temperatures at which collisions occur at energies in excess of those accessible in the laboratory. Beyond that, our investigations become more speculative, but not necessarily less rewarding. In fact, cosmology may be regarded as one testing ground for theories of physics. Any constraints that present-day observations can put on the very early universe, however weak, are important because that era is the only time and place in the history of the universe where such very high-energy nuclear reactions have occurred. The limit of our confidence at present is a thermal energy of about 100 GeV, up to which the unified Weinberg–Salaam model of the electroweak interaction has been tested. Even at these energies, the interactions of quarks dominate the physics of the cosmological fluid and cause considerable uncertainty. As the theory improves, one might hope to calculate such numbers as the expected density of free quarks in the universe (Wagoner 1980*b*).

Earlier still. At earlier times and higher energies, physics becomes still more speculative, but the problems one can hope to solve are even more fascinating. Modern gauge theories give reason to hope that the strong interaction will be unified with the electroweak interaction in a 'grand unified theory', which might be expected to explain why the entropy of the universe is what it is and why the number of baryons is very near to the closure density, but not equal to it (or is it?). This could take us to energies as high as 10^{14} GeV, at which all the interactions could have comparable strength. In fact, such theories make even the vacuum a dynamical entity and may explain how chaotic initial conditions evolved into the homogeneous universe we see today. A recent model for this is the 'inflationary universe'. But even at energies of 10^{14} GeV the geometry can still be regarded as classical. Not until we reach the 'Planck mass' of 10^{19} GeV (see Exer. 17, § 12.6) do we expect that quantum gravity will play a role. Perhaps it, too, will one day be part of a unified theory of all physical interactions. Perhaps then we will learn whether it is meaningful to ask what came earlier than the big bang.

Open or closed? The cosmological question which seems to generate the most interest is whether our universe is open or closed. The reason this question excites such debate is not scientific; the value of q_0 is probably not crucial to an understanding of the main features of the early universe, and the different futures predicted by Einstein's equations for the different models will have no practical influence on humanity. Rather, the interest is philosophical: some scientists prefer infinite universes lasting infinite times, while others prefer finite but unbounded ones that may cycle through an unending sequence of 'big bangs'. Both philosophical points of view depend not only on the assumed value of q_0, but also on the cosmological principle: no local measurement can tell us directly whether our universe is finite or not or, indeed, anything at all about regions outside our present particle horizon.

The evidence is scanty. Direct measurements of dH/dt are not reliable yet. Observations of the mean density of the universe fall well short of the critical density, Eq. (12.35), which is 8×10^{-54} m^{-2} or 1×10^{-26} kg m^{-3}. The evidence of the deuterium abundance discussed suggests 6×10^{-28} kg m^{-3}. Other measures, however, argue for somewhat larger numbers, up to 3×10^{-27} kg m^{-3}. These include the evidence that there is considerable dark matter ('missing mass') in galaxies and clusters, as deduced from their dynamical behavior. At the present time there is no strong evidence for a 'closure density' of 10^{-26} kg m^{-3}. So the universe appears to be open. Perhaps the most interesting question is: Why is the universe so near to the critical density?

12.5 Bibliography

Standard cosmology is treated in great detail in Weinberg (1972) and updated by Zel'dovich (1979). Cosmological models become somewhat more complex when the assumption of isotropy is dropped, but they retain the same overall features: the Big Bang, open vs. closed. See Ryan & Shepley (1975). A well-balanced introduction to cosmology is Heidmann (1980).

Astrophysics makes use of cosmological models in studying, for example, galactic evolution, the cosmic microwave background radiation, and galaxy formation. These are discussed in Peebles (1980), Liang & Sachs (1980), and Balian *et al.* (1980).

Much cosmological research is directed at deciding whether our universe is open or closed, i.e., at determining the present mass density. See Gott *et al.* (1974) for one view, Ostriker *et al.* (1974) for another.

An important current research area is into inhomogeneous cosmologies. See MacCallum (1979). Another subject closely allied to theoretical cosmology is singularity theory: Geroch & Horowitz (1979), Tipler *et al.* (1980). See also the stimulating article by Penrose (1979) on time asymmetry in cosmology.

12.6 Exercises

1 Use the metric of a two-sphere to prove the statement associated with Fig. 12.1, that the rate of increase of the distance between any two

points as the sphere expands (as measured *on* the sphere!) is proportional to the distance between them.

2 The astronomer's distance unit, the parsec, is defined to be the distance from the Sun to a star whose parallax is exactly one second of arc. (The parallax of a star is half the maximum change in its angular position as measured from Earth as Earth orbits the Sun.) Given that the radius of Earth's orbit is $1 \, AU = 10^{11}$ m, calculate the length of one parsec.

3 Show that if $h_{ij}(t_1) \neq f(t_1, t_0) h_{ij}(t_0)$ for all i and j in Eq. (12.3), then distances between galaxies would increase anisotropically: the Hubble law would have to be written as

$$v^i = H^i{}_j x^j \tag{12.53}$$

for a matrix $H^i{}_j$ not proportional to the identity.

4 Show that if galaxies are assumed to move along the lines $x^i = $ const., and to see the local universe as homogeneous, then g_{0i} in Eq. (12.5) must vanish.

5(a) Prove the statement leading to Eq. (12.8), that we can deduce G_{ij} of our three-spaces by setting Φ to zero in Eqs. (10.15)–(10.17).

(b) Derive Eqs. (12.9) and (12.10).

6 Show that the metric, Eq. (12.7), is not locally flat at $r = 0$ unless $A = 0$ in Eq. (12.12).

7(a) Find the coordinate transformation leading to Eq. (12.20).

(b) Show that the intrinsic geometry of a hyperbola $t^2 - x^2 - y^2 - z^2 = $ const. > 0 in Minkowski spacetime is identical with that of Eq. (12.20) in appropriate coordinates.

(c) Use the Lorentz transformations of Minkowski space to prove that the $k = -1$ universe is homogeneous and isotropic.

8 Show from Eq. (12.24) that if the radiation has a black-body spectrum of temperature T, then T is inversely proportional to R.

9 Prove Eq. (12.25).

10 With the cosmological constant Λ, Eq. (12.27) becomes

$$(\dot{R}/R)^2 = -k/R^2 + 8\pi\rho/3 + \Lambda/3. \tag{12.54}$$

Repeat the qualitative analysis of the analogue of Fig. 12.3. What qualitatively new cosmological behaviors can you find? Be sure to consider both signs of Λ.

11 (Parts of this exercise are suitable only for students with access to a computer or programmable calculator.) Construct a more realistic equation of state for the universe as follows.

(a)   Assume that, today, the matter density is $\rho_m = m \times 10^{27}$ kg m^{-3} (where m is of order 1) and that the cosmic radiation has black-body temperature 2.7 K. Find the ratio $\varepsilon = \rho_r / \rho_m$, where ρ_r is the energy density of the radiation. Find the number of photons per baryon, $\sim \varepsilon m_p c^2 / kT$.

(b)   Find the general form of the energy-conservation equation, $T^{0\mu}{}_{,\mu} = 0$, in terms of $\varepsilon(t)$ and $m(t)$.

(c)   Numerically integrate this equation and Eq. (12.27) back in time from the present, assuming $\dot{R}/R = 75$ km s^{-1} Mpc^{-1} today, and assuming there is no exchange of energy between matter and radiation. Do the integration for $m = 0.3$, 1.0, and 3.0. Stop the integration when the radiation temperature reaches $E_i/26.7k$, where E_i is the ionization energy of hydrogen (13.6 eV). This is roughly the temperature at which there are enough photons to ionize all the hydrogen: there is roughly a fraction 2×10^{-9} photons above energy E_i when $kT = E_i/26.7$, and this is roughly the fraction needed to give one such photon per H atom. For each m, what is the value of $R(t)/R_0$ at that time, where R_0 is the present scale factor? Explain this result. What is the value of t at this epoch?

(d)   Determine whether the pressure of the matter is still negligible compared to that of the radiation. (You will need the temperature of the matter, which equals the radiation temperature now because the matter is ionized and therefore strongly coupled to the radiation.)

(e)   Integrate the equations backwards in time from the decoupling time, now with the assumption that radiation and matter exchange energy in such a way as to keep their temperatures equal. In each case, how long ago was the time at which $R = 0$, the Big Bang?

12(a)   Show that a photon which propagates on a radial null geodesic of the metric, Eq. (12.14), has energy $-p_0$ inversely proportional to $R(t)$.

(b)   Show from this that a photon emitted at time t_e and received at time t_r by observers at rest in the cosmological reference frame is redshifted by

$$1 + z = R(t_r)/R(t_e). \tag{12.55}$$

(c)   Calculate the redshift of decoupling by assuming that the cosmic microwave radiation has temperature 2.7 K today and had the temperature $E_i/20k$ at decoupling, where $E_i = 13.6$ eV is the energy needed to ionize hydrogen (see Exer. 11c).

(d)   From Eq. (12.14) show that for light which travels a short distance between emission and absorption, z is proportional to r, independent of k.

13   If Hubble's constant is 75 km s^{-1} Mpc^{-1}, what is the minimum present density for a $k = +1$ universe?

14(a)   Prove Eq. (12.40) and deduce Eq. (12.41) from it.

(b)   Fill in the indicated steps leading to Eq. (12.52).

15 Astronomers usually do not speak in terms of intrinsic luminosity and flux. Rather, they use absolute and apparent magnitude. The (bolometric) *apparent magnitude* of a star is defined by its flux F relative to a standard flux F_s:

$$m = -2.5 \log_{10}(F/F_s), \tag{12.56}$$

where $F_s = 3 \times 10^{-8}$ J m^{-2} s^{-1} is roughly the flux of visible light at Earth from the brightest stars in the night sky. The *absolute magnitude* is defined as the apparent magnitude the object would have at a distance of 10 pc:

$$M = -2.5 \log_{10}[L/4\pi(10 \text{ pc})^2 F_s]. \tag{12.57}$$

Rewrite Eq. (12.52) in astronomer's language as:

$$m - M = 5 \log_{10}(z/10 \text{ pc } H_0) + 1.09(1 - q_0)z. \tag{12.58}$$

For objects of the same absolute magnitude M, then, the deviation of a plot of m vs. $\log z$ from a straight line measures q_0. If q_0 is 0.1, to what redshift must one be able to observe in order to measure q_0, if magnitudes are uncertain to ± 0.5? If evolution causes M to be a function of z, how does this change one's ability to measure q_0 from this equation?

16 The angular diameter distance d_A to an object is defined as the ratio of the actual diameter D of the object to its apparent angular diameter θ. (In flat space this would, of course, be the proper distance to the object.)

(a) Put the observer at the center of the coordinates and the object at the coordinate r_0. If it emitted its light at time t_e when the radius was $R_e = R(t_e)$, show that $D = R_e r_0 \theta$.

(b) From this, show that $d_A = R_e r_0 = (1+z)^{-2} d_L$.

(c) Show that the analogue of Eq. (12.52) is

$$H_0 d_A = z - \tfrac{1}{2}(3 + q_0)z^2 + \cdots.$$

(d) Show that the angular diameter of an object can actually *increase* as z increases.

17 Estimate the times earlier than which our uncertainty about the laws of physics prevents us drawing firm conclusions about cosmology as follows.

(a) Deduce from Eqs. (12.27) and (12.32) that, in the radiation-dominated early universe, where k is negligible, the temperature T behaves as

$$T = \beta t^{-1/2}, \qquad \beta = (45\hbar^3/32\pi^3)^{1/4}k^{-1}.$$

(b) Assuming that our knowledge of particle physics is uncertain for $kT > 10^3$ GeV, find the earliest time t at which we can have confidence in the physics.

(c) Quantum gravity is probably important when a photon has enough energy kT to form a black hole within one wavelength ($\lambda = h/kT$). Show that this gives $kT \sim \hbar^{1/2}$. This is the *Planck temperature*. At what time t is this an important worry?

Appendix A
Summary of linear algebra

For the convenience of the student we collect those aspects of linear algebra that are important in our study. We hope that none of this is new to the reader.

Vector space. A collection of elements $V = \{A, B, \ldots\}$ forms a *vector space* over the real numbers if and only if they obey the following axioms (with a, b real numbers).

(1) V is an abelian group with operation $+$ $(A + B = B + A \in V)$ and identity 0 $(A + 0 = A)$.

(2) Multiplication of vectors by real numbers is an operation which gives vectors and which is

(i) distributive over vector addition, $a(A + B) = a(A) + a(B)$;

(ii) distributive over real number addition, $(a + b)(A) = a(A) + b(A)$;

(iii) Associative with real number multiplication, $(ab)(A) = a(b(A))$;

(iv) consistent with the real number identity, $1(A) = A$.

This definition could be generalized to vector spaces over complex numbers or over any field, but we shall not need to do so.

A set of vectors $\{A, B, \ldots\}$ is said to be *linearly independent* if and only if there do not exist real numbers $\{a, b, \ldots, f\}$ such that

$$aA + bB + \cdots + fF = 0.$$

The dimension of the vector space is the largest number of *linearly independent* vectors one can choose. A basis for the space is any linearly independent set of vectors $\{A_1, \ldots, A_n\}$, where n is the dimension of the space. Since for any B the set $\{B, A_1, \ldots, A_n\}$ is linearly dependent, it follows that B can be written as a

linear combination of the basis vectors:
$$B = b_1 A_1 + b_2 A_2 + \cdots + b_n A_n.$$
The numbers $\{b_1, \ldots, b_n\}$ are called the components of B on $\{A_1, \ldots, A_n\}$.

An *inner product* may be defined on a vector space. It is a rule associating with any pair of vectors, A and B, a real number $A \cdot B$, which has the properties:

(1) $\quad A \cdot B = B \cdot A$,

(2) $\quad (aA + bB) \cdot C = a(A \cdot C) + b(B \cdot C)$.

By (1), the map $(A, B) \rightarrow (A \cdot B)$ is symmetric; by (2), it is bilinear. The inner product is called positive-definite if $A \cdot A > 0$ for all $A \neq 0$. In that case the *norm* of the vector A is $|A| \equiv (A \cdot A)^{1/2}$. In relativity we deal with inner products that are indefinite: $A \cdot A$ has one sign for some vectors and another for others. In this case the norm, or magnitude, is often defined as $|A| \equiv |A \cdot A|^{1/2}$. Two vectors A and B are said to be orthogonal if and only if $A \cdot B = 0$.

It is often convenient to adopt a set of basis vectors $\{A_1, \ldots, A_n\}$ that are *orthonormal*: $A_i \cdot A_j = 0$ if $i \neq j$ and $|A_k| = 1$ for all k. This is not necessary, of course. The reader unfamiliar with nonorthogonal bases should try the following. In the two-dimensional Euclidean plane with Cartesian (orthogonal) coordinates x and y and associated Cartesian (orthonormal) basis vectors e_x and e_y, define A and B to be the vectors $A = 5e_x + e_y$, $B = 3e_y$. Express A and B as linear combinations of the nonorthogonal basis $\{e_1 = e_x, e_2 = e_y - e_x\}$. Notice that, although e_1 and e_x are the same, the 1 and x components of A and B are *not* the same.

Matrices. A matrix is an array of numbers. We shall only deal with square matrices, e.g.

$$\begin{pmatrix} 1 & 2 \\ 3 & 1 \end{pmatrix} \quad \text{or} \quad \begin{pmatrix} 1 & 2 & 5 \\ -6 & 3 & 18 \\ 10^5 & 0 & 0 \end{pmatrix}.$$

The *dimension* of a matrix is the number of its rows (or columns). We denote the elements of a matrix by A_{ij}, where the value of i denotes the row and that of j denotes the column; for a 2×2 matrix we have

$$\mathbf{A} = \begin{pmatrix} A_{11} & A_{12} \\ A_{21} & A_{22} \end{pmatrix}.$$

A column vector W is a set of numbers W_i, for example $\begin{pmatrix} W_1 \\ W_2 \end{pmatrix}$ in two dimensions. (Column vectors form a vector space in the usual way.) The following rule governs multiplication of a column vector by a matrix to give a column vector $V = \mathbf{A} \cdot W$:

$$\begin{pmatrix} V_1 \\ V_2 \end{pmatrix} = \begin{pmatrix} A_{11} & A_{12} \\ A_{21} & A_{22} \end{pmatrix} \begin{pmatrix} W_1 \\ W_2 \end{pmatrix} = \begin{pmatrix} A_{11} W_1 + A_{12} W_2 \\ A_{21} W_1 + A_{22} W_2 \end{pmatrix}.$$

In index notation this is clearly

$$V_i = \sum_{j=1}^{2} A_{ij} W_j.$$

For n-dimensional matrices and vectors, this generalizes to

$$V_i = \sum_{j=1}^{n} A_{ij} W_j.$$

Notice that the sum is on the *second* index of **A**.

Matrices form a vector space themselves, with addition and multiplication by a number defined by:

$$\mathbf{A} + \mathbf{B} = \mathbf{C} \Rightarrow C_{ij} = A_{ij} + B_{ij}.$$
$$a\mathbf{A} = \mathbf{B} \Rightarrow B_{ij} = aA_{ij}.$$

For $n \times n$ matrices, the dimension of this vector space is n^2. A natural inner product may be defined on this space:

$$\mathbf{A} \cdot \mathbf{B} = \sum_{i,j} A_{ij} B_{ij}.$$

One can easily show that this is positive-definite. More important than the inner product, however, for our purposes, is *matrix multiplication*. (A vector space with multiplication is called an algebra, so we are now studying the matrix algebra.) For 2×2 matrices, the product is

$$\mathbf{AB} = \mathbf{C} \Rightarrow \begin{pmatrix} C_{11} & C_{12} \\ C_{21} & C_{22} \end{pmatrix}$$
$$= \begin{pmatrix} A_{11} & A_{12} \\ A_{21} & A_{22} \end{pmatrix} \begin{pmatrix} B_{11} & B_{12} \\ B_{21} & B_{22} \end{pmatrix}$$
$$= \begin{pmatrix} A_{11}B_{11} + A_{12}B_{22} & A_{11}B_{12} + A_{12}B_{22} \\ A_{21}B_{11} + A_{22}B_{21} & A_{21}B_{12} + A_{22}B_{22} \end{pmatrix}$$

In index notation this is

$$C_{ij} = \sum_{k=1}^{2} A_{ik} B_{kj}.$$

Generalizing to $n \times n$ matrices gives

$$C_{ij} = \sum_{k=1}^{n} A_{ik} B_{kj}.$$

Notice that the index summed on is the second of A and the first of B. Multiplication is associative but not commutative; the identity is the matrix whose elements are δ_{ij}, the Kronecker delta symbol ($\delta_{ij} = 1$ if $i = j$, 0 otherwise).

The *determinant* of a 2×2 matrix is

$$\det \mathbf{A} = \det \begin{pmatrix} A_{11} & A_{12} \\ A_{21} & A_{22} \end{pmatrix}$$
$$= A_{11}A_{22} - A_{12}A_{21}.$$

Given any $n \times n$ matrix **B** and an element B_{lm} (for fixed l and m), we call $\mathbf{S}_{lm}$ the $(n-1) \times (n-1)$ submatrix defined by excluding row l and column m from **B**, and we call D_{lm} the determinant of $\mathbf{S}_{lm}$. For example, if $\mathbf{B}$ is the 3×3 matrix

$$\mathbf{B} = \begin{pmatrix} B_{11} & B_{12} & B_{13} \\ B_{21} & B_{22} & B_{23} \\ B_{31} & B_{32} & B_{33} \end{pmatrix},$$

then the submatrix $\mathbf{S}_{12}$ is the 2×2 matrix

$$\mathbf{S}_{12} = \begin{pmatrix} B_{21} & B_{23} \\ B_{31} & B_{33} \end{pmatrix}$$

and its determinant is

$$D_{12} = B_{21}B_{33} - B_{23}B_{31}.$$

Then the determinant of $\mathbf{B}$ is defined as

$$\det(\mathbf{B}) = \sum_{j=1}^{n} (-1)^{i+j} B_{ij} D_{ij} \quad \text{for any } i.$$

In this expression one sums only over j for fixed i. The result is independent of which i was chosen. This enables one to define the determinant of a 3×3 matrix in terms of that of a 2×2 matrix, and that of a 4×4 in terms of 3×3, and so on.

Because matrix multiplication is defined, it is possible to define the multiplicative inverse of a matrix, which is usually just called its inverse;

$$(\mathbf{B}^{-1})_{ij} = (-1)^{i+j} D_{ji} / \det(\mathbf{B}).$$

The inverse is defined if and only if $\det(\mathbf{B}) \neq 0$.

Appendix B
Hints and solutions to selected exercises

Chapter 1

1 3.7×10^{-24} kg m^{-1}; 3.5×10^{-43} kg m; 10^{-7}; 10^{-4} kg; 1.1×10^{-12} kg m^{-3}; 10^3 kg m^{-3}; 3.7×10^{-16} kg m^{-3}.

2 3×10^6 m s^{-1}; 9×10^{35} N m^{-2}; 3.3×10^9 s; 9×10^{16} J m^{-3}; 9×10^{17} m s^{-2}.

6 Write out all the terms.

7 $M_{00} = \mu^2 - \alpha^2$, $M_{01} = \mu\nu - \alpha\beta$, $M_{11} = \nu^2 - \beta^2$, $M_{22} = a^2$, $M_{33} = b^2$, $M_{02} = M_{03} = M_{12} = M_{13} = M_{23} = 0$.

8 (c) Use various specific choices of Δx^i; e.g. $\Delta x = 1$, $\Delta y = 0$, $\Delta z = 0 \Rightarrow M_{11} = -M_{00}$.

10 Null; spacelike; timelike; null.

13 The principle of relativity implies that if time dilation applies to one clock (like one based on light travel times over known distances, which is effectively the sort we use to calibrate our time coordinate), then it applies to all (like the pion half-life). Algebra gives the result.

14 (d) For (a), 3.7×10^{-5}.

16 (a) In Fig. 1.14, we want the ratio $\bar{t}/t$ for event $\mathcal{B}$ (these are its time-coordinate differences from $\mathcal{A}$). The coordinates of $\mathcal{B}$ in $\mathcal{O}$ are (t, vt). The first line of Eq. (1.12) implies $\bar{t} = t(1 - v^2)^{1/2}$.

17 (a) 12 m; (b) 1.25×10^{-8} s = 3.75 m; 211 m²: spacelike; (c) 9 m; 20 m; (d) No: spacelike separated events have no unique time ordering; (e) If one observer saw the door close, all observers must have seen it close. The finite speed of transmission ($< c$) of the shock wave along the pole prevents it behaving rigidly.

18 (b) $\tanh[N \tanh^{-1}(0.9)] \approx 1 - 2(19)^{-N/2}$.

Chapter 2

1 (a) -4; (b) 7, 1, 26, 17; (c) same as (b); (d) $-15, 27, 30, -2$; (e) $A^0 B_0 = 0$, $A^2 B_3 = 0$, $A^3 B_1 = 12$, etc.; (f) -4; (g) the subset of (e) with indices drawn from $(1, 2, 3)$ only.

2 (c) γ, α free; μ, λ dummy; 16 equations.

10 Choose each of the basis vectors for $\vec{A}$, getting the four equations in Eq. (2.13) in turn.

12 (b) $(35/6, -37/6, 3, 5)$.

14 (a) $-(0.75/1.25) = -0.6$ in the z direction.

15 (a) $(\gamma, \gamma v, 0, 0)$; (b) $(\gamma, \gamma v^x, \gamma v^y, \gamma v^z)$ with $\gamma = (1 - v \cdot v)^{-1/2}$. (c) $v^x = U^x / U^0 = 0.5$, etc.

18 (a) Given $\vec{a} \cdot \vec{a} > 0$, $\vec{b} \cdot \vec{b} > 0$, $\vec{a} \cdot \vec{b} = 0$; then $(\vec{a} + \vec{b}) \cdot (\vec{a} + \vec{b}) = \vec{a} \cdot \vec{a} + 2\vec{a} \cdot \vec{b} + \vec{b} \cdot \vec{b} > 0$.
(b) If $\vec{a}$ is timelike, use the frame in which $\vec{a} \rightarrow (a, 0, 0, 0)$.

19 (b) $v = \alpha t(1 + \alpha^2 t^2)^{-1/2}$, $\alpha = 1.1 \times 10^{-16}$ m^{-1}; $t = 2.0 \times 10^{17}$ m = 6.7×10^8 s = 22 yr.
(c) $v = \tanh(\alpha \tau)$, $x = \alpha^{-1}[(1 + \alpha^2 t^2)^{1/2} - 1]$, 10 yr.

22 (a) 4 kg, 3.7 kg, 0.25 $(e_x + e_y)$; (b) $(3, -\frac{1}{2}, 1, 0)$ kg, 3 kg, 2.8 kg, $-\frac{1}{6}e_x + \frac{1}{3}e_y$, 0.2 e_y.

23 $|v|^2 = 2/3$.

24 Work in the CM frame.

25 (a) Lorentz transformation of $\vec{p}$.

27 The cooler, because ratio of the rest masses is $1 + 3.3 \times 10^{-16}$.

33 $E_{max} = 8 \times 10^5 \, m_p$, above the γ-ray band.

Chapter 3

4 (b) $\vec{p} \rightarrow (-1/4, -3/8, 15/8, -23/8)$; (d) yes.

6 (a) Consider $\vec{p} = \tilde{\omega}^0$, an element of the basis dual to $\{\vec{e}_\alpha\}$; (b) $(1, 0, 0, 1)$.

9 $\eth T(\mathcal{P}) \to (-15, -15)$; $\eth T(\mathcal{Q}) \to (0, 0)$.

10 (a) A partial derivative *wrt* x^β holds all x^α fixed for $\alpha \neq \beta$, so $\partial x^\alpha / \partial x^\beta = 0$ if $\alpha \neq \beta$. Of course, if $\alpha = \beta$ then $\partial x^\alpha / \partial x^\beta = 1$.

12 (a) By definition, $\tilde{n}(\vec{v}) = 0$ if $\vec{v}$ is tangent to S, so if $\vec{v}$ is not tangent, then $\tilde{n}(\vec{v}) \neq 0$.

(b) If $\vec{v}$ and $\vec{w}$ point to the same side of S then there exists a *positive number* α such that $\vec{w} = \alpha \vec{v} + \vec{T}$, where $\vec{T}$ is tangent to S. Then $\tilde{n}(\vec{w}) = \alpha \tilde{n}(\vec{v})$ and both have the same sign.

(c) On the Cartesian basis, the components of $\tilde{n}$ are $(0, 0, \beta)$ for some β. Thus any $\tilde{n}$ is a multiple of any other.

16 (e) 10 and 6 in four dimensions.

18 (b) $\tilde{q} \to (-1, -1, 1, 1)$.

20 (a) In matrix language, $\Lambda^{\bar{\alpha}}{}_\beta A^\beta$ is the product of the matrix $\Lambda^{\bar{\alpha}}{}_\beta$ with the column vector A^β, while $\Lambda^\alpha{}_{\bar{\beta}} p_\alpha$ is the product of the *transpose* of $\Lambda^\alpha{}_{\bar{\beta}}$ with the column vector p_α. Since $\Lambda^\alpha{}_{\bar{\beta}}$ is inverse to $\Lambda^{\bar{\alpha}}{}_\beta$, these are the same transformation if $\Lambda^\alpha{}_{\bar{\beta}}$ equals the transpose of its inverse.

21 (a) The associated vectors for $t = 0$ and $t = 1$ point *inwards*.

24 (b) No.

25 Use the inverse property of $\Lambda^\alpha{}_{\bar{\beta}}$ and $\Lambda^{\bar{\alpha}}{}_\beta$.

26 (a) $A^{\alpha\beta} B_{\alpha\beta} = -A^{\beta\alpha} B_{\alpha\beta} = -A^{\beta\alpha} B_{\beta\alpha} = -A^{\mu\nu} B_{\mu\nu} = -A^{\alpha\beta} B_{\alpha\beta}$. Therefore $2A^{\alpha\beta} B_{\alpha\beta} = 0$. The justification for each of the above steps: antisymmetry, symmetry, relabel dummy indices, relabel dummy indices.

28 Arbitrariness of $\vec{U}$.

30 (a) Since $\vec{D} \cdot \vec{D} = -x^2 + 25t^2x^2 + 2t^2 \neq -1$, $\vec{D}$ is not a four-velocity field; (f) $5t$; (h) $-5t$ because of (e); (i) the vector gradient of ρ has components $\{\rho^{,\alpha}\} = (-2t, 2x, -2y, 0)$; (j) $\nabla_{\vec{U}} \vec{D} \to [t^2, 5t^3 + 5x(1 + t^2), \sqrt{2}(1 + t^2), 0]$.

33 (d) Given any matrix (A) in $0(3)$, let (Λ) be the 4×4 matrix

$$\begin{pmatrix} 1 & 0 & 0 & 0 \\ 0 & & & \\ 0 & & (A) & \\ 0 & & & \end{pmatrix}.$$

Show that this is in $L(4)$ and that the product of any two such matrices is one of the same type, so that they form a subgroup. These are pure rotations of the spatial axes (relative velocities of the two frames are zero). Transformations like Eq. (1.12) are pure boosts (spatial axes aligned, relative velocity nonzero). The most general Lorentz transformation involves both boost and rotation.

34 (c) $g_{uu} = g_{vv} = 0$, $g_{uv} = -1/2$, $g_{uy} = g_{uz} = g_{yz} = 0$, $g_{yy} = g_{zz} = 1$; (e) $\tilde{d}u = \tilde{d}t - \tilde{d}x$, $\tilde{d}v = \tilde{d}t + \tilde{d}x$, $\mathbf{g}(\vec{e}_u, \ ) = -\tilde{d}v/2$, $\mathbf{g}(\vec{e}_v, \ ) = -\tilde{d}u/2$. Notice that the basis dual to $\{\vec{e}_u, \vec{e}_v\}$, which is $\{\tilde{d}u, \tilde{d}v\}$, is not the same as the set of one-forms mapped by the metric from the basis vectors.

Chapter 4

2 Particles contributing to this flux need not be moving exclusively in the x direction. Moreover, consider a change to the non-orthogonal coordinates $(t, x, y' \equiv x + y, z)$. A surface of constant x is unchanged, so the flux across it is unchanged, but the 'x direction', which is the direction in which now t, y', and z are constant, is in the old $\vec{e}_x + \vec{e}_y$ direction. Is the unchanged flux now to be regarded as a flux in this new direction as well? The loose language carries an implicit assumption of orthogonality in it.

3 (a) In Galilean physics $\mathbf{p}$ changes when we change frames, but in relativity $\vec{p}$ does not: only its components change.
(b) If in Galilean physics we define a four-vector $(m, \mathbf{p})$ then the Galilean transformation changes this to $(m, \mathbf{p} - m\mathbf{v})$, where $\mathbf{v}$ is the relative velocity of the two frames. This is an approximation to the relativistic one (see Eq. (2.21)) in which terms of order v^2 are neglected.

4 The required density is, by definition, N^0 in the frame in which $\vec{U} \to (1, 0, 0, 0)$. In this frame $\vec{N} \cdot \vec{U} = -N^0$.

8 (a) Consider a two-dimensional space whose coordinates are, say, p and T, each point of which represents an equilibrium state of the fluid for that p and T. In such a space the $\mathrm{d}\rho$ of Eq. (4.25) is just $\langle \tilde{d}\rho, \vec{\Delta} \rangle$ where $\vec{\Delta}$ is whatever vector points from the old state to the new one, the change in state contemplated in Eq. (4.25). Since we want Eq. (4.25) to hold for arbitrary $\vec{\Delta}$, it must hold in the one-form version. See Schutz (1980b).
(b) If $\tilde{\Delta}q = \tilde{d}q$, then $T\partial S/\partial x^i = \partial q/\partial x^i$, where x^i is either p or T. The identity $\partial^2 q/\partial T \partial p = \partial^2 q/\partial p \partial T$ implies $(\partial T/\partial p)(\partial S/\partial T) = \partial S/\partial p$, which will almost never be true.

11 (a) By definition of 'rotation': see Exer. 20b, § 2.9.
(b) Suppose M has the property $O^{\mathrm{T}}MO = M$ for any orthogonal matrix O. Consider the special case of a rotation about x^3, where $O_{11} = \cos\theta$, $O_{12} = \sin\theta$, $O_{21} = -\sin\theta$, $O_{22} = \cos\theta$, $O_{33} = 1$, all other elements zero. Then $O^{\mathrm{T}}MO = M$ for arbitrary θ implies $M_{13} = M_{23} = M_{31} = M_{32} = 0$, $M_{11} = M_{22}$, and $M_{21} = -M_{12}$. By relabeling, a rotation about x implies $M_{12} = M_{21} = 0$, $M_{33} = M_{22}$. Therefore M is proportional to I.

13 $U_\alpha = \eta_{\alpha\gamma}U^\gamma$; $\eta_{\alpha\gamma}$ is constant, so $\eta_{\alpha\gamma,\beta} = 0$; $U^\alpha{}_{,\beta}U^\gamma\eta_{\alpha\gamma} = U^\gamma{}_{,\beta}u^\alpha\eta_{\gamma\alpha}$ (relabeling) $= U^\gamma{}_{,\beta}U^\alpha\eta_{\alpha\gamma}$ (symmetry of η).

14 Since $\vec{U} \to_{MCRF} (1, 0, 0, 0)$, multiplying by U_α and summing on α picks out the zero component in the MCRF.

16 There is no guarantee that the MCRF of one element is the same as that of its neighbor.

20 (b) F^0 is the rate of generation of energy per unit volume, and F^i is the ith component of the force. F^α is the only self-consistent generalization of the concept of force to relativity.

21 (a) $T^{\alpha\beta} = \rho_0 U^\alpha U^\beta$, $\vec{U} \to \gamma(1, \beta, 0, 0)$, $\gamma = (1 - \beta^2)^{-1/2}$.

(b) At any point on the ring, the particles have speed ωa. At position (x, y) we have $U^\alpha \to \gamma(1, -\omega y, \omega x, 0)$, $\gamma = (1 - \omega^2 a^2)^{-1/2}$. In the inertial frame their number density is $N[2\pi^2(\delta a)^2]^{-1}$, which equals nU^0, where n is the number density in their rest frame. Therefore $n = N[2\pi^2\gamma (\delta a)^2]^{-1}$ and $T^{\alpha\beta} = mnU^\alpha U^\beta$.

(c) Add (b) to itself with $\omega \to -\omega$. For example, $T^{0x} = 0$ and $T^{xx} = 2mn\omega^2 y^2 \gamma^2$.

22 No bias means that T^{ij} is invariant under rotations. By Exer. 11 $T^{ij} = p\delta^{ij}$ for some p. Since $T^{0i} = 0$ in the MCRF, Eq. (4.36) holds. Clearly $\rho = \gamma nm$, where $\gamma = (1 - v^2)^{-1/2}$. The contributions of each particle to T^{zz}, say, will be the momentum flux it represents. For a particle with speed v in the direction (θ, ϕ), this is a z component of momentum $m\gamma v \cos\theta$ carried across a $z = $ const. surface at a speed $v \cos\theta$. If there are n particles per unit volume, with random velocities, then $T^{zz} = n(m\gamma v)(v)$ times the average value of $\cos^2\theta$ over the unit sphere. This is $\frac{1}{3}$, so $T^{zz} = p = \gamma nmv^2/3$. Thus, $p/\rho = v^2/3 \to \frac{1}{3}$ as $v \to 1$. (In this limit $m\gamma$ remains finite, the energy of each photon.)

24 (c) $R/x = [(1 - v)/(1 + v)]^{1/2}$.

25 (h) $E^{\bar{x}} = E^x$; let $E^y e_y + E^z e_z$ be called $E_\perp$, the part of E perpendicular to v. Similarly, let $E'_\perp = E^{\bar{y}} e_{\bar{y}} + E^{\bar{z}} e_{\bar{z}}$. Then $E'_\perp = \gamma(E_\perp + v \times B)$.

Chapter 5

3 (b) (i) Good except at origin $x = y = 0$: usual polar coordinates. (ii) Undefined for $x < 0$, fails at $x = 0$, good for $x > 0$: maps the right-hand plane of (x, y) onto the whole plane of (ξ, η). (iii) Good except at origin and infinity: an inversion of the plane through the unit circle.

4 Compute slope of curve.

5 (a) and (b) have same path, $x^2 + y^2 = 1$.

7 $\Lambda^{2'}_{\ 1} = -y/r^2$; $\Lambda^1_{\ 2'} = -y$.

8 (a) $V^r = r^2(\cos^3\theta + \sin^3\theta) + 6r(\sin\theta\cos\theta)$, $W^\theta = (\cos\theta - \sin\theta)/r$; (b) $(\tilde{d}f)_r = 2r(1 + \sin 2\theta)$; (c) $(\tilde{W})_\theta = r(\cos\theta - \sin\theta)$.

11 (b) Although it is possible to do this using matrices, the straightforward expansion of the transformation equation is less error prone. Thus, the r–r component of $\nabla\vec{v}$ is

$$\Lambda^{1'}{}_\alpha \Lambda^\beta{}_{1'} V^\alpha{}_{,\beta} = \Lambda^{1'}{}_1 \Lambda^1{}_{1'} V^1{}_{,1} - \Lambda^{1'}{}_2 \Lambda^1{}_{1'} V^2{}_{,1} + \Lambda^{1'}{}_1 \Lambda^2{}_{1'} V^1{}_{,2} + \Lambda^{1'}{}_2 \Lambda^2{}_{1'} V^2{}_{,2}$$
$$= 2r(\cos^3\theta + \sin^3\theta) + 6\sin\theta\cos\theta.$$

(c) This gives the same as (b): $V^r{}_{;r} \equiv V^{1'}{}_{;1'}$ is as above, and the others are: $V^r{}_{;\theta} = 2r^2\sin\theta\cos\theta(\sin\theta - \cos\theta) + 3r(\cos^2\theta - \sin^2\theta)$, $V^\theta{}_{;r} = 2\sin\theta\cos\theta(\sin\theta - \cos\theta) + 3(\cos^2\theta - \sin^2\theta)/r$, $V^\theta{}_{;\theta} = 2r\sin\theta\cos\theta(\sin\theta + \cos\theta) - 6\sin\theta\cos\theta$.
(d) $2(x + y)$.
(e) $2r(\sin\theta + \cos\theta)$, same as (d).

12 (c) These functions, which are the same as in (b), are related to the answers given for Exer. 11c as follows: $p_{r;r} = V^r{}_{;r}, p_{r;\theta} = V^r{}_{;\theta}, p_{\theta;r} = r^2 V^\theta{}_{;r}, p_{\theta;\theta} = r^2 V^\theta{}_{;\theta}$. It happens that $p_{\theta;r} = p_{r;\theta}$ for this one-form field. This is not generally true, but happens only because $\tilde{p}$ is the gradient of a function.

14 $A^{r\theta}{}_{;\theta} = r(1 + \cos\theta - \tan\theta)$; $A^{rr}{}_{;r} = 2r$.

15 Of the first-derivative components, only $V^\theta{}_{;\theta} = 1/r$ is nonzero. Of the second-derivative components, the only nonzero ones are $V^\theta{}_{;r;\theta} = -1/r^2$, $V^r{}_{;\theta;\theta} = -1$, and $V^\theta{}_{;\theta;r} = -1/r^2$.

17 $\partial\vec{e}_\mu/\partial x^{\nu'} = [\partial(\Lambda^\alpha{}_{\mu'}\vec{e}_\alpha)/\partial x^\beta]\Lambda^\beta{}_{\nu'} = \Lambda^\alpha{}_{\mu'}\Lambda^\beta{}_{\nu'}\partial\vec{e}_\alpha/\partial x^\beta + \Lambda^\alpha{}_{\mu',\beta}\Lambda^\beta{}_{\nu'}\vec{e}_\alpha$
$\Rightarrow \Gamma^{\lambda'}{}_{\mu'\nu'} = \Lambda^\alpha{}_{\mu'}\Lambda^\beta{}_{\nu'}\Lambda^{\lambda'}{}_\gamma\Gamma^\gamma{}_{\alpha\beta} + \Lambda^\alpha{}_{\mu',\beta}\Lambda^\beta{}_{\nu'}\Lambda^{\lambda'}{}_\alpha$

20 $\Gamma^\nu{}_{\alpha\beta} = \tfrac{1}{2}g^{\nu\mu}(g_{\mu\alpha,\beta} + g_{\mu\beta,\alpha} - g_{\alpha\beta,\mu} + c_{\alpha\mu\beta} + c_{\beta\mu\alpha} - c_{\mu\alpha\beta})$.

21 (a) Compute the vectors $(dt/d\lambda, dx/d\lambda)$ and $(dt/da, dx/da)$ and show they are orthogonal.
(b) For arbitrary a and λ, x and t obey the restriction $|x| > |t|$. The lines $\{x > 0, t = \pm x\}$ are the limit of the $\lambda = $ const. hyperbolae as $a \to 0^+$, but for any finite λ the limit $a \to 0^+$ takes an event to the origin $x = t = 0$. To reach $x = t = 1$, for example, one can set $(\lambda = \ln(2/a)$, which sends $\lambda \to \infty$ as $a \to 0$.
(c) $g_{\lambda\lambda} = -a^2$, $g_{aa} = 1$, $g_{a\lambda} = 0$, $\Gamma^\lambda{}_{a\lambda} = 1/a$, $\Gamma^a{}_{\lambda\lambda} = a$, all other Christoffel symbols zero. Note the close analogy to Eqs. (5.3), (5.30), and (5.44). Note also that $g_{\lambda\lambda}$ (the time–time component of the metric) vanishes on the null lines $|x| = |t|$, another property we will see again when we study black holes.

22 Use Eq. (5.68).

Chapter 6

1 (a) Yes, singular points depend on the system.

(b) Yes: if we define $r = (x^2 + y^2)^{1/2}$, the map $\{X = (x/r)\tan(\pi r/2),$ $Y = (y/r)\tan(\pi r/2)\}$ shows that, as a manifold, the interior of the unit circle $(r < 1)$ is indistinguishable from the whole plane $(X, Y$ arbitrary). This map distorts distances, but the metric is not part of the definition of the manifold.

(c) No, discrete.

(d) This consists of the unit circle and the coordinate axes. It has the structure of a one-dimensional manifold everywhere except at the five intersection points, such as $(1, 0)$.

4 (a) Symmetry on (γ', μ') means there are $\frac{1}{2}n(n+1) = $ ten independent pairs (γ', μ'). Since α can assume four values independently, there are 40 coefficients.

(b) The number of symmetric combinations $(\lambda', \gamma', \mu')$ is $\frac{1}{6}n(n+1) \times (n+2) = 20$, times four for α, gives 80.

(c) Two symmetric pairs of ten independent combinations each give 100.

6 $g^{\alpha\beta}g_{\beta\mu,\alpha} = g^{\beta\alpha}g_{\beta\mu,\alpha}$ (symmetry of metric) $= g^{\alpha\beta}g_{\alpha\mu,\beta}$ (relabeling dummy indices) $= g^{\alpha\beta}g_{\mu\alpha,\beta}$ (symmetry of metric). This cancels the second term in brackets.

9 For polar coordinates in two-dimensional Euclidean space, $g = r^2 \sin^2\theta$. In three dimensions it is $g = r^4 \sin^2\theta$. Notice in these cases $g > 0$, so formulae like Eqs. (6.40) and (6.42), which are derived from Eq. (6.39), should have $-g$ replaced by g.

10 The vector maintains the angle it makes with the side of the triangle as it moves along. On going around a corner the angle with the new side exceeds that with the old by an amount which depends only on the interior angle at that corner. Summing these changes gives the result.

15 Consider a spacelike curve $\{x^\alpha(\lambda)\}$ parametrized by λ, and going from $x^\alpha(a)$ to $x^\alpha(b)$. Its length is $\int_a^b [g_{\alpha\beta}(dx^\alpha/d\lambda)(dx^\beta/d\lambda)]^{1/2}/d\lambda$. Assume that λ is chosen so that the integrand is constant along the given curve. Now change the curve to $\{x^\alpha(\lambda) + \delta x^\alpha(\lambda)\}$ with $\delta x^\alpha(a) = \delta x^\alpha(b) = 0$. The first-order change in the length is, after an integration by parts, $\int_a^b [\frac{1}{2}g_{\alpha\beta,\gamma}U^\alpha U^\beta - d(g_{\gamma\alpha}U^\alpha)\,d\lambda]\delta x^\gamma\,d\lambda$, where $U^\alpha \equiv dx^\alpha/d\lambda$. For a geodesic, the term in square brackets vanishes.

18 (b) Each pair $\alpha\beta$ or $\mu\nu$ is antisymmetric, so has six independent combinations for which the component need not vanish. Each pair can be chosen independently from among these six, but the component is symmetric under the exchange of one pair with the other. (c) Eq. (6.69) allows us to write Eq. (6.70) as $R_{\alpha[\beta\mu\nu]} = 0$. There are only $n(n-1) \times$

$(n-2)/6 =$ four independent choices for the combination $\beta\mu\nu$, by anti-symmetry. In principle α is independent, but if α equals any one of β, μ, and ν then Eq. (6.70) reduces to one of Eq. (6.69). So Eq. (6.70) is at most four equations determined only by, say, the value α. However, Eq. (6.69) allows us also to change Eq. (6.70) to $R_{\beta[\alpha\mu\nu]} = 0$. Thus, if we had earlier taken, say, $\alpha = 1$ and $\beta = 2$, then that equation would have been equivalent to the one with $\alpha = 2$ and $\beta = 1$: all values of α give the same equation.

25 Use Eqs. (6.69) and (6.70).

29 $R_{\theta\phi\theta\phi} = \sin^2\theta$.

32 Compare with Exer. 34, § 3.10.

35 The nonvanishing algebraically independent Christoffel symbols are:
$\Gamma^t{}_{tr} = -\Phi'$, $\Gamma^r{}_{tt} = -\Phi'\exp(2\Phi - 2\Lambda)$, $\Gamma^r{}_{rr} = \Lambda'$, $\Gamma^r{}_{\theta\theta} = -r\exp(-2\Lambda)$, $\Gamma^r{}_{\phi\phi} = -r\sin^2\phi\,\exp(-2\Lambda)$, $\Gamma^\theta{}_{r\theta} = \Gamma^\phi{}_{r\phi} = r^{-1}$, $\Gamma^\theta{}_{\phi\phi} = -\sin\theta\cos\theta$, $\Gamma^\phi{}_{\theta\phi} = \cot\theta$. Here primes denote r derivatives. (Compare these with the Christoffel symbols you calculated in Exer. 29.) As explained in the solution to Exer. 18, in calculating $R_{\alpha\beta\mu\nu}$ we should concentrate on the pairs $(\alpha\beta)$ and $(\mu\nu)$, choosing each from the six possibilities $(tr, t\theta, t\phi, r\theta, r\phi, \theta\phi)$. Because $R_{\alpha\beta\mu\nu} = R_{\mu\nu\alpha\beta}$, we do not need to calculate, say, $R_{\theta\phi r\phi}$ after having calculated $R_{r\phi\theta\phi}$. This gives 21 independent components. Again following Exer. 18, one of the components with four distinct indices, say $R_{tr\theta\phi}$, can be calculated from others. We catalog, therefore, the following 20 algebraically independent components:

$R_{trtr} = [\Phi'' - (\Phi')^2 - \Phi'\Lambda']\exp(2\Phi)$, $\qquad R_{tr t\theta} = R_{tr t\phi} = R_{tr r\theta} = R_{tr r\phi} = 0$,

$R_{t\theta t\theta} = -r\Phi'\exp(2\Phi - 2\Lambda)$, $\qquad R_{t\theta t\phi} = R_{t\theta r\theta} = R_{t\theta r\phi} = R_{t\theta\theta\phi} = 0$,

$R_{t\phi t\phi} = -r\Phi'\sin^2\theta\exp(2\Phi - 2\Lambda)$, $\qquad R_{t\phi r\theta} = R_{t\phi r\phi} = R_{t\phi\theta\phi} = 0$,

$R_{r\theta r\theta} = r\Lambda'$, $\qquad R_{r\theta r\phi} = R_{r\theta\theta\phi} = 0$, $\qquad R_{r\phi r\phi} = r\sin^2\theta\Lambda'$,

$R_{r\phi\theta\phi} = 0$, $\qquad R_{\theta\phi\theta\phi} = r^{-2}\sin^2\theta[1 - \exp(-2\Lambda)]$.

See if you can use the spherical symmetry and time independence of the metric to explain why certain of these components vanish. Also compare $R_{\theta\phi\theta\phi}$ with the answer to Exer. 29 and see if you can explain why they are different.

36 Since $\phi = 0$ is already inertial, we can look for a coordinate transformation of the form $x^{\alpha'} = (\delta^\alpha{}_\beta + L^\alpha{}_\beta)x^\beta$, where $L^\alpha{}_\beta$ is of order ϕ. The solution to Exer. 17, § 5.9, gives $\Gamma^{\lambda'}{}_{\mu'\nu'}$, which must vanish at P. Since $\Lambda^{\alpha'}{}_\beta = \delta^\alpha{}_\beta + L^\alpha{}_\beta + L^\alpha{}_{\mu,\beta}x^\mu$, we find $L^\alpha{}_{(\beta,\nu)} = \frac{1}{2}\Gamma^\alpha{}_{\beta\nu}$ at P. The antisymmetric part, $L^\alpha{}_{[\beta,\nu]}$, is undetermined, and represents a Lorentz transformation of order ϕ. Since we are only looking for *an* inertial system, we can set $L^\alpha{}_{[\beta,\nu]} = 0$. Calculating $\Gamma^\alpha{}_{\beta\nu}$ at P (as in Exer. 3, § 7.6, below) gives the new coordinates. In particular, the equation $x^{i'} = 0$ gives the motion of the origin of the new frame, whose acceleration is $d^2x^i/dt^2 = -\Gamma^i{}_{tt} = -\phi_{,i}$. We shall interpret this in the next chapter, where we identify ϕ as the Newtonian gravitational potential and see that this acceleration expresses the equivalence principle.

37 (b) The ranges of the coordinates must be deduced: $0 < \chi < \pi, 0 < \theta < \pi$, $0 < \phi < 2\pi$. Then the volume is $2\pi^2 r^3$.

Chapter 7

1 Consider a fluid at rest, where $U^i = 0$ and $U^0 = 1$ in a local inertial frame. Then $\partial n/\partial t = R$, so Eq. (7.3) implies creation (or destruction) of particles by the curvature.

2 $g^{00} = -(1 - 2\phi)$, $g^{ij} = \delta^{ij}(1 + 2\phi)$.

3 $\Gamma^0{}_{00} = \dot\phi$, $\Gamma^0{}_{0i} = \phi_{,i}$, $\Gamma^0{}_{ij} = -\dot\phi\delta_{ij}$, $\Gamma^i{}_{00} = \phi_{,i}$, $\Gamma^i{}_{0j} = -\dot\phi\delta_{ij}$, $\Gamma^i{}_{jk} = \delta_{jk}\phi_{,i} - \delta_{ij}\phi_{,k} - \delta_{ik}\phi_{,j}$.

9 (a) $R_{0i0j} = \phi_{,ij} + \delta_{ij}\phi_{,00}$, $R_{0ijk} = \phi_{,0j}\delta_{ik} - \phi_{,0k}\delta_{ij}$, $R_{ijkl} = \delta_{ik}\phi_{,jl} + \delta_{jl}\phi_{,ik} - \delta_{il}\phi_{,jk} - \delta_{jk}\phi_{,il}$.
(c) The 'acceleration', in Newtonian language, is $-\phi_{,i}$. The difference between the accelerations of nearby particles separated by ξ^j is therefore $-\xi^j\phi_{,ij}$.

16 (b) In terms of a Lorentz basis, the following vector fields are Killing fields: $\vec{e}_t, \vec{e}_x, \vec{e}_y, \vec{e}_z, x\vec{e}_y - y\vec{e}_x, y\vec{e}_z - z\vec{e}_y, z\vec{e}_x - x\vec{e}_z, t\vec{e}_x + x\vec{e}_t, t\vec{e}_y + y\vec{e}_t, t\vec{e}_z + z\vec{e}_t$. Any linear combination with constant coefficients of solutions to Eq. (7.45) is a solution to Eq. (7.45), which means that the analogous fields to these in another Lorentz frame are derivable from these.

Chapter 8

2 (c) (i) 7.425×10^{-11} m^{-2}; (ii) 8.261×10^{-12} m^{-2}; (iii) 1.090×10^{-16} m^{-1}; (iv) 2.756×10^{-12}.
(d) $m_{PL} = 2.176 \times 10^{-8}$ kg; $t_{PL} = 5.390 \times 10^{-44}$ s. Typical elementary-particle lifetimes are 10^{-24} s or greater. The heaviest known particles are less than 10^{-29} kg.

3 (a) -2.122×10^{-6}; (b) -9.873×10^{-9}; (c) -6.961×10^{-10}; (d) 9.936×10^{-5}.

17 (a) $M = Rv^2 = (c^3/2\pi P^2) \sim 1.3$ km ~ 1 M$_\odot$. (b) 100 km ~ 68 M$_\odot$.

18 (a) $|\Lambda| \leqslant 4 \times 10^{-35}$ m^{-2}.

19 (a) $T^{00} = \rho$, $T^{01} = -\rho\Omega x^2$, $T^{02} = \rho\Omega x^1$, $T^{03} = 0$. The components T^{ij} are not fully determined by the given information, but they must be of order $\rho v^i v^j$, i.e. of order $\rho\Omega^2 R^2$.

(b) Since $\nabla^2 \bar{h}^{00} = -16\pi\rho$, $\bar{h}^{00}$ is just minus four times the Newtonian potential, $\bar{h}^{00} = 4M/r$ exactly. For $\bar{h}^{0i}$ we have $\bar{h}^{0i} = -4\rho\Omega \int y^2 |x - y|^{-1} d^3y$. Use the binomial expansion $|x - y|^{-1} = r^{-1}[1 + x\cdot y/r^2 + 0(R/r)^2]$. By symmetry, $\int y^i d^3y = 0$, $\int y^i y^j d^3y = 0$ if $i \neq j$, and $\int y^1 y^1 d^3y = \int y^2 y^2 d^3y = \int y^3 y^3 d^3y = (4\pi/15)R^5$. This implies $\bar{h}^{01} = -(16\pi/15)\rho R^5 \Omega x^2/r^3$. In terms of the angular momentum J, we find $\bar{h}^{01} = -2Jx^2/r^3$, $\bar{h}^{02} = 2Jx^1/r^3$, $\bar{h}^{03} = 0$. These fall off as r^{-2} and are correct to order r^{-3}. A more careful study of the properties of the solutions of $\nabla^2 f = g$ would show that these are in fact exact: the higher-order terms all vanish. The components $\bar{h}^{ij}$ are small compared to $\bar{h}^{00}$ and $\bar{h}^{0i}$ because T^{ij} is small. Therefore Eq. (8.31) gives $h^{\mu\nu}$ and the metric $g_{00} = -1 + 2M/r + 0(\Omega^2 R^2)$, $g_{01} = 2Jx^2/r^3 + 0(\Omega^3 R^3)$, $g_{02} = -2Jx^1/r^3 + 0(\Omega^3 R^3)$, $g_{03} = 0$, $g_{ij} = \delta_{ij}(1 + 2M/r) + 0(\Omega^2 R^2)$. Compare this with Eq. (7.8). In standard spherical coordinates, g_{00} is the same, $g_{0r} = g_{0\theta} = 0$, $g_{0\phi} = -2J/r$, spatial line element $dl^2 = [1 + 2M/r - 0(\Omega^2 R^2)](dr^2 + r^2 d\theta^2 + r^2 \sin^2\theta d\phi^2)$.

(c) Such a particle obeys the geodesic equation with $p_0 \equiv -E$ and $p_\phi \equiv L$ constant, p^r and p^θ zero. To this order, the normalization $\vec{p}\cdot\vec{p} = -m^2$ implies $E = m(1 - M/r + L^2/2m^2 r)$, just as in Newtonian theory (except for the rest mass). The r component of the geodesic equation implies, again to lowest order, $L = m(Mr)^{1/2}$, again as in Newtonian theory. One orbit, $\Delta\phi = 2\pi$, will take a time $\Delta t = (dt/d\phi)\Delta\phi$. Now $dt/d\phi = (dt/d\tau)/(d\phi/d\tau) = U^0/U^\phi = p^0/p^\phi$, and this can be expressed in terms of E, L, and the metric; a straightforward calculation gives $(\Delta t)_{\text{prograde}} - (\Delta t)_{\text{retrograde}} = -8\pi J/M$, independently of r. In principle, this allows measurement of a body's angular momentum by the study of particle orbits far from it.

(d) 0.16 ms.

Chapter 9

7 A solution of Eq. (9.22) is *uniquely* determined by the initial position and U^α. The function $U^\alpha = \delta^\alpha{}_0$ satisfies Eq. (9.22) for all time (by virtue of Eq. (9.23)), and so must be the unique solution for initial data in which $U^\alpha = \delta^\alpha{}_0$.

8 No.

9 For the light beam, $ds^2 = 0 \Rightarrow dt/dx = (g_{xx}/|g_{tt}|)^{1/2} \approx 1 + \frac{1}{2}h_{xx}^{\text{TT}}(t)$. Therefore $\Delta t \approx (2 + \langle h_{xx}^{\text{TT}}\rangle)\varepsilon$, where $\langle h_{xx}^{\text{TT}}\rangle$ is some mean value of h_{xx}^{TT} during the time of flight of the photon. Since $\langle h_{xx}^{\text{TT}}\rangle$ changes with time while ε does not, free particles do see accelerations relative to their neighbors.

10 (a) For example, $R_{tyxz} = -\frac{1}{2}\omega^2 h_{yz}$; (b) $\xi_i = -\frac{1}{2}B(x - t)^2$, $\xi_i = 0$.

20   In the TT coordinates of the wave, choose the wave's direction to be z, the ellipse's principal axes to be x and y, and the masses' separation to be along the unit vector $s \to (\sin\theta\cos\phi, \sin\theta\sin\phi, \cos\theta)$ in the usual spherical coordinates of the frame. Then in Eq. (9.42) we replace h_{xx}^{TT} by $h_{ss}^{TT} \equiv s^\alpha s^\beta h_{\alpha\beta}^{TT} = \sin^2\theta(\cos 2\phi + ia\sin 2\phi)h_{xx}^{TT}$. There will be no driving term if $\theta = 0$, i.e. if the masses lie along the direction of the wave's propagation (compare with Exer. 12).

27 $I^{lm} = \sum_A m_{(A)}x_{(A)}^i x_{(A)}^j$.

28   (a) $(4\pi/3)\delta^{ij}\int \rho r^4 dr; 0$. (b) Result of (a) $+ Ma^i a^j$; $M(a^i a^j - \frac{1}{3}\delta^{ij}a^2)$.
(c) $I^{xx} = a^2 M/5$, $I^{yy} = b^2 M/5$, $I^{zz} = c^2 M/5$, $I^{xx} = (2a^2 - b^2 - c^2)M/15$, $I^{yy} = (2b^2 - a^2 - c^2)M/15$, $I^{zz} = (2c^2 - a^2 - b^2)M/15$, all other components zero.

   (d)
$$I^{xx} = [a^2\cos^2\omega t + b^2\sin^2\omega t)M/5,$$
$$I_{yy} = (b^2\cos^2\omega t + a^2\sin^2\omega t)M/5, \qquad I^{zz} = c^2 M/5,$$
$$I^{xy} = \cos\omega t\sin\omega t(a^2 - b^2)M/5,$$
$$I^{xx} = (a^2(3\cos^2\omega t - 1) + b^2(3\sin^2\omega t - 1) - c^2]M/15,$$
$$I^{yy} = [b^2(3\cos^2\omega t - 1) + a^2(3\sin^2\omega t - 1) - c^2]M/15,$$
$$I^{zz} = (2c^2 - a^2 - b^2)M/15, \qquad I^{xy} = I^{xy},$$

others zero.
(e) $I^{xx} = 2ma^2 = I^{yy}$, others zero; $I^{xx} = 2ma^2/3 = I^{yy}$, $I^{zz} = -4ma^2/3$, others zero.
(f) Same as (e).
(g) $I^{xx} = 2m(A^2\cos^2\omega t + Al_0\cos\omega t + l_0^2/4)$, others zero; $I^{xx} = 2I^{xx}/3$, $I^{yy} = -I^{xx}/3 = I^{zz}$, others zero.
(h) $I^{xx} = (m + M)(A^2\cos^2\omega t + Al_0\cos\omega t + l_0^2/4)$, other zero; $I^{xx} = 2I^{xx}/3$, $I^{yy} = -I^{xx}/3 = I^{zz}$, and $\omega^2 = k/\mu$, where $\mu = mM/(M + m)$ is the reduced mass.

32   There is no radiation in (a)–(c) and (e), and no *quadrupole* radiation in (f). For (d) first put the time dependence of I_{ij} from Exer. 28 into complex form, e.g., $I_{xx} = \frac{1}{10}M(a^2 - b^2)\exp(-2i\omega t)$. Then use Eqs. (9.75)–(9.77) with $\Omega = 2\omega$ and correct permutations of indices. Along x axis, $\bar{h}_{zz}^{TT} = -\bar{h}_{yy}^{TT} = -\frac{2}{5}M\omega^2(a^2 - b^2)(\exp[-2i\omega(t - r)]/r$, $\bar{h}_{zy}^{TT} = 0$. Along y axis, $\bar{h}_{zz}^{TT} = -\bar{h}_{xx}^{TT} = \frac{2}{5}M\omega^2(a^2 - b^2)\exp[-2i\omega(t - r)]/r$, $\bar{h}_{xz}^{TT} = 0$. Along z axis, $\bar{h}_{xx}^{TT} = -\bar{h}_{yy}^{TT} = -\frac{4}{5}M\omega^2(a^2 - b^2)\exp[-2i\omega(t - r)]/r$, $\bar{h}_{xy}^{TT} = i\bar{h}_{xx}^{TT}$. This means that the radiation is linearly polarized in the equatorial plane, circularly polarized along the z axis. Notice there is no radiation if the body is axially symmetric, i.e. if $a = b$. For (h), on y axis $\bar{h}_{zz}^{TT} = -(m + M)\omega^2\{2A^2\exp[-2i\omega(t - r)] + Al_0\exp[-i\omega(t - r)]\}/r = -\bar{h}_{yy}^{TT}$, $\bar{h}_{xy}^{TT} = 0$. On x axis, no radiation.

33   Without loss of generality choose the wave to be moving in the x–y plane. Then $P_{xx} = \sin^2\theta$, $P_{yy} = \cos^2\theta$, $P_{zz} = 1$, $P_{xy} = -\sin\theta\cos\theta$, others zero. By matrix multiplication or otherwise we find $\bar{h}_{xx}^{TT} = f\sin^2\theta$, $\bar{h}_{yy}^{TT} = f\cos^2\theta$, $\bar{h}_{xy}^{TT} = -f\sin\theta\cos\theta$, $\bar{h}_{zz}^{TT} = -f$, others zero, where $f =$

$m\omega^2 \sin^2\theta \{2A^2 \exp[-2i\omega(t-r)] + Al_0 \exp[-i\omega(t-r)]\}/r$. If we let l^i be $(-\sin\theta, \cos\theta, 0)$, then $\vec{l}$ and $\vec{e}_z$ are two orthogonal vectors perpendicular to the motion of the wave. We find $\bar{h}_{ll}^{TT} \equiv l^i l^j \bar{h}_{ij}^{TT} = f$. Recalling $\bar{h}_{zz}^{TT} = -f$, we see that the wave is always 100% linearly polarized with the ellipse's axes in the x–y plane and parallel to $\vec{e}_z$: the component rotated by 45° is absent.

35   Let the wave be traveling in the x–z plane at an angle θ with the z axis. Let $\vec{e}_y$ and $\vec{l} \to (\cos\theta, 0, -\sin\theta)$ be orthogonal vectors in the plane of polarization. Then $\bar{h}_{yy}^{TT} = (1 - \frac{1}{2}\sin^2\theta)f$, $\bar{h}_{ly}^{TT} \equiv l^i \bar{h}_{iy}^{TT} = -i\cos\theta\, f$, $\bar{h}_{ll}^{TT} \equiv l^i l^j \bar{h}_{ij}^{TT} = -\bar{h}_{yy}^{TT}$, where $f = 2ml_0^2\omega^2 \exp[-2i\omega(t-r)]/r$. From Exer. 14b we see that the wave is elliptically polarized with principal axes $\vec{l}$ and $\vec{e}_y$. The percentage of circular polarization is $|\bar{h}_{ly}^{TT}/\bar{h}_{yy}^{TT}|^2$, which ranges from zero at the equator to 100% at the poles.

36   (a) $P^2 = -\pi^2(M+m)^2\mu^3/2E^3$, $e^2 = 1 + 2EL^2/(M+m)^2\mu^3$, $a = -mM/2E$ (recall $E < 0$), where $\mu = mM/(m+M)$.

37   Bring out the R^{-1} as r^{-1}, as in Eq. (9.94), but expand $t-R$ about $t-r$. Eq. (9.95) and its consequences still follow. The first term is Eq. (9.96) but higher terms depend on $(d/dt)^n \int \rho y_j y_k \ldots y_l y_m \, d^3y$, where there are n factors of y_j. If ρ is spherical, then the integrations give indices of the form $\delta_{jk} \cdots \delta_{lm}$ (see Exer. 42) and permutations. The TT projection eliminates traces and so completely eliminates these terms.

Chapter 10

5   Such a transformation must leave all quantities invariant, but it sends $t \to -t$ and hence $dr/dt \to -dr/dt$. Only if it vanishes can it be invariant.

8   (a) Use reasoning similar to that in Exer. 5.

10   (b) $\exp(\Phi) = 0.999\,997, 0.760$; $z = 3 \times 10^{-6}, 0.315$.

15 $\rho = \rho_c + \rho_1 r^2$, $p = p_c(1 + \rho_1\Gamma_c/\rho_c r^2)$, $m = 4\pi r^3\rho_c/3 + 4\pi r^5\rho_1/5$, with $\rho_1 = -2\pi\rho_c(\rho_c + p_c)(\rho_c + 3p_c)/(3\Gamma_c p_c)$. Estimate the error in neglecting the next term to be the square of the first correction, i.e., choose r such that the corrections are $<3\% \Rightarrow r^2 \lesssim |\rho_c/30\rho_1\Gamma_c|$.

19   (a) 1.12×10^{42} kg m² s⁻¹; (b) 1.4×10^4 s⁻¹; yes, by 50%; (c) 2.7×10^{-4}; (d) 5×10^9 Gauss.

Chapter 11

7   (a) Use Eq. (11.21) in $ds^2 = g_{00}dt^2 + g_{\phi\phi}d\phi^2$ to get $\tau = 20\pi\sqrt{7}\,M$.
     (b) Same as coordinate time interval, Eq. (11.22): $20\pi\sqrt{10}\,M$.
     (c) Integrate $ds^2 = g_{00}dt^2$ over the time in (b): $40\pi\sqrt{2}\,M$.
     (d) 6.4 ms: innermost stable orbit.
     (e) $40/\sqrt{2}$ yr, independent of M.

9 (a) Turning points are at $r = 12.5\ M$ and $5.47\ M$. (Notice that the high $\tilde{L}$ enables the particle to turn around *inside* $6\ M$.) The orbit changes by $\Delta\phi = 7.4$ rad between these points. A full orbit has $\Delta\phi = 14.8$ rad, or a perihelion shift of 8.5 rad. The approximation Eq. (11.34) gives only 2.3 rad. This shows that highly noncircular orbits can have much greater shifts.

10 (a) $(\tilde{E}/\tilde{L})^2 = [1 + 36 M^2/\tilde{L}^2 + (1 - 12 M^2/\tilde{L}^2)^{3/2}]/(54\ M^2)$.

14 In arc sec per orbit and per year: Venus $(0.052, 0.085)$, Earth $(0.038, 0.038)$, Mars $(0.025, 0.013)$.

22 Approx. $10^5\ M_\odot$.

25 Estimate the total breakup force to be M/R^2, giving $R_T \sim (M_H/\rho)^{1/3}$. About $10^6\ M_\odot$.

34 (a) $(2 - \sqrt{2})M$, (b) $E \leqslant m_1 + m_2 - \sqrt{(m_1^2 + m_2^2)} \approx m_1(1 - \tfrac{1}{2} m_1/m_2)$.

Chapter 12

7 (c) A Lorentz transformation leaves the metric and hence the hyperbola unchanged, but changes the origin of spatial coordinates. Any point on the hyperbola can be made into this origin by choosing the correct transformation.

8 The energy density of black-body radiation is proportional to T^4.

10 To V_M in Eq. (12.29) must be added the term $-\Lambda R^3/3$. If $\Lambda > 0$, then the new V_M reaches a negative maximum. If this maximum value exceeds -1 then there exist $k = +1$ models which collapse from $R = \infty$ to a finite R and re-expand. If the maximum is less than -1 then the $k = +1$ model expands from $R = 0$ to infinite radius. If the maximum is only slightly less than 1, this model will 'hesitate' or 'coast' for a long time near the value of R at which the maximum occurs. If $\Lambda < 0$ the new V_M increases monotonically from $-\infty$ at $R = 0$ to $+\infty$ at $R = \infty$. All models expand from $R = 0$ to finite radii and re-collapse. There is no qualitative difference between the matter-dominated and radiation-dominated cases.

11 (a) For black-body radiation $\rho_r = a_s T^4/c^2$, where $a_s = 7.56 \times 10^{-16}\ \mathrm{J\ m^{-3}\ deg^{-4}\ K^{-4}}$ is Stephan's constant. This give $\varepsilon = 4.5 \times 10^{-4}\ m$, and about $2m \times 10^9$ photons per baryon.

12 (a) Because of homogeneity, p^r is a constant. The result follows from null-ness of $\vec{p}$.

References

Abraham, R. & Marsden, J. E. (1978) *Foundations of Mechanics*, 2nd edn (Benjamin/Cummings, Reading, Mass.).

Adler, R., Bazin, M. & Schiffer, M. (1975) *Introduction to General Relativity*, 2nd edn (McGraw-Hill, N.Y.).

Alpher, R. A. & Herman, R. C. (1949) *Phys. Rev.*, **75**, 1089.

Anderson, J. L. (1967) *Principles of Relativity Physics* (Academic Press, N.Y.).

Arzeliès, H. (1966) *Relativistic Kinematics* (Pergamon Press, London). Translation of a 1955 French edn.

Ashtekar, A. (1980) 'Asymptotic structure of the gravitational field at spatial infinity'. In Held (1980*b*), p. 37.

Bahcall, J. N. (1978) *Ann. Rev. Astron. and Astrophys.*, **16**, 241.

Balian, R., Audouze, J. & Schramm, D. N. (1980) *Physical Cosmology*. Proceedings of 1979 Les Houches summer school (North-Holland, Amsterdam).

Benenti, S. & Francaviglia, M. (1980) 'The theory of separability of the Hamilton–Jacobi equation and its applications to general relativity'. In Held (1980*a*), p. 393.

Bertotti, B. (1974) *Experimental Gravitation* (Academic Press, N.Y.). Lectures at the course LVI of the Enrico Fermi summer school, Varenna, 1972.

Bertotti, B. (1977) *Experimental Gravitation* (Academica Nazionale dei Lincei, Rome). Proceedings of a symposium in Pavia, 1976.

Bishop, R. L. & Goldberg, S. I. (1968) *Tensor Analysis on Manifolds* (Macmillan, London).

Blandford, R. & Teukolsky, S. A. (1976) *Astrophys. J.*, **205**, 580.

Blandford, R. & Thorne, K. S. (1979) 'Black hole astrophysics'. In Hawking & Israel (1979), p. 454.

Bohm, D. (1965) *The Special Theory of Relativity* (W. A. Benjamin, N.Y.).

Borel, E. (1960) *Space and Time* (Dover, N.Y.).

Boriakoff, V., Ferguson, D. C., Haugan, M. P., Terzian, Y. & Teukolsky, S. A. (1982) *Astrophys. J.*, **261**, L101.

Bowler, M. G. (1976) *Gravitation and Relativity* (Pergamon, Oxford).

Branch, D. (1977). In Schramm, D. N. *Supernovae* (Reidel, Dordrecht), p. 21.

Buchdahl, H. A. (1959) *Phys. Rev.*, **116**, 1027.

Buchdahl, H. A. (1981) *Seventeen Simple Lectures on General Relativity Theory* (Wiley, N.Y.).

Burke, W. L. (1980) *Spacetime, Geometry, Cosmology* (University Science Books, Mill Valley, California).

Burke, W. L. (1981) *Astrophys. J.*, **244**, L1.

Butterworth, E. M. & Ipser, J. R. (1976) *Astrophys. J.*, **204**, 200.

Canuto, V. (1974) *Ann. Rev. Astron. and Astrophys.*, **12**, 167.

Canuto, V. (1975) *Ann. Rev. Astron. and Astrophys.*, **13**, 335.

Carr, B. J. (1977) *Mon. Not. Roy. Astron. Soc.*, **181**, 293.

Cartan, E. (1923) *Ann. École Norm. Sup.*, **40**, 325.

Carter, B. (1969) *J. Math. Phys.*, **10**, 70.

Carter, B. (1973) In DeWitt, B. & DeWitt, C. (eds.). *Black Holes* (Gordon and Breach, N.Y.).

Carter, B. (1979) 'The general theory of the mechanical, electromagnetic, and thermodynamic properties of black holes'. In Hawking & Israel (1979), p. 294.

Caves, C. M., Thorne, K. S., Drever, R. W. P., Sandberg, V. D. & Zimmerman, M. (1980) *Rev. Mod. Phys.*, **52**, 341.

Chandrasekhar, S. (1957) *Stellar Structure* (Dover, N.Y.).

Chandrasekhar, S. (1979) 'An introduction to the theory of the Kerr metric and its perturbations'. In Hawking & Israel (1979), p. 370.

Chandrasekhar, S. (1980) In Wayman, P. *Highlights of Astronomy* (Reidel, Dordrecht). Proceedings of the 1979 IAU meeting in Montreal, p. 45.

Chandrasekhar, S. (1983) *The Mathematical Theory of Black Holes* (Oxford University Press).

Choquet-Bruhat, Y., DeWitt-Morette, C. & Dillard-Bleick, M. (1977) *Analysis, Manifolds, and Physics* (North-Holland, Amsterdam).

Choquet-Bruhat, Y. & York, J. W. (1980) 'The Cauchy problem'. In Held (1980*a*), p. 99.

Clarke, C. (1979) *Elementary General Relativity* (Edward Arnold, London).

Clayton, D. (1968) *Principles of Stellar Evolution and Nucleosynthesis* (McGraw-Hill, N.Y.).

Comins, N. & Schutz, B. F. (1978) *Proc. Roy. Soc. A*, **364**, 211.

Cowling, S. (1983), Ph. D. Thesis, University College Cardiff, U.K.

Davies, P. C. W. (1980) 'Quantum fields in curved space'. In Held (1980*a*), p. 255.

Deruelle, N. & Piran, T. (1983) *Gravitational Radiation* (North-Holland, Amsterdam). Lectures at the 1982 Les Houches summer school.

DeWitt, B. S. (1979) 'Quantum gravity: the new synthesis'. In Hawking & Israel (1979), p. 680.

Dicke, R. H. (1964) In Dewitt, B. & Dewitt, C. *Relativity, Groups and Topology* (Gordon and Breach, N.Y.). Lectures at the 1963 Les Houches summer school.

Dixon, W. G. (1978) *Special Relativity, The Foundation of Macroscopic Physics* (Cambridge University Press).

Dodson, C. T. J. & Poston, T. (1977) *Tensor Geometry* (Pitman, London).

Douglass, D. H. & Braginsky, V. B. (1979) 'Gravitational radiation experiments'. In Hawking & Israel (1979), p. 90.

Durisen, R. H. (1975) *Astrophys. J.*, **199**, 179.

Durrell, C. V. (1960) *Readable Relativity* (Harper and Bros. N.Y.) Reprint of a 1926 ed.

Eardley, D. M. & Press, W. H. (1975) *Ann. Rev. Astron. and Astrophys.*, **13**, 381.

Ehlers, J., Rosenblum, A., Goldberg, J. & Havas, P. (1976) *Astrophys. J.*, **208**, 77.

Epstein, R. (1977) *Astrophys. J.*, **216**, 92.

Fermi, E. (1956) *Thermodynamics* (Dover, N.Y.).

Fischer, A. E. & Marsden, J. E. (1979) 'The initial-value problem and the dynamical formulation of general relativity'. In Hawking & Israel (1979), p. 138.

Fock, V. (1964) *The Theory of Space, Time, and Gravitation*, 2nd edn (Pergamon Press, Oxford).

Foster, J. & Nightingale, J. D. (1979) *A Short Course in General Relativity* (Longman, London).

Frankel, T. (1979) *Gravitational Curvature, an Introduction to Einstein's Theory* (W. H. Freeman, San Francisco).

French, A. P. (1968) *Special Relativity* (W. W. Norton, N.Y.).

Friedman, J. L. (1978) *Commun. Math. Phys.*, **63**, 243.

Friedman, J. L. & Schutz, B. F. (1978) *Astrophys. J.*, **222**, 281.

Futamase, T. (1983) *Phys. Rev. D*, **28**, 2373.

Futamase, T. & Schutz, B. F. (1983) *Phys. Rev. D*, **28**, 2363.

Gamow, G. (1948) *Nature*, **162**, 680.

Geroch, R. (1978) *General Relativity from A to B*, (University of Chicago Press).

Geroch, R. & Horowitz, G. T. (1979) 'Global structure of spacetimes'. In Hawking & Israel (1979), p. 212.

Gibbons, G. W. (1979) 'Quantum field theory in curved spacetime'. In Hawking & Israel (1979), p. 639.

Goldberg, J. N. (1980) 'Invariant transformations, conservation laws, and energy-momentum'. In Held (1980a), p. 469.

Gott, J. R., Gunn, J. E., Schramm, D. N. & Tinsley, B. M. (1974), *Astrophys. J.*, **194**, 543.

Grishchuk, L. P. & Polnarev, A. G. (1980) 'Gravitational waves and their interactions with matter and fields'. In Held (1980b), p. 393.

Hagedorn, R. (1963) *Relativistic Kinematics* (Benjamin, N.Y.).

Harrison, B. K., Thorne, K. S., Wakano, M. & Wheeler, J. A. (1965) *Gravitation Theory and Gravitational Collapse* (University of Chicago Press).

Hartle, J. B. (1978) *Phys. Rep.*, **46**, 201.

Hawking, S. W. (1972) *Commun. Math. Phys.*, **25**, 152.

Hawking, S. W. (1975) *Commun. Math. Phys.*, **43**, 199.

Hawking, S. W. (1979) 'The path-integral approach to quantum gravity'. In Hawking & Israel (1979), p. 746.

Hawking, S. W. & Ellis, G. F. R. (1973) *The Large-Scale Structure of Space–Time* (Cambridge University Press).

Hawking, S. W. & Israel, W. (1979) *General Relativity: An Einstein Centenary Survey* (Cambridge University Press).

Heidmann, J. (1980) *Relativistic Cosmology* (Springer, Berlin).

Held, A. (1980*a*) *General Relativity and Gravitation* (Plenum, London), vol. 1.

Held, A. (1980*b*) *General Relativity and Gravitation* (Plenum, London), vol. 2.

Hermann, R. (1968) *Differential Geometry and the Calculus of Variations* (Academic Press, N.Y.).

Hicks, N. J. (1965) *Notes on Differential Geometry* (D. Van Nostrand, N.Y.).

d'Inverno, R. A. (1980) 'A review of algebraic computing in general relativity'. In Held (1980*a*), p. 491.

Isenberg, J. & Nester, J. (1980) 'Canonical gravity'. In Held (1980*a*), p. 23.

Israel, W. & Stewart, J. (1980) 'Progress in relativistic thermodynamics and electrodynamics of continuous media'. In Held (1980*b*), p. 491.

Jackson, J. D. (1975) *Classical Electrodynamics* (Wiley, N.Y.).

Jammer, M. (1969) *Concepts of Space*, 2nd edn (Harvard University Press, Cambridge, Mass.)

Kilmister, C. W. (1970) *Special Theory of Relativity* (Pergamon, Oxford).

Kirschner, R. (1976) *Scientific American*, December, p. 88.

Kobayashi, S. & Nomizu, K. (1963) *Foundations of Differential Geometry* (Interscience, N.Y.), vol. I

Kobayashi, S. & Nomizu, K. (1969) *Foundations of Differential Geometry* (Interscience, N.Y.), vol II.

Kramer, D., Stephani, H., MacCallum, M. & Herlt, E. (1981) *Exact Solutions of Einstein's Field Equations* (Cambridge University Press).

Landau, L. D. & Lifshitz, E. M. (1959) *Fluid Mechanics* (Pergamon, Oxford).

Landau, L. D. & Lifshitz, E. M. (1962) *The Classical Theory of Fields* (Pergamon, Oxford).

Lerch, F. J. (1978) *Geophys. Res. Lett.*, **5**, 1031.

Liang, E. P. T. & Sachs, R. K. (1980) 'Cosmology'. In Held (1980*b*), p. 329.

Lightman, A. P., Press, W. H., Price, R. H. & Teukolsky, S. A. (1975) *Problem Book in Relativity and Gravitation* (Princeton University Press).

Lovelock, D. & Rund, H. (1975) *Tensors, Differential Forms, and Variational Principles* (Wiley, N.Y.).

MacCallum, M. A. H. (1979) 'Anisotropic and inhomogeneous relativistic cosmologies'. In Hawking & Israel (1979), p. 533.

Mathews, J. & Walker, R. L. (1965) *Mathematical Methods of Physics* (Benjamin, N.Y.)

McCrea, W. H. (1979), *Q. J. Roy. Astron. Soc.*, **20**, 251.

McVittie, G. C. (1965) *General Relativity and Cosmology* (University of Illinois Press).

Marder, L. (1971) *Time and the Space-Traveller* (George Allen and Unwin, London).

Mehra, J. (1974) *Einstein, Hilbert, and the Theory of Relativity* (Reidel, Dordrecht).

Miller, J. C. & Sciama, D. W. (1980) 'Gravitational collapse to the black hole state'. In Held (1980*b*), p. 359.

Misner, C. W. (1964) In DeWitt, B. & DeWitt, C. (eds.), *Relativity, Groups, and Topology* (Gordon and Breach, N.Y.), p. 883.

Misner, C. W., Thorne, K. S. & Wheeler, J. A. (1973) *Gravitation* (Freeman, San Francisco).

Møller, C. (1972) *The Theory of Relativity* (Clarendon Press, Oxford).

Muirhead, H. (1973) *The Special Theory of Relativity* (Macmillan, London).

Novikov, I. D. & Thorne, K. S. (1973) In DeWitt, B. & DeWitt C. (eds.), *Black Holes* (Gordon and Breach, N.Y.).

Ostriker, J. P. (1979) 'Astrophysical sources of gravitational radiation'. In Smarr (1979a), p. 461.

Ostriker, J. P., Peebles, P. J. E. & Yahil, A. (1974), *Astrophys. J.*, **193**, L1.

Pais, A. (1982) '*Subtle is the Lord . . .*', *The Science and the Life of Albert Einstein* (Oxford University Press).

Papapetrou, A. (1974) *Lectures in General Relativity* (Reidel, Dordrecht).

Peebles, P. J. E. (1971) *Physical Cosmology* (Princeton University Press).

Peebles, P. J. E. (1980) *Large Scale Structure of the Universe* (Princeton University Press).

Penrose, R. (1979) In Hawking & Israel (1979), p. 581.

Penzias, A. A. (1979) *Rev. Mod. Phys.*, **51**, 425.

Peters, P. C. (1964) *Phys. Rev.*, **B 136**, 1224.

Pound, R. V. & Rebka, G. A. (1960) *Phys. Rev. Lett.*, **4**, 337.

Pound, R. V. & Snider, J. L. (1965) *Phys. Rev.*, **B 140**, 788.

Press, W. H. & Thorne, K. S. (1972), *Ann. Rev. Astron. and Astrophys.*, **10**, 335.

Price, R. H. (1972a) *Phys. Rev. D*, **5**, 2419.

Price, R. H. (1972b) *Phys. Rev. D*, **5**, 2439.

Rees, M. J. (1977) In *Proceedings of the 8th Texas Symposium on Relativistic Astrophysics*, *Ann. N.Y. Acad. Sci.*, **302**, 613.

Rindler, W. (1969) *Essential Relativity* (Van Nostrand, N.Y.).

Robertson, H. P. & Noonan, T. W. (1968) *Relativity and Cosmology* (W. B. Saunders, Philadelphia).

Robinson, D. C. (1975) *Phys. Rev. Lett.*, **34**, 905.

Ryan, M. P. & Shepley, L. C. (1975) *Homogeneous Relativistic Cosmologies* (Princeton University Press).

Sachs, R. K. & Wu, H. (1977) *General Relativity for Mathematicians* (Springer, Berlin).

Schild, A. (1967) In Ehlers, J. (ed.) *Relativity Theory and Astrophysics: I. Relativity and Cosmology* (American Math. Soc., Providence, R.I.).

Schmidt, B. G. (1973) In Israel, W. (ed.) *Relativity, Astrophysics and Cosmology* (Reidel, Dordrecht).

Schouten, J. A. (1954) *Ricci-Calculus* (Springer, Berlin).

Schrödinger, E. (1950) *Space-Time Structure* (Cambridge University Press).

Schutz, B. F. (1980a) *Phys. Rev.*, **D 22**, 249.

Schutz, B. F. (1980b) *Geometrical Methods of Mathematical Physics* (Cambridge University Press).

Schutz, B. F. (1984) *Am. J. Phys.*, **52**, 412.

Sears, F. W. & Brehme, R. W. (1968) *Introduction to the Theory of Relativity* (Addison-Wesley, Reading, Mass.).

Sexl, R. U. & Sexl, H. (1979) *White Dwarfs–Black Holes* (Academic Press, N.Y.).

Shapiro, I. I. (1980) 'Experimental tests of the general theory of relativity'. In Held (1980b), p. 469.

Shapiro, S. L. & Teukolsky, S. A. (1983) *Black Holes, White Dwarfs, and Neutron Stars* (Wiley, N.Y.).

Smarr, L. L. (1979a) (ed.) *Sources of Gravitational Radiation* (Cambridge University Press).

Smarr, L. L. (1979b) 'Gauge conditions, radiation formulae, and the two-black-hole collision'. In Smarr (1979a), p. 245.

Smith, F. G. (1977) *Pulsars* (Cambridge University Press).

Smith, J. H. (1965) *Introduction to Special Relativity* (Benjamin, Reading, Mass.)

Smoot, G. F. & Lubin, P. M. (1979) *Astrophys. J.*, **234**, L83.

Spivak, M. (1979) *A Comprehensive Introduction to Differential Geometry*, (Publish or Perish, Boston), vols. 1–5.

Stephani, H. (1982) *General Relativity, An Introduction to the Theory of the Gravitational Field* (Cambridge University Press).

Synge, J. L. (1960) *Relativity: The General Theory* (North-Holland, Amsterdam).

Synge, J. L. (1965) *Relativity: The Special Theory* (North-Holland, Amsterdam).

Tammann, G. A., Sandage, A. & Yahil, A. (1980) In Balian *et al.* (1980), p. 53.

Taylor, E. F. & Wheeler, J. A. (1966) *Spacetime Physics* (Freeman, San Francisco).

Taylor, J. H. & Weisberg, J. M. (1982) *Astrophys. J.*, **253**, 908.

Terletskii, Y. P. (1968) *Paradoxes in the Theory of Relativity* (Plenum, N.Y.).

Teukolsky, S. A. (1972) *Phys. Rev. Lett.*, **29**, 114.

Thirring, W. (1978) *Classical Field Theories* (Springer, Vienna).

Thorne, K. S. (1967) In DeWitt, C. W., Schatzman, E. & Veron, P. (eds.), *High Energy Astrophysics* (Gordon and Breach, N.Y.).

Thorne, K. S. (1980*a*) *Rev. Mod. Phys.*, **52**, 285.

Thorne, K. S. (1980*b*) *Rev. Mod. Phys.*, **52**, 299.

Thorne, K. S., Caves, C. M., Sandberg, V. D. & Zimmerman, M. (1979) 'The quantum limit for gravitational-wave detectors and methods of circumventing it'. In Smarr (1979*a*), p. 49.

Tipler, F. J., Clarke, C. J. S. & Ellis, G. F. R. (1980), 'Singularities and horizons – a review article'. In Held (1980*b*), p. 97.

Tonnelat, M.-A. (1964) *Les Vérifications Expérimentales de la Relativité Générale* (Masson et Cie, Paris).

Trautman, A. (1962) In Witten, L. (ed.) *Gravitation: An Introduction to Current Research* (Wiley, N.Y.).

Tyson, J. A. & Giffard, R. P. (1978), *Ann. Rev. of Astron. and Astrophys.*, **16**, 521.

de Vaucouleurs, G. (1982) *Observatory*, **102**, 178

Wagoner, R. V. (1980*a*) In Balian *et al.* (1980), p. 180.

Wagoner, R. V. (1980*b*). In Balian *et al.* (1980), p. 395.

Walker, M. & Will, C. M. (1980) *Phys. Rev. Lett.* **45**, 1741.

Ward, W. R. (1970) *Astrophys. J.*, **162**, 345.

Weber, J. (1961) *General Relativity and Gravitational Waves* (Interscience, N.Y.).

Weber, J. (1980) 'The search for gravitational radiation'. In Held (1980*b*), p. 345.

Weinberg, S. (1972) *Gravitation and Cosmology* (Wiley, N.Y.).

Weinberg, S. (1977) *The First Three Minutes* (Deutsch, London).

Weinberg, S. (1979) 'Ultraviolet divergences in quantum theories of gravitation'. In Hawking & Israel (1979), p. 790.

Will, C. M. (1972) *Physics Today*, **25**, 23 (October 1972).

Will, C. M. (1979) 'The confrontation between gravitation theory and experiment'. In Hawking & Israel (1979), p. 24.

Will, C. M. (1981) *Theory and Experiment in Gravitational Physics*, (Cambridge University Press).

Williams, L. P. (1968) *Relativity Theory: Its Origins and Impact on Modern Thought* (Wiley, N.Y.).

Wilson, R. W. (1979) *Rev. Mod. Phys.*, **51**, 433.

Winicour, J. (1980) 'Angular momentum in general relativity'. In Held (1980*b*), p. 71.

Yano, K. (1955) *The Theory of Lie Derivatives and its Applications* (North-Holland, Amsterdam).

Young, P., Gunn, J. E., Kristian, J., Oke, J. B. & Westphal, J. A. (1981) *Astrophys. J.*, **244**, 736.

Zel'dovich, Ya. B. (1979) 'Cosmology and the early universe'. In Hawking & Israel (1979), p. 518.

Zel'dovich, Ya. B. & Novikov, I. D. (1971) In Thorne, K. S & Arnett, W. D. (eds.), *Stars and Relativity* (*Relativistic Astrophysics Vol. I*) (University of Chicago Press).

Index

accelerated observer, in SR, 150
acceleration, 44, 51, 244, 278
  absolute, 3
  uniform, 56, 150
active galactic nuclei, 310, 311
adiabatic, 109
angular diameter, in cosmology, 333
angular diameter distance, 341
angular momentum, 298, 317
  as source of metric, 213
  conservation of, 191
  in collapse, 296
  in gravitational waves, 250
  in Kerr metric orbit, 301, 302
  of a black hole, 305
  of star, 265
  of the Sun, 274
  total, 193
angular velocity, 280
  of inertial frames, 301
  of zero-angular-momentum particle, 298
anisotropy, 320
  dipole, 335, 336
  quadrupole, 336
aphelion, 282
area theorem, *see* Hawking's area theorem
assumption of mediocrity, 322
asymptotic flatness, 254

baryon, 106
basis

non-coordinate, 144, 146, 147
orthonormal, 144, 147, 315
transformation of, 40
Bianchi identities, 173, 175, 200, 257
  contracted, 174
  twice-contracted, 174
big bang, 326, 327, 328, 337, 338, 340
binary pulsar system, 284, 314
  as source of gravitational waves, 232,
    241
binary star system, 265
  gravitational wave effects on, 250
  orbit of, 249
Birkhoff's theorem, 258
black body, 339
  luminosity of, 317
black hole, 106, 212, 226, 241, 253, 254,
    260, 264, 266, 271, 288, 294, 315, 318
  angular momentum of, 305
  as a black body, 308, 309
  entropy of, 310
  formed by collapse, 295
  in modern astrophysics, 310
  Kerr, *see* Kerr solution
  lifetime of, 308
  luminosity of, 308
  rotational energy of, 305
  Schwarzschild, *see* Schwarzschild
    solution
  small, 308
  stationary, 296
  temperature of, 308